Developmental Neurobiology of the Autonomic Nervous System

Contemporary Neuroscience

Developmental Neurobiology of the Autonomic Nervous System, edited by
   Phyllis M. Gootman
Neurobiology of the Trace Elements, edited by **Ivor E. Dreosti** and **Richard
   M. Smith**
   Volume 1: *Trace Element Neurobiology and Deficiencies*
   Volume 2: *Neurotoxicology and Neuropharmacology*

Developmental Neurobiology of the Autonomic Nervous System

Edited by

Phyllis M. Gootman

Edited by

Humana Press • Clifton, New Jersey

Library of Congress Cataloging-in-Publication Data

Main entry under title:

Developmental neurobiology of the autonomic nervous system.

(Contemporary neuroscience)
Includes bibliographies and index.
    1. Nervous system, Autonomic—Growth. I. Gootman, Phyllis M.,
1938– . II. Series.   [DNLM: 1. Autonomi
Nervous System—embryology.   2. Autonomic Nervous
System—growth & development.   3. Neurobiology.
WL 600 D4895]
QP368.D48 1986   612'.64018   86-3077
ISBN 0-89603-080-6

DEDICATION

I dedicate this book to my husband, Dr. Norman Gootman, and to my children, Sharon Hillary and Craig Seth, for their understanding and help.

Preface

For many years there were few studies concerning the role of the autonomic nervous system in visceral regulation, but examination of abstracts recently published in *Federation Proceedings* shows the rapid change that has occurred over the past fifteen years. Topics of special interest among the scientists abstracted include not just autonomic function in the adult, but also the maturation of autonomic function. The growing availability of sophisticated electrophysiological, tissue culture, and histochemical techniques has engendered a surge in the amount of experimental information available on autonomic maturation such that a symposium published in 1981 (Ciba Symposium #83, "Development of the Autonomic Nervous System," edited by G. Burnstock, Pittman Medical Ltd., NY, 1981) is already severely dated. Thus, the field is ready for another somewhat more detailed examination of aspects of the recent findings on autonomic nervous system development. This book is intended to bring scientists in related fields up to date on those recent advances.

The editor would particularly like to thank Dr. Nancy M. Buckley for her help in preparation of this book.

Phyllis M. Gootman

Introduction

This book, covering many key aspects of autonomic nervous system maturation, was suggested by the success of a symposium on the developing autonomic nervous system held at the Spring 1982 meeting of the Federation of American Scientists for Experimental Biology (*Federation Proceedings* 1983, **42,** 1609). It was obvious from the FASEB symposium that there is increasing interest in the developing autonomic nervous system, particularly with respect to its role in regulating visceral function. Some additional topics that were not covered in the FASEB symposium are also included in this book.

The editor feels that the readers of this volume are, in all probability, already cognizant of the state of knowledge of the adult autonomic nervous system. Therefore, a review of classical autonomic physiology, pharmacology, and neuroanatomy is not provided.

For a recent detailed discussion of the ontogeny and phylogeny of the developing nervous system, I would recommend the book published not long ago by D. Purves and J. W. Lichtman, *Principles of Neural Development* (Sinauer, Sunderland, MA, 1985). Another recent book, *Autonomic Nerve Funtion in the Vertebrates* by F. Nilsson (Springer-Verlag, New York, 1984), presents a comparative examination of autonomic nervous system function in vertebrates. For a summary of recent advances in the many aspects of catecholamines as they bear on autonomic nervous system research, I would recommend the series of three books edited by E. Usdin, A. Carlsson, A. Dahlstrom, and J. Engel; *Catecholamines,* Part A: *Basic and Peripheral Mechanisms;* Part B: *Neuropharmacology and Central Nervous System—Theoretical Aspects;* Part C: *Neuropharmacology and the Central Nervous System—Therapeutic Aspects* (Liss, NY, 1984).

It is obvious from the table of contents of our book that there are areas of autonomic function that have not been covered. For example, central transmitters and pathways are not discussed because of very limited information and the fact that most chapters are concerned with the autonomic nervous system, which is, of course, a peripheral nervous system. Nevertheless, the editor feels that the level of information available in this volume, supplied by experts from around the world, will be of considerable interest to the reader.

A chapter is included by Dr. Julian Smith on the ontogeny of the autonomic nervous system and its very earliest divergence and differentiation. Since the field of development of parasympathetic function is limited, we are fortunate to have a chapter by Dr. Ezio Giacobini on the development of peripheral parasympathetic neurons and synapses. Doctor Theodore A. Slotkin covers the development of sympathetic-adrenal medullary aspects.

Because of the amount of information available with respect to autonomic innervation of the developing myocardium, four chapters have been devoted to various aspects of autonomic regulation of the heart. The chapter by Drs. Margaret L. Kirby and Donald E. Stewart is concerned with the development of autonomic innervation of the avian heart, and Dr. Howard L. Cohen's chapter is concerned primaily with the development of autonomic innervation of the mammalian heart. Doctor Theodore A. Slotkin examines an aspect of sympathetic innervation of the developing heart, i.e., endocrine control of synaptic development. Doctors Robert F. Reder, Ofer Binah, and Peter Danilo, Jr. cover the electrophysiology of the developing myocardium.

Development of the autonomic innervation of regional circulations is covered in the chapters by Drs. Nancy M. Buckley, Phyllis M. Gootman, Geoffrey Burnstock, and Tim Cowen, and, to some degree, by Dr. Peter G. Smith.

There is one chapter that at first glance might appear to belong in a book on respiratory function. Though control of respiration is a somatic function (i.e., skeletal muscles are innervated by alpha-motoneurons), the close interrelationship between circulatory regulation and respiration convinced the editor to include such a chapter. Thus, information concerned with the state of knowledge of development of respiratory regulation is available from Dr. Andrew M. Steele's chapter on development of central respiratory function.

Contributors

OFER BINAH • Rappaport Family Institute for Research in the Medical Sciences, Technion-Israel Institute of Technology, Faculty of Medicine, Haifa, Israel

NANCY M. BUCKLEY • Department of Physiology and Biophysics, Albert Einstein College of Medicine, New York, New York

GEOFFREY BURNSTOCK • Department of Anatomy and Embryology and Centre for Neuroscience, University College London, London, UK

HOWARD L. COHEN • Department of Physiology, Downstate Medical Center, Brooklyn, New York

TIM COWEN • Department of Anatomy and Embryology and Center for Neuroscience, University College London, London UK

PETER DANILO, Jr. • Department of Pharmacology, College of Physicians and Surgeons of Columbia University, New York, New York

EZIO GIACOBINI • Department of Pharmacology, Southern Illinois University School of Medicine, Springfield, Illinois

PHYLLIS M. GOOTMAN • Department of Physiology, Downstate Medical Center, Brooklyn, New York

MARGARET L. KIRBY • Department of Anatomy, Medical College of Georgia, Augusta, Georgia

ROBERT F. REDER • Department of Medical Affairs, Knoll Pharmaceutical Company, Whippany, New Jersey

THEODORE A. SLOTKIN • Department of Pharmacology, Duke University Medial Center, Durham, North Carolina

JULIAN SMITH • Institut d'Embryologie du CNRS et du Collège de France, Nogent-sur-Marne, France

PETER G. SMITH • Department of Pharmacology, Duke University Medical Center, Durham, North Carolina

ANDREW M. STEELE • Schneider Children's Hospital of Long Island Jewish Medical Center, New Hyde Park, New York

DONALD E. STEWART • Department of Anatomy, Medical College of Georgia, Augusta, Georgia

Contents

Chapter 1
ONTOGENY OF THE AUTONOMIC NERVOUS SYSTEM: CELL LINE DIVERGENCE AND DIFFERENTIATION
Julian Smith

xiii

Chapter 2
DEVELOPMENT OF PERIPHERAL PARASYMPATHETIC NEURONS AND SYNAPSES
Ezio Giacobini

Chapter 3
DEVELOPMENT OF THE SYMPATHOADRENAL AXIS
Theodore A. Slotkin

Chapter 4
ENDOCRINE CONTROL OF SYNAPTIC DEVELOPMENT IN THE SYMPATHETIC NERVOUS SYSTEM: THE CARDIAC–SYMPATHETIC AXIS
Theodore A. Slotkin

Chapter 5
DEVELOPMENT OF ANS INNERVATION TO THE AVIAN HEART
Margaret L. Kirby and Donald E. Stewart

Chapter 6

DEVELOPMENT OF AUTONOMIC INNERVATION IN MAMMALIAN MYOCARDIUM

Howard L. Cohen

Chapter 7
AUTONOMIC EFFECTS IN THE DEVELOPING HEART
Robert F. Reder, Ofer Binah, and Peter Danilo, Jr.

Chapter 8
DEVELOPMENT, AGING, AND PLASTICITY OF PERIVASCULAR AUTONOMIC NERVES
Tim Cowen and Geoffrey Burnstock

Chapter 9
REGULATION OF REGIONAL VASCULAR BEDS BY THE DEVELOPING AUTONOMIC NERVOUS SYSTEM
Nancy M. Buckley

Chapter 10
RELATIONSHIPS BETWEEN THE SYMPATHETIC NERVOUS SYSTEM AND FUNCTIONAL DEVELOPMENT OF SMOOTH MUSCLE END ORGANS
Peter G. Smith

Chapter 11
DEVELOPMENT OF CENTRAL AUTONOMIC REGULATION OF CARDIOVASCULAR FUNCTION
Phyllis M. Gootman

Chapter 12
DEVELOPMENTAL CHANGES IN NEURAL CONTROL
OF RESPIRATION
Andrew M. Steele

Chapter 1

Ontogeny of the Autonomic Nervous System

Cell Line Divergence and Differentiation

Julian Smith

All elements of the vertebrate autonomic nervous system (ANS) originate from the neural crest, a short-lived embryonic structure that arises at the neurula stage from the lateral ridges of the neural primordium. As the neural tube closes the crest cells migrate, often over relatively long distances, to specific sites in the developing embryo where their progeny differentiate into the various categories of cells that compose the ANS—and also into many other cell types, including sensory neurons and glia, skeletal and mesenchymal tissues, endocrine cells, adrenal medulla, and melanocytes. Embryologists have long been fascinated by this transformation of a relatively undifferentiated and apparently homogeneous neural crest cell population into a multiplicity of distinct cell lines, each ultimately possessing characteristic phenotypic features (for an extensive review of the literature, *see* ref. 47). The last decade or so has witnessed an intensification of research efforts in this field, and this chapter will consider some of the recent advances that have been made, with particular reference to the events underlying the conversion of neural crest to autonomic

1

ganglion cells. Even confined to this relatively limited domain, the problem of the generation of phenotypic diversity from a unique source remains posed, for apart from the neuronal/glial dichotomy considerable morphological and chemical heterogeneity exists within the autonomic neuron population itself. Increased understanding of the mechanisms whereby such variety is engendered will help cast new light on plastic phenomena in the ANS at later stages of development.

1. Origin of Autonomic Ganglion Cells

The majority of experiments that will be described here concern avian embryos, which for obvious practical reasons have been and continue to be the most popular experimental material for the study of neural crest development.

Although initially the object of some controversy, the origin of autonomic ganglion cells from the neural crest has been recognized for many years, primarily as a result of classical ablation studies (33). On the other hand, the establishment of a precise correspondence between different axial levels of the neural primordium and given derivatives of the neural crest had to await the introduction of a suitable method for labeling neural crest cells, thus allowing them to be distinguished from the tissues through which they migrate and in which they differentiate. In this respect, quail-chick chimeras (45,46) have been of inestimable service in permitting a particularly fine analysis of neural crest migration and differentiation in vivo.

The construction of mosaic embryos was initially used to determine with exactitude the site of origin along the neural axis of the different peripheral ganglia. By replacing sequential portions of the neural primordium of the chick embryo with the equivalent region from a quail embryo at an identical stage of development, Le Douarin and Teillet (52,53), after histological analysis of the resulting chimeras (which are viable and develop normally), were able to compose a fate map of crest derivatives. Concerning the origin of autonomic ganglia, these workers showed that neural crest cells from the vagal (opposite somites 1–7) and lumbosacral (beyond somites 28) levels invade the splanchnopleure and give rise to the myenteric plexuses, whereas ganglion cell precursors from the trunk level (somites 6–28) are restricted in their migration to the dorsal mesenchyme where they produce the sympa-

thetic chains, aortic and adrenal plexuses, and the adrenomedullary cords. Subsequent studies enabled the origin of
various other autonomic structures to be localized to specific
points on the neuraxis (66,67,72,94).

Besides providing invaluable data of a topographical nature by
revealing an axial heterogeneity of the crest cell population with
respect to its ultimate differentiated fate, these experiments raised
the important question of the limitation of crest potentialities at
different levels of the neural primordium. Further experiments
conducted to examine this problem demonstrated that changing
the position of crest cells along the neural axis prior to their migration did not lead to phenotypic anomalies in the ANS of the
operated embryo. Thus, trunk crest grafted at the vagal level colonized the gut and developed into physiologically normal
cholinergic ganglia (50); conversely, it was shown that the mesen/
rhombencephalic crest could, when transplanted to the midtrunk
level, form adrenergic sympathetic ganglia and adrenomedullary
cells (54). Even more strikingly, the prosencephalic neural crest,
which does not normally give rise to neurons, can do so if grafted
to an appropriate heterotopic position (2,72).

Apparently, then, all regions of the neural crest can provide the
entire spectrum of autonomic ganglion cells. Any local restriction
of these potentialities during normal development must therefore
be the result of the operation of extrinsic factors during the steps
leading to gangliogenesis.

One of the keys to explaining the developmental divergence of
autonomic neuron precursors most probably lies in differences in
crest cell migration pathways at various levels of the embryo.
Some of these have now been plotted in detail, both by immuno-
cytochemically staining fibronectin, one of their major matrix
constituents, and by marking the crest cells themselves with a
selective monoclonal antibody (17,98–100). Interestingly, the
migration routes differ strikingly according to the axial level considered.

At the vagal level of the neuraxis most of the enteric ganglion
precursors migrate via a pathway situated immediately below the
ectoderm to the pharynx, which they invade posteriorly,
colonizing it up to the cloaca and forming the plexuses of
Meissner and Auerbach. In contrast, at the trunk level neural precursors take either of two available dorsoventral pathways. One
leads crest cells between the neural tube and the somitic mesenchyme, and the other conducts them through the space between

adjacent somites. Cells taking the latter path become distributed along the roots of the dorsal aorta where they condense to produce the primary sympathetic ganglion chains dorsolateral to the notochord. The cells that pass between somite and neural tube form the dorsal root ganglia (DrG).

In addition to yielding information on the mechanisms that govern the migration process, studies such as these which designate the cellular and acellular environment encountered by the crest cells, aid in identifying potential sources of the factors that influence the early stages of differentiation of ANS ganglion cell precursors. This point will be taken up in a subsequent section.

2. Cell Line Segregation and Plasticity

Whether the pluripotentiality of the neural crest denotes the existence within the population of individual cells possessing the ability to generate a gamut of cell types, or whether it signifies the presence of predetermined cell lineages is a question that is currently exciting much interest. A number of experimental approaches have been made to try to distinguish between these two possibilities (which are not necessarily mutually exclusive). They include attempts to identify markers of differentiation on neural crest cells at very early stages, analysis of the behavior of clones, examination of the relationship between differentiation and cell division, and studies on the restriction of developmental potential with time. Because of these efforts, it is becoming increasingly apparent that at least certain crest cells are limited in their capabilities and are partly or totally committed to a particular phenotype before leaving the neural primordium. Even the results of the heterotopic transplantation studies described above revealed that crest cells are not wholly equivalent in terms of their neurogenic possibilities at every region of the neural axis (54, 56,72), and data obtained from in vivo culture experiments led to the same conclusion (69,91). The only experimental support for the existence of plasticity at the level of the neural crest cell itself is provided by the results of clonal analysis, which suggest the existence of a bipotential precursor of melanocytes and catecholamine (CA)-containing cells (5,15,86). On the other hand, direct proof of the heterogeneity of premigratory crest in the neural primordium is still lacking and suggestions that it contains predetermined cell lineages are based on indirect evidence.

2.1. A Neuronal Precursor Line in the Neural Crest

Evidence derived from in vitro studies of crest cell development may be taken to support the idea that the neural crest population is initially heterogeneous. Neuronal cells, identifiable on the basis of varied criteria, can develop in cultured neural crest (see subsequent section). However, even under the most favorable conditions that have been found, only a small fraction of the cells expresses this phenotype, indicating that the crest cannot be validly considered to be a uniform population of naive cells, all of which respond in an identical manner to a given set of environmental conditions.

In a comprehensive study of neuronal differentiation in cultures of neural crest grown in a medium of defined chemical composition, Ziller et al. (104) noted the rapid appearance of neuronal markers (neurofilament proteins and tetanus toxin binding sites) in a small subset of neural crest cells that were morphologically recognizable as neurons within 24 h of culture. Particularly significant was the demonstration that this transformation occurred without cell division. Although alternative interpretations may be possible, it can be suggested that these results reflect the fact that precursors of at least a certain type of neuron are segregated early (before or at the start of the migration phase) from the remainder of the crest population. The finding that a small fraction of crest cells (isolated in vitro) possesses antigens common to a class of peripheral neurons (3,8) points to a similar conclusion.

2.2. Segregation of Autonomic and Sensory Cell Lines

As described earlier, transplanting neural crest from one level of the embryo to another has yielded pertinent information on the developmental capabilities of crest cells; similarly, manipulation of the environment in which differentiating ganglion cells evolve has provided valuable insight into the restriction of these potentialities during embryogenesis and, consequently, into the divergence of cell lines and its reversibility.

In experiments performed in our laboratory, various developing quail peripheral ganglia were transplanted into the neural crest cell migration pathways of younger chick embryo hosts.

Fragments of autonomic (ciliary, Remak, or sympathetic chain) or sensory (DrG) ganglia were removed from quail embryos at 4.5–15 d of incubation and inserted at the adrenomedullary level between the somites and neural tube of a 2-d chick embryo (i.e., before the onset of crest cell migration) and the behavior and fate of the transplanted ganglion cells were studied at various times after the operation. The grafted fragment quickly dissociated into individual cells that, within a few days, and after dividing actively (*18*), became distributed definitively among various neural crest derivatives of the host embryo (*49,55,56*). The patterns of localization of cells derived from the different types of peripheral ganglia were essentially similar, with one striking exception: whereas parasympathetic, sympathetic, and sensory ganglion cells all colonized host sympathetic ganglia, plexuses, and the adrenal medulla, in which they developed histochemically demonstrable CA, DrG alone contained cells able to contribute neurons and glia to the host spinal ganglia. Furthermore, this ability was lost in DrG older than 7 d of incubation, a time at which all sensory neuron precursors in quail spinal ganglia have ceased dividing (*85*).

These results reveal that the tissue environment can still exert a profound effect on neural crest-derived cells even at a relatively advanced stage of gangliogenesis, and demonstrate that all types of embryonic peripheral ganglia contain cells capable of expressing an autonomic phenotype. On the other hand, autonomic ganglia, unlike the neural crest cells from which they derive, never possess sensory potentialities, indicating an early and irreversible segregation of the autonomic and sensory lineages. Sensory precursors are present exclusively in the spinal ganglia where they are apparently to be found only among the still mitotically active elements of their population. A model for the development of sensory and autonomic neurons and glia, taking into account the above data, has been presented (*48,49,56*). It proposes that separate pools of precursors for sensory and autonomic cell lines exist in the neural crest, but that sensory neurons can only develop in very close proximity to the central nervous system. If during the course of migration they reach more distal positions, they die or at any rate lose their sensory potentialities; autonomic precursors do not have such stringent survival requirements and are to be found in a wide range of peripheral ganglia, including a great many in which they may never normally express the autonomic phenotype. Although speculative, this model is compatible with a number of other experimental observations. Thus, cultured neural crest is able after transplantation in vivo to differentiate into autonomic neurons, but not into sensory ganglion cells (22), suggesting that the sensory precursors are short-lived in the absence of appropriate factors. Concerning the autonomic potential of sensory ganglia, it has been demonstrated that cells with "autonomic" features (i.e., the ability to synthesise CA) differentiate when sensory ganglion rudiments (70) or dissociated 9- to 15-d DrG (Z.G. Xue, J. Smith, and N.M. Le Douarin, *Proc. Natl. Acad. Sci. USA,* in press) are cultured in vitro. Indeed, there are recent indications that in mammalian species the expression of catecholaminergic properties is a normal feature of a few cells within certain spinal or cranial ganglia (36,41,80).

2.3. Neuronal Potentialities in a Satellite Cell Precursor Population

In all the previously discussed experiments involving transplantation and "redifferentiation" of embryonic peripheral ganglion cells it was not possible to determine the initial phenotypic status of those cells that expressed novel neuronal features in their new embryonic microenvironment. Thus, it remained a distinct possibility that the cells responding to new developmental cues were not in fact neurons, but belonged to the abundant population of small ganglionic cells most of which are probably normally precursors of satellite cells. Relatively undifferentiated in ganglia at the ages used for transplantation, these cells would seem to be plausible candidates for phenotypic reprogramming. To verify this hypothesis, however, a suitable cell-type-specific marker is required. Fortuitously, the mixed embryological origin of a number of cranial sensory ganglia enabled the quail/chick chimera system to provide the appropriate specificity of labeling.

In the nodose ganglion, for example, the neurons derive from the placodal ectoderm, whereas the satellite cells arise from the neural crest (68). Grafting a quail rhombencephalic neural primordium isotopically into a chick embryo therefore results in a chimeric ganglion in which the satellite cells and the neurons are of quail and chick origin, respectively. Fragments of mosaic ganglia constructed in this way were removed at 5.5–9 d of incubation and grafted into 2-d chick embryos (2). Subsequent histological analysis showed that the crest-derived (satellite) population of the nodose ganglion provided cells able to differentiate into enteric and sympathetic neurons and adrenomedullary cells—but not into DrG neurons (as a distal cranial ganglion, the nodose, according to the theory described in the previous section, would not be a source of precursor cells of the sensory lineage). Experiments involving chimeric nodose ganglia in which the neuronal population was selectively labeled with the quail marker demonstrated that nodose neurons all degenerated within a few days of grafting.

It can be deduced from these results that autonomic neuronal potentialities are present within the population of nonneuronal cells of the developing nodose ganglion and can be expressed in appropriate intraembryonic locations. Although it has not been confirmed that the extrapolation of these conclusions to all other types of peripheral ganglia is justified, the demonstration (18) that ciliary ganglion neurons (labeled with [^{3}H]thymidine) also die shortly after transplantation into a second embryo indicates that here, too, it is exclusively the nonneuronal cell population that undergoes differentiation in response to host-derived signals.

It is to be noted that results obtained with very different experimental systems also lead to the conclusion that neuronal potentialities exist in the nonneuronal population of peripheral ganglia. Thus, modification of the ionic composition of the culture medium caused the conversion of cells that were morphologically unrecognizable as neurons in 10-d chick embryo DrG into neurons in vitro (6). More direct evidence is to be found in the recent report that a large proportion of nonneuronal cells (defined by a glia-specific monoclonal antibody) in short-term cultures of 6–14-d chick embryo ciliary ganglia also possess what are usually considered to be neuronal properties, i.e., NGF receptors and a high-affinity norepinephrine uptake system (82).

2.4. Plasticity of Chemical Differentiation in Autonomic Ganglia

If the experiments detailed above provide evidence of neural crest plasticity, in the sense that cells probably destined to differentiate into a particular phenotypic category can be compelled to develop in an alternative direction, they do not constitute examples of the plasticity of the differentiated state. There is no indication, for example, that the cells of the "satellite" lineage, which acquire neuronal properties in transplanted nodose ganglia, possess any recognizable phenotypic features characteristic of terminally differentiated glial cells prior to their tardy expression of neuronal traits. Cases of what one might call true phenotypic switching, in which cells display properties specific first for one particular cell type and subsequently, in response to a modification of their environment, for another, are few. However, an unambiguous illustration of a phenomenon of this sort does exist and concerns the expression of neurotransmitter-related features in cultures of primary sympathetic neurons of the superior cervical ganglion (SCG) of the newborn rat (75).

The development in vitro of SCG principal neurons, which are already adrenergic at birth, depends strongly on the culture conditions used. Thus, when dissociated and grown in the absence of nonneuronal cells, postmitotic SCG neurons continue their adrenergic maturation (59,60). In contrast, if they are grown with certain types of nonneuronal cells or conditioned media, the neurons evolve along cholinergic lines, synthesizing acetylcholine (ACh), forming cholinergic synapses, and accumulating numerous clear cytoplasmic vesicles (35,43,76–78). Electrophysiological, ultrastructural, and biochemical studies on individual neurons showed this transformation to be a phenotypic change at the single cell level (27,42,81). Subsequent experiments showed that a dual expression of adrenergic and cholinergic properties can be an intermediate stage in the transition from adrenergic to cholinergic function (79). However, under highly cholinergic culture conditions, CA accumulation and tyrosine hydroxylase (TH) activity drop markedly (103).

Although no other system has been subject to such intensive scrutiny, there are persuasive indications that similar adrenergic/cholinergic switching can occur in cultured rat adrenal chromaffin cells (74).

Two lines of evidence suggest that chemical plasticity of this kind may be displayed by certain neurons of the ANS during the course of normal development. The first concerns a population of cells, presumably neural crest derivatives, containing endogenous CA and the enzymes necessary for their synthesis, that appear in the developing gut of the rat embryo during the 12th day of gestation (10,11,97). The CA-synthesizing activity is expressed only temporarily and is lost within 3–4 d. The cells themselves do not disappear immediately since a CA uptake system can be demonstrated when all CA enzyme immunoreactivity has vanished (37). Although their ultimate phenotype has not been identified, it is tempting to conclude that the transient adrenergic cells are enteric neuron precursors that subsequently switch to producing an alternative neurotransmitter. More strongly convincing evidence for plasticity has been provided by the results of a comprehensive study of the ontogeny of the cholinergic sympathetic innervation of sweat glands in the footpads of the rat. The data are wholly consistent with the hypothesis that the glands are initially innervated by noradrenergic axons that lose their endogenous CA and become functionally cholinergic during the first three postnatal weeks (44). Also in this case, CA uptake can be observed after other demonstrable adrenergic properties have disappeared.

In contrast to the amply documented adrenergic/cholinergic switch, examples of the opposite transformation have so far proved elusive. Cells from transplanted quail ciliary ganglia were indeed able to colonize the sympathetic ganglia or adrenal medulla of the chick host where they developed adrenergic properties, as shown by histofluorescence of endogenous CA (55). However, the results discussed in the preceding section indicate that the cells concerned were in all probability not recruited from the population of cholinergic neurons. Interestingly, in the case of SCG neurons in vitro, once the cholinergic phenotype has become firmly established reversion to adrenergic function cannot be brought about by withdrawing the conditioned medium (78).

3. Establishment of Chemical Diversity in the Autonomic Neuron Population

Although previously presented evidence suggests that what might be described as a broad autonomic determination is acquired relatively early in neurogenesis by a category of neural crest progeny, a large spectrum of developmental options is nonetheless left open to the cells initially set apart in this way. For autonomic neurons display a considerable degree of morphological and, particularly, biochemical heterogeneity and this fact, in addition to posing the problem of the integrated function of these diverse elements, raises interesting questions concerning their ontogeny. One of the more striking manifestations of this biochemical diversity—and an obvious basis for categorizing autonomic neuronal phenotypes—is the wide variety of neuroactive substances produced, stored, released, and in some cases degraded or taken up by nerve cells. The number of identified molecules of this type (some of which may coexist within the same cell) has long since reached double figures and it is hardly surprising that the development of different neurotransmitter-related properties has attracted an increasing amount of attention over recent years, both from phenomenological and deterministic standpoints (4,47,51,75,90). Although the picture is far from complete at the present time, the results of some of these studies have enabled progress to be made in the identification of embryonic sources of factors—as yet ill-defined—that play a part in the chemical differentiation of autonomic derivatives of the neural crest.

3.1. Chemical Differentiation of Neural Crest Cells During Development

A consideration of the chronology of the appearance of various neurotransmitters and neuropeptides, together with a knowledge of the pathways followed by peripheral ganglion progenitor cells, can help to provide insight into the respective differentiation processes.

A logical first step in the study of events leading to the expression of differentiated properties by neural crest derivatives is to define the status of the neural crest itself with respect to these same features. Within the limits of the histochemical and immunocytochemical techniques currently available to analyze CA-associated properties at the cellular level, it has been clearly demonstrated that no adrenergic traits are present in neural crest cells before or during their migration to the sites of sympathetic gangliogenesis (1,13). Confirming this cytological data, Fauquet et al. (23) showed that freshly dissected neural crest is unable to convert radioactive tyrosine to CA. The first detectable appearance of the catecholaminergic phenotype in sympathetic ganglioblasts in fact coincides with their aggregation to form the primary sympathetic chains, at about 3.5 d of incubation in the chick and quail embryo (1,13,20,83, and our own unpublished results).

In contrast, a number of properties characteristic of the cholinergic phenotype are already present in premigratory and migrating neural crest. Thus, microsurgically excised mesencephalic neural crest can be shown by means of a highly sensitive radiochemical technique to convert [^{3}H]choline to ACh via a choline acetyltransferase (CAT)-mediated reaction (93). A similar activity is also associated with neural crest cells from other axial levels (*90*). In addition to its ability to produce ACh, neural crest also possesses the catabolic enzyme acetylcholinesterase (AChE) which is present in presumptive crest cells soon after closure of the neural tube (*9,16*). Subsequently, it is specifically found in ~90% of the migrating crest cell population at both cephalic and trunk levels (*9*). From these findings, it can be concluded that key constituents of the cholinergic phenotype are present, albeit at very low levels, in neural crest as soon as it is formed. The role of such a system at extremely early stages of development remains to be explained, as does its relationship to the definitive cholinergic neuronal phenotype. As might be expected, cholinergic properties can easily be detected in parasympathetic and enteric ganglia in the earliest stages at which they can be examined (*7,55,91*).

Recently, a number of studies has been carried out on the ontogeny of neuropeptide-producing cells. Like CA, none of the neuropeptides that have been examined so far in early embryos has been detected in crest-derived cells at stages prior to peripheral ganglion formation. However, somatostatin appears early in the developing sympathetic nervous system, a short time after the first adrenergic cells can be demonstrated in the sympathetic chains (*28,61*), aortic plexuses, and adrenal medulla (*28*). In the avian gut, evidence has been obtained that vasoactive intestinal polypeptide and substance P appear several days after the arrival of neural crest cells in this location (*25*) and a sequential development of various neurotransmitters and peptides in the intestine has been proposed (*29,84*). However, these findings have been contested (*21*) and it is not clear to what extent this apparent succession corresponds to a real developmental pattern and to what degree it reflects differences in the sensitivity of the immunocytochemical techniques used.

3.2. Environmental Factors Involved in the Chemical Differentiation of Autonomic Neurons In Vivo

Over the last decade, efforts to elucidate the interactions involved in the early phases of the chemical differentiation of autonomic neuron precursors have been largely devoted to a study of adrenergic cell differentiation, in great part because of the ease with with CA-containing cells can be identified by histofluorescence. As has already been implied, sympathoblast precursors receive the stimulus to acquire their adrenergic properties during the 36 h or so that separate the onset of migration from the aggregation of the primary sympathetic ganglia. The neural tube, notochord, and somites (structures lining or bordering the migratory paths) have all been considered potential candidates for a role in the initiation of adrenergic differentiation in crest cells, and results from a variety of experimental studies have in fact implicated all three tissues. When various combinations of tissues were grafted on the chorioallantoic membrane of the chick, adrenergic cells were found to develop from the neural crest only if the neural tube and somitic mesenchyme were both present (13). Experiments involving organ culture extended these results; it was concluded that somitic mesenchyme alone is able to promote the differentiation of sympathoblasts provided that it has received a prior conditioning stimulus from the neural tube and notochord (73). The role of the notochord was revealed by studies in which aneural gut mesenchyme (taken from a 5-d chick embryo, i.e., before the arrival of neural crest-derived enteric neuron precursors) was cultured with neural primordium (tube plus associated crest) on the chorioallantoic membrane. Crest cells migrated into the gut and developed into enteric ganglia in which adrenergic cells normally never appeared. They did do so, however, if the notochord was included in the tissue association (12,95).

The complexity of the interactions between crest, neural tube, notocord, and somites has been strikingly demonstrated by the results of experiments designed to examine the effects of surgical excision of the neural tube and/or notochord on peripheral neurogenesis in vivo (96). If both axial organs were ablated just after migration of the neural crest cells, the somitic cells degenerated as did the neural crest cells associated with the somitic mesenchyme at that time. The degeneration of both somite and neural crest could be prevented by the presence of the neural tube *or* the notochord: In either case, adrenergic ganglion-like structures developed. Qualitatively at least, neural tube and notochord seem to play analogous roles in adrenergic differentiation and both structures may influence crest cells only indirectly via a common effect on somitic cells.

In mammalian systems, pre- and postganglionic influences and humoral factors play an important part in the maturation of young sympathetic neurons once they have received their initial adrenergic stimulus (4). The examples of chemical plasticity quoted earlier suggest that the adrenergic phenotype is relatively unstable and may require continual reinforcement. Depolarization (101), calcium ions (102), and glucocorticoids (26,65) have all been shown to aid in maintaining sympathetic neurons in an adrenergic state. It is reasonable to suppose that similar mechanisms operate in the avian embryo, although to date virtually no attention has been paid to this aspect of adrenergic differentiation in birds.

Information available on cholinergic neuron development is scanty. Although evidence presented in the previous section suggests that enteric neuroblasts do not require an inductive stimulus in order to trigger their chemical differentiation, it is most probable that target-derived factors are required for the stimulation and consolidation of their initially feeble expression of cholinergic properties. Gut mesenchyme is an environment that allows excellent cholinergic differentiation of quail neural crest cells on the chorioallantoic membrane (50,91). It should be noted that enteric neuron differentiation within the splanchnic mesenchyme in vivo appears to be a remarkably autonomous process, occurring apparently normally in the total absence of neural tube and notochord (96).

Very little is known concerning the tissue origin of factors responsible for the development of neuropeptide-producing neurons. The chronological and topical coincidence of somatostatin- and CA-containing cells during the ontogeny of the sympathetic nervous system (28,61) suggests a priori that the expression of the two phenotypes is triggered by the same environmental factors, and evidence for coregulation has indeed been obtained in vitro (see next section). However, no data obtained from experimentation in vivo are as yet available.

As concerns other neuropeptides, the results of interspecific heterotopic grafts have provided yet another example of the elicitation of latent phenotypes in an appropriate embryonic situation, demonstrating that the tissue environment plays a key role in the differentiation of VIP- and substance P-containing enteric neurons (24). However, the same experiments also indicated that information relating to the chronology of the development of these peptidergic neurons is intrinsic to the crest cells.

3.3. *Neural Crest Differentiation Studied in Culture In Vitro*

Cultures of neural crest are being increasingly used as models for studying chemical and morphological differentiation of peripheral neurons. An in vitro approach should theoretically facilitate access to interactive mechanisms that govern these developmental processes. In particular, it provides a unique opportunity for testing the effects of tissue extracts and other soluble factors. One should bear in mind, however, that a number of hard-to-control variables are also inseparable from culture techniques and, consequently, results obtained in vitro should be interpreted with the necessary prudence. Despite this reservation, studies with cultured crest have confirmed several findings obtained using other experimental systems and have yielded information that would be difficult or impossible to obtain by other means.

The first experiments showed that cholinergic and adrenergic cells can develop in cultured neural crest from the cephalic and trunk regions of the embryo, respectively (14,32). Later, it was confirmed that both these levels of the neural crest are bipotential with respect to their ability to give rise to cholinergic and adrenergic derivatives (23,38,63). Such differentiation occurs in the presence of conventional tissue culture medium supplemented with horse or calf serum. No requirement for exogenous growth factors has ever been noted, although since these would most probably be provided by the complex components of the medium, this fact is not highly conclusive.

The ACh-synthesizing ability that can be detected in crest explants at the time they are put into culture (93) generally increases over the culture period (38,63), but can be modulated by varying the composition of the medium or by coculturing the neural crest cells with various young embryonic tissues (23). Interestingly, CAT activity can be selectively stimulated in cultured neural crest by adding heart-conditioned medium (87). The active constituent is possibly the same as the factor(s) obtained from a variety of nonneuronal cells (including heart) that increases cholinergic properties in cultures of diverse central and peripheral neurons at later developmental stages (30,31,40,64,71,77).

The fact that adrenergic cells can develop in crest cultures grown without notochord or somitic tissue (14,38,63)— or even in the absence of continuous cellular interactions of any kind (86)—appears to contradict evidence from other sources, including data obtained in our laboratory with cultured crest. Using an admittedly different methodology, we found that very little CA was produced by cultures of surgically isolated neural crest unless other types of cells were also present (23); somitic mesenchyme and notochord were quantitatively the most effective inducers of adrenergic differentiation, although other tissues also possessed stimulatory activity. The discrepancies between our results and those of others could be caused by a number of factors, including differences in the developmental stages of the embryos or in the techniques employed for isolating the crest. However, it has recently been shown that the explanation of the disagreement resides principally in the quantity of 10-d chick embryo extract routinely added to the medium. If the concentration, normally 2% in our studies, is increased to 10% (that used by other investigators) or, even more effectively, to 15%, then substantial adrenergic differentiation occurs in the absence of other cell types (34 and C. Ziller, M. Fauquet, J. Smith, and N. M. Le Douarin, in preparation). The requirement for heterologous tissue interactions can therefore be replaced by one or more soluble factors present in the young embryo. Their nature has not been elucidated. They may consist of extracellular matrix material, certain components of which can exert a stimulatory action on CA production (57,89). On the other hand, they are almost certainly not glucocorticoid hormones, which can markedly enhance adrenergic differentiation in neural crest, but only in the presence of other cell types (92).

It is worth emphasizing that adrenergic cells differentiate in cultures of neural crest alone only after a certain delay: CA production can first be detected at 4–6 d of culture (14,63), which, taking into account the age of the crest at the time of explantation (2 d), is 2–4 d later than it should if the in vivo chronology were respected. In fact, division of adrenergic cell precursors for at least 3 d appears to be mandatory for the expression of the phenotype in vitro (39). However, neural crest cells can undergo extensive adrenergic differentiation in culture with a time-course virtually identical to that of sympathoblast development *in situ*. This occurs when "migrating" neural crest taken from a 3-d embryo is cultured in the presence of the sclerotomal component of the somite, a tissue with which it is intimately associated at this stage of development and that constitutes the normal microenvironment for sympathetic neuron precursor differentiation in vivo. In this type of culture, adrenergic properties that are undetectable when the tissue is first explanted appear within 24 h and increase markedly over the next few days (23). Similarly, somatostatin is first observable in these cultures after 2 d, a time-point that corresponds quite well to its appearance in vivo (28), whereas it can be detected in cultures of "premigratory" neural crest alone only after 5 d in vitro (62). Explanation of these differences in the kinetics of development in the two types of system might well provide an important clue to understanding the mechanism of sympathoblast differentiation.

Although not particularly suitable for attempting to unravel the initial steps of the adrenergic differentiation process (the relevant triggering events have almost certainly already occurred when the explants are put into culture), the neural crest/sclerotome system has proved to be a useful in vitro model for examining—and quantifying—some of the effects of environmental factors such as steroid hormones on the early development of sympatho-adrenal precursors (92).

An important point raised by all of these studies on neural crest development in vitro concerns the lack of uniformity of the response of crest cells to a given set of environmental conditions. Only a small percentage of the cells that develop over a period of several days possesses cytochemically demonstrable properties that can be related to neuronal differentiation. Thus, CA-containing cells were estimated to represent approximately 0.2% of the population in 9-d neural crest cultures (63). Although this would be explicable if neuronal and nonneuronal lineages were already distinct in the premigratory crest, as has been suggested (see previous section), the neuronal phenotypes arising in culture also display considerable chemical heterogeneity. In addition to ACh, found in every type of neural crest culture examined (23), and CA and somatostatin, which appear under the conditions described above, serotonin-containing neurons have also been demonstrated in quail neural crest cultures (88). Furthermore, the neurons that develop rapidly in serum-free conditions (104) almost certainly constitute an additional phenotypic class since their neurotransmitter has so far escaped identification.

All this diversity could simply be a reflection of an equivalent number of already committed precursor cell types. Alternatively, it may be suggested that, although bathed in a homogeneous culture medium, all neural crest cells, even if they are potentially plastic with respect to neurotransmitter choice, are not in an identical situation vis-à-vis their neighbors, and that locally different inductive cell–cell interactions could well occur. The true explanation lies probably somewhere between these two extremes. The situation is additionally complicated by the possible coexistence within the same cell of two or more neurotransmitters/neuropeptides that might be coregulated during development. Such is apparently the case for the somatostatin- and CA-related phenotypes that occur together in neuron-like cells differentiating from neural crest in sclerotome-associated cultures and whose expression is modulated in an analogous manner in response to various modifications of the liquid medium (J. García-Arrarás, M. Fauquet, M. Chanconie and J. Smith, *Dev. Biol.*, in press). It has also been argued, although the evidence is admittedly sketchy, that ACh and CA coexist in developing avian neuroblasts (*90*). If such colocalization were the general rule, then the number of neuronal cell types actually arising in culture would be smaller than it appears to be.

4. Conclusions

When the now classical neural primordium transplantation experiments of Le Douarin and her colleagues first provided firm evidence for the pluripotentiality of the neural crest population, it was enticing to confer the same property on individual crest cells by extrapolation. However, in the light of experimental evidence that has accumulated since then—much of it obtained by other types of transplantation experiments in vivo by the same group—it is becoming increasingly likely that, at least in their broad outlines, important developmental options have already been taken by subsets of crest cells when (or even before) they begin to leave the neural tube. This is not to deny that a number of "secondary" choices (that may nonetheless be of prime significance) could be left undecided until a later stage of development. For instance, the acquisition of neurotransmitter-related characteristics may be dictated to a cell already possessing an "autonomic" imprint by appropriate environmental factors encountered before, during, or after ganglion formation. It should be emphasized that this notion is largely inspired by the demonstration of neurotransmitter plasticity in mammalian sympathetic ganglion cells. However, the universality of this process remains unproven and no such plasticity has so far been shown to occur in avian systems. It cannot therefore yet be discounted that cells committed to a particular neurotransmitter-related phenotype are present in the autonomic neuron precursor population in the neural crest.

The detection of predetermined cell lineages in premigratory neural crest is currently the object of much attention in several laboratories and a clearer picture will no doubt emerge presently. However, regardless of the outcome of these investigations the importance of factors extrinsic to the neural crest in the establishment of the definitive autonomic neuronal phenotype remains uncontested. External cues play a key role in the orientation of neural crest cells along specific developmental pathways and, even when migration has ended, crest cell derivatives can remain responsive to such signals for a considerable time. The awakening of dormant autonomic neuronal potentialities in nonneuronal peripheral ganglion cells at a relatively late stage of development is a forceful illustration of this point.

The role of environmental factors will, of course, be different according to whether or not the neural crest proves to be a mosaic of already determined cells: in the first case the environment would exert a selection, and in the second it would impose a program of differentiation. A close study of the mechanism of action of the factors responsible for various aspects of neuronal differentiation could therefore constitute a useful approach to the problem of crest cell determination. Experiments designed to characterize and identify factors influencing the chemical differentiation of neural crest have begun to bring some light to bear on this question.

Although the importance of external cues has been particularly well demonstrated with regard to the choice of neurotransmitter-related properties, much remains to be explained in this domain. If differences in crest cell migration pathways could reasonably account for some cases of phenotypic divergence, it is less immediately clear how diversity arises within a ganglion in which the constituent cells have presumably followed the same route and thus been subjected to the same environmental agents. Sympathetic neurons are a striking example: various categories can be recognized according to the neurotransmitter (19,44) and the combination of neurotransmitter and neuropeptide (58) that they contain. Although this complexity could be considered an argument for the existence of predetermined progenitors, such heterogeneity may alternatively be the result of local intervention of external factors at the level of the developing ganglion. All cells within a ganglion are clearly not in an equivalent topographical situation and may be subject selectively to intra- or extra-ganglionic influences causing them to acquire (or lose) the ability to synthesize certain neurotransmitters or neuropeptides. A discriminatory effect of target cells in maintaining or modifying the chemical properties of neurons that synapse upon them can equally be envisaged.

It is obvious from many of the preceding remarks that the ontogeny of the ANS from the neural crest is still far from being completely understood. Despite the undeniable progress that has been made in the last few years, blank spots remain in a number of important areas. To fill them even partially requires continued application of the approved mixture of inspiration and perspiration.

Acknowledgments

The author wishes to thank Prof. N. M. Le Douarin for many stimulating discussions and Mrs. E. Bourson for typing the manuscript. Financial support was provided by the Centre National de le Recherche Scientifique, Research Grant RO1 DEO 42 57 04 CBY from the US National Institutes of Health and by Basic Research Grant 1-866 from March of Dimes Birth Defects Foundation.

References

1. Allan, I. J. and D. F. Newgreen (1977) Catecholamine accumulation in neural crest cells and the primary sympathetic chain. *Am. J. Anat.* **149**, 413–421.
2. Ayer-Le Lièvre, C. S. and N. M. Le Douarin (1982) The early development of cranial sensory ganglia and the potentialities of their component cells studied in quail-chick chimeras. *Dev. Biol.* **94**, 291–310.
3. Barald, K. D. (1982) Monoclonal Antibodies to Embryonic Neurons: Cell-Specific Markers for Chick Ciliary Ganglion, In *Neuronal Development* (N. Spitzer, ed.) Plenum, New York, pp. 110–119.
4. Black, I. B. (1982) Stages of neurotransmitter development in autonomic neurons. *Science* **215**, 1198–1204.
5. Bronner-Fraser, M., M. Sieber-Blum, and A. M. Cohen (1980) Clonal analysis of the avian neural crest: migration and matura-

tion of mixed neural crest clones injected into host chicken embryos. *J. Comp. Neurol.* **193**, 423–434.

6. Chalazonitis, A. and G. D. Fischbach (1980) Elevated potassium induces morphological differentiation of dorsal root ganglionic neurons in dissociated cell culture. *Dev. Biol.* **78**, 173–183.

7. Chiappinelli, V., E. Giacobini, G. Pilar, and H. Uchimura (1976) Induction of cholinergic enzymes in chick ciliary ganglion and in muscle cells during synapse formation. *J. Physiol.* **257**, 749–766.

8. Ciment, G. and J. A. Weston (1982) Early appearance in neural crest and crest-derived cells of an antigenic determinant present in avian neurons. *Dev. Biol.* **93**, 355–367.

9. Cochard, P. and P. Coltey (1983) Cholinergic traits in the neural crest: acetylcholinesterase in crest cells of the chick embryo. *Dev. Biol.* **98**, 221–238.

10. Cochard, P., M. Goldstein, and I. B. Black (1978) Ontogenetic appearance and disappearance of tyrosine hydroxylase and catecholamines in the rat embryo. *Proc. Natl. Acad. Sci. USA* **75**, 2986–2990.

11. Cochard, P., M. Goldstein, and I. B. Black (1979) Initial development of the noradrenergic phenotype in autonomic neuroblasts of the rat embryo in vivo. *Dev. Biol.* **71**, 100–114.

12. Cochard, P. and N. M. Le Douarin (1982) Development of intrinsic innervation of the gut. *Scand. J. Gastroenterol.* **17**, Suppl. 71, 1–14.

13. Cohen, A. M. (1972) Factors directing the expression of sympathetic nerve traits in cells of neural crest origin. *J. Exp. Zool.* **179**, 167–182.

14. Cohen, A. M. (1977) Independent expression of the adrenergic phenotype by neural crest cells in vitro. *Proc. Natl. Acad. Sci. USA* **74**, 2899–2903.

15. Cohen, A. M. and I. B. Konigsberg (1975) A clonal approach to the problem of neural crest determination. *Dev. Biol.* **46**, 262–280.

16. Drews, U. (1975) Cholinesterase in embryonic development. *Prog. Histochem. Cytochem.* **7**, 1–52.

17. Duband, J. L. and J. P. Thiery (1982) Distribution of fibronectin in the early phase of avian cephalic neural crest migration. *Dev. Biol.* **93**, 308–323.

18. Dupin, E. (1984) Cell division in the ciliary ganglion of quail embryos *in situ* and after back-transplantation into the neural crest migration pathways of chick embryos. *Dev. Biol.* **105**, 288–299.

19. Edgar, D., Y.-A. Barde, and H. Thoenen (1981) Subpopulations of cultured chick sympathetic neurons differ in their requirement for survival factors. *Nature* **289**, 294–295.

20. Enemar, A., B. Falck, and R. Håkanson (1965) Observations on the appearance of norepinephrine in the sympathetic nervous system of the chick embryo. *Dev. Biol.* **11**, 268–283.

21. Epstein, M. L., J. Hudis, and J. L. Dahl (1983) The development of peptidergic neurons in the foregut of the chick. *J. Neurosci.* **3,** 2431–2447.

22. Erickson, C.A., K. W. Tosney, and J. A. Weston (1980) Analysis of migratory behavior of neural crest and fibroblastic cells in embryonic tissues. *Dev. Biol.* **77,** 142–156.

23. Fauquet, M., J. Smith, C. Ziller, and N. M. Le Douarin (1981) Differentiation of autonomic neuron precursors in vitro: cholinergic and adrenergic traits in cultured neural crest cells. *J. Neurosci.* **1,** 478–492.

24. Fontaine-Pérus, J., M. Chanconie, and N. M. Le Douarin (1982) Differentiation of peptidergic neurons in quail–chick chimeric embryos. *Cell Differentiation* **11,** 183–193.

25. Fontaine-Pérus, J., M. Chanconie, J. M. Polak, and N. M. Le Douarin (1981) Origin and development of VIP- and substance P-containing neurons in the embryonic avian gut. *Histochemistry* **71,** 313–323.

26. Fukada, K. (1980) Hormonal control of neurotransmitter choice in sympathetic neuron cultures. *Nature* **287,** 553–555.

27. Furshpan, E. J., P. R. MacLeish, P. H. O'Lague, and D. D. Potter (1976) Chemical transmission between rat sympathetic neurons and cardiac myocytes developing in microcultures: evidence for cholinergic, adrenergic, and dual-function neurons. *Proc. Natl. Acad. Sci. USA* **73,** 4225–4229.

28. García-Arrarás, J., M. Chanconie, and J. Fontaine-Pérus (1984) In vivo and in vitro development of somatostatin-like immunoreactivity in the peripheral nervous system of quail embryos. *J. Neurosci.* **4,** 1549–1558.

29. Gershon, M. D., G. Teitelman, and T. P. Rothman (1981) Development of enteric neurons from nonrecognizable precursor cells. *CIBA Foundation Symp.* **83,** Pitman Medical, London, pp. 51–69.

30. Giller, E. L., B. K. Schrier, A. Shainberg, H. R. Fisk, and P. G. Nelson (1973) Choline acetyltransferase activity is increased in combined cultures of spinal cord and muscle cells from mice. *Science* **182,** 588–589.

31. Godfrey, E. W., B. K. Schrier, and P. G. Nelson (1980) Source and target specificities of a conditioned medium factor that increases choline acetyltransferase activity in cultured spinal cord cells. *Dev. Biol.* **77,** 100–119

32. Greenberg, J. H. and B. K. Schrier (1972) Development of choline acetyltransferase activity in chick cranial neural crest cells in culture. *Dev. Biol.* **61,** 86–93.

33. Hörstadius, S. (1950) *The Neural Crest; Its Properties and Derivatives in the Light of Experimental Research.* Oxford University Press, London.

34. Howard, M. J., M. Bronner-Fraser, and T. Tomosky-Sykes (1982) Adrenergic differentiation of neural crest cells in vitro without the neural tube. *Soc. Neurosci. Abstr.* **8,** 257.
35. Johnson, M. I., C. D. Ross, M. Meyers, E. L. Spitznagel, and R. P. Bunge (1980) Morphological and biochemical studies on the development of cholinergic properties in cultured sympathetic neurons. I. Correlative changes in choline acetyltransferase and synaptic vesicle cytochemistry. *J. Cell Biol.* **84,** 680–691.
36. Jonakait, G. M., K. A. Markey, M. Goldstein, and I. B. Black (1984) Transient expression of selected catecholaminergic traits in cranial sensory and dorsal root ganglia of the embryonic rat. *Dev. Biol.* **101,** 51–60.
37. Jonakait, G. M., J. M. Wolf, P. Cochard, M. Goldstein, and I. B. Black (1979) Selective loss of noradrenergic phenotype in neuroblasts of the rat embryo. *Proc. Natl. Acad. Sci. USA* **76,** 4683–4686.
38. Kahn, C. R., J. T. Coyle, and A. M. Cohen (1980) Head and trunk neural crest in vitro: autonomic neuron differentiation. *Dev. Biol.* **77,** 340–348,
39. Kahn, C. R. and M. Sieber-Blum (1983) Cultured quail neural crest cells attain competence for terminal differentiation into melanocytes before competence to terminal differentiation into adrenergic neurons. *Dev. Biol.* **95,** 232–238.
40. Kato, A. C. and M. J. Rey (1982) Chick ciliary ganglion in dissociated cell culture. I. Cholinergic properties. *Dev. Biol.* **94,** 121–130.
41. Katz, D. M., K. A. Markey, M. Goldstein, and I. B. Black (1983) Expression of catecholaminergic characters by primary sensory neurons in the normal adult rat in vivo. *Proc. Natl. Acad. Sci. USA* **80,** 3526–3530.
42. Landis, S. C. (1976) Rat sympathetic neurons and cardiac myocytes developing in microcultures: correlation of the fine structure of endings with neurotransmitter function in single neurons. *Proc. Natl. Acad. Sci. USA* **73,** 4220–4224.
43. Landis, S. C. (1980) Developmental changes in the neurotransmitter properties of dissociated sympathetic neurons: a cytochemical study of the effects of medium. *Dev. Biol.* **77,** 349–361.
44. Landis, S. C. and D. Keefe (1983) Evidence of neurotransmitter plasticity in vivo: developmental changes in properties of cholinergic sympathetic neurons. *Dev. Biol.* **98,** 349–372.
45. Le Douarin, N. M. (1969) Particularitiés du noyau interphasique chez la caille japonaise (Coturnix coturnix japonica). Utilisation de ces particularités comme "marquage biologique" dans les recherches sur les interactions tissulaires et les migrations cellulaires au cours de l'ontogenèse. *Bull. Biol. Fr. Belg.* **103,** 435–452.

46. Le Douarin, N. M. (1973) A biological cell labeling technique and its use in experimental embryology. *Dev. Biol.* **30**, 217–222.
47. Le Douarin, N. M. (1982) *The Neural Crest.* Cambridge University Press, London.
48. Le Douarin, N. M. (1984) A Model for Cell Line Divergence in the Ontogeny of the Peripheral Nervous System, in *Cellular and Molecular Biology of Neuronal Development.* (I. Black, ed.) Plenum, New York, pp. 3–28.
49. Le Douarin, N. M., C. S. Le Lièvre, G. Schweizer, and C. M. Ziller (1979) An Analysis of Cell Line Segregation in the Neural Crest, in *Cell Lineage, Stem Cells, and Cell Determination.* INSERM Symp. **10** (N. M. Le Douarin, ed.) Elsevier/North Holland, Amsterdam, pp. 353–365.
50. Le Douarin, N. M., D. Renaud, M. A. Teillet, and G. H. Le Douarin (1975) Cholinergic differentiation of presumptive adrenergic neuroblasts in interspecific chimeras after heterotopic transplantation. *Proc. Natl. Acad. Sci. USA* **72**, 728–732.
51. Le Douarin, N. M., J. Smith, and C. Le Lièvre (1981) From the neural crest to the ganglia of the peripheral nervous system. *Ann. Rev. Physiol.* **43**, 653–671.
52. Le Douarin, N. M. and M. A. Teillet (1971) Localisation, par la méthode des greffes interspécifiques du territoire neural dont dérivent les cellules adrénales surrénaliennes chez l'embryon d'oiseau. *C. R. Acad. Sci. Paris* **272**, 481–484.
53. Le Douarin, N. M. and M. A. Teillet (1973) The migration of neural crest cells to the wall of the digestive tract in avian embryo. *J. Embryol. Exp. Morphol.* **30**, 31–48.
54. Le Douarin, N. M. and M. A. Teillet (1974) Experimental analysis of the migration and differentiation of neuroblasts of the autonomic nervous system and of neurectodermal mesenchymal derivatives, using a biological cell marking technique. *Dev. Biol.* **41**, 162–184.
55. Le Douarin, N. M., M. A. Teillet, C. Ziller, and J. Smith (1978) Adrenergic differentiation of cells of the cholinergic ciliary and Remak ganglia in avian embryo after in vivo transplantation. *Proc. Natl. Acad. Sci. USA* **75**, 2030–2034.
56. Le Lièvre, C. S., G. Schweizer, C. M. Ziller, and N. M. Le Douarin (1980) Restrictions of developmental capabilities in neural crest cell derivatives as tested by in vivo transplantation experiments. *Dev. Biol.* **77**, 362–378.
57. Loring, J., B. Glimelius, and J. A. Weston (1982) Extracellular matrix materials influence quail neural crest cell differentiation in vitro. *Dev. Biol.* **90**, 165–174.
58. Lundberg, J. M., T. Hökfelt, A. Änggård, L. Terenius, R. Elde, K. Markey, M. Goldstein, and J. Kimmel (1982) Organizational

principles in the peripheral sympathetic nervous system: subdivision by coexisting peptides (somatostatin-, avian pancreatic polypeptide-, and vasoactive intestinal polypeptide-like immunoreactive materials). *Proc. Natl. Acad. Sci. USA* **79,** 1303–1307.

59. Mains, R. E. and P. H. Patterson (1973) Primary cultures of dissociated sympathetic neurons. I. Establishment of long-term growth in culture and studies of differentiated properties. *J. Cell Biol.* **59,** 329–345.

60. Mains, R. E. and P. H. Patterson (1973) Primary cultures of dissociated sympathetic neurons. II. Initial studies on catecholamine metabolism. *J. Cell Biol.* **59,** 346–360.

61. Maxwell, G. D., P. D. Sietz, and P. H. Chenard (1984) Development of somatostatin-like immunoreactivity in embryonic sympathetic ganglia. *J. Neurosci.* **4,** 576–584.

62. Maxwell, G. D., P. D. Sietz, and S. Jean (1984) Somatostatin-like immunoreactivity is expressed in neural crest cultures. *Dev. Biol.* **101,** 357–366.

63. Maxwell, G. F., P. D. Sietz, and C.E. Rafford (1982) Synthesis and accumulation of putative neurotransmitters by cultured neural crest cells. *J. Neurosci.* **2,** 879–888.

64. McLennan, I. S. and I. A. Hendry (1980) Influence of cardiac extracts on cultured ciliary ganglia. *Dev. Neurosci.* **3,** 1–10.

65. McLennan, I. S., C. E. Hill, and I. A. Hendry (1980) Glucocorticoids modulate transmitter choice in developing superior cervical ganglion. *Nature* **283,** 206–207.

66. Narayanan, C. H. and Y. Narayanan (1978) Determination of the embryonic origin of the mesencephalic nucleus of the trigeminal nerve in birds. *J. Embryol. Exp. Morphol.* **43,** 85–105.

67. Narayanan, C. H. and Y. Narayanan (1978) On the origin of the ciliary ganglion in birds studied by the method of interspecific transplantation of embryonic brain regions between quail and chick. *J. Embryol. Exp. Morph.* **47,** 137–148.

68. Narayanan, C. H. and Y. Narayanan (1980) Neural crest and placodal contributions in the development of the glossopharyngeal–vagal complex in the chick. *Anat. Rec.* **196,** 71–82.

69. Newgreen, D. F., I. Jahnke, I. J. Allan, and I. L. Gibbins (1980) Differentiation of sympathetic and enteric neurons of the fowl embryo in grafts to the chorioallantoic membrane. *Cell Tiss. Res.* **208,** 1–20.

70. Newgreen, D. F. and R. O. Jones (1975) Differentiation in vitro of sympathetic cells from chick embryo sensory ganglia. *J. Embryol. Exp. Morphol.* **33,** 43–56.

71. Nishi, R. and D. K. Berg (1981) Two components from eye tissue

that differentially stimulate the growth and development of ciliary ganglion neurons in cell culture. *J. Neurosci.* **1**, 505–513.

72. Noden, D. M. (1978) The control of avian cephalic neural crest cytodifferentiation. II. Neural tissues. *Dev. Biol.* **67**, 313–329.

73. Norr, S. C. (1973) In vitro analysis of sympathetic neuron differentiation from chick neural crest cells. *Dev. Biol.* **34**, 16–38.

74. Ogawa, M., T. Ishikawa, and A. Irimajiri (1984) Adrenal chromaffin cells form functional cholinergic synapses in culture. *Nature* **307**, 66–68.

75. Patterson, P. H. (1978) Environmental determination of autonomic neurotransmitter functions. *Ann. Rev. Neurosci.* **1**, 1–17.

76. Patterson, P. H. and L. L. Y. Chun (1974) The influence of nonneuronal cells on catecholamine and acetylcholine synthesis and accumulation in cultures of dissociated neurons. *Proc. Natl. Acad. Sci. USA* **71**, 3607–3610.

77. Patterson, P. H. and L. L. Y. Chun (1977) The induction of acetylcholine synthesis in primary cultures of dissociated rat sympathetic neurons. I. Effects of conditioned medium. *Dev. Biol.* **56**, 263–280.

78. Patterson, P. H. and L. L. Y. Chun (1977) The induction of acetylcholine synthesis in primary cultures of dissociated rat sympathetic neurons. II. Developmental aspects. *Dev. Biol.* **60**, 473–481.

79. Potter, D. D., S. C. Landis, and E. J. Furshpan (1981) Adrenergic–cholinergic dual function in cultured sympathetic neurons of the rat. *CIBA Foundation Symp.* **83**, Pitman Medical, London, pp. 123–138.

80. Price, J. and A. W. Mudge (1983) A subpopulation of rat dorsal root ganglion neurons is catecholaminergic. *Nature* **304**, 241–243.

81. Reichardt, L. F. and P. H. Patterson (1977) Neurotransmitter synthesis and uptake by individual rat sympathetic neurons developing in microcultures. *Nature* **270**, 147–151.

82. Rohrer, H. and I. Sommer (1983) Simultaneous expression of neuronal and glial properties by chick ciliary ganglion cells during development. *J. Neurosci.* **3**, 1683–1693.

83. Rothman, R. P., M. D. Gershon, and H. Holtzer (1978) The relationship of cell division to the acquisition of adrenergic characteristics by developing sympathetic ganglion cell precursors. *Dev. Biol.* **65**, 322–341.

84. Saffrey, M. J., J. M. Polak, and G. Burnstock (1982) Distribution of vasoactive intestinal polypeptide-, substance P-, enkephalin-, and neurotensin-like immunoreactive nerves in the chicken gut during development. *Neurosci.* **7**, 279–293.

85. Schweizer, G., C. Ayer-Le Lièvre and N. M. Le Douarin (1983) Restrictions of developmental capacities in the dorsal root ganglia during the course of development. *Cell Differentiation* **13**, 191–200.

86. Sieber-Blum, M. and A. M. Cohen (1980) Clonal analysis of quail neural crest cells: they are pluripotent and differentiate in vitro in the absence of noncrest cells. *Dev. Biol.* **80,** 96–106.

87. Sieber-Blum, M. and C. R. Kahn (1982) Suppression of catecholamine and melanin synthesis and promotion of cholinergic differentiation of quail neural crest cells by heart cell conditioned medium. *Stem Cells* **2,** 344–353.

88. Sieber-Blum, M., W. Reed, and H. G. W. Lidov (1983) Serotoninergic differentiation of quail neural crest cells in vitro. *Dev. Biol.* **99,** 352–359.

89. Sieber-Blum, M., F. Sieber, and K. M. Yamada (1981) Cellular fibronectin promotes adrenergic differentiation of quail neural crest cells in vitro. *Exp. Cell Res.* **133,** 285–295.

90. Smith, J. (1983) Early Events in Autonomic Neuron Development: The Cholinergic/Adrenergic Choice, in *Dale's Principle and Communication Between Neurons.* (N. N. Osborne, ed.) Pergamon, Oxford, pp. 143–159.

91. Smith, J., P. Cochard, and N. M. Le Douarin (1977) Development of choline acetyltransferase and cholinesterase activities in enteric ganglia derived from presumptive adrenergic and cholinergic levels of the neural crest. *Cell Differentiation* **6,** 199–216.

92. Smith, J. and M. Fauquet (1984) Glucocorticoids stimulate adrenergic differentiation in cultures of migrating and premigratory neural crest. *J. Neurosci.* **4,** 2160–2172.

93. Smith, J., M. Fauquet, C. Ziller, and N. M. Le Douarin (1979) Acetylcholine synthesis by mesencephalic neural crest in the process of migration in vivo. *Nature* **282,** 852–855.

94. Teillet, M. A. (1978) Evolution of the lumbo-sacral neural crest in the avian embryo: origin and differentiation of the ganglionated nerve of Remak studied in interspecific quail-chick chimeras. *W. Roux's Arch. Dev. Biol.* **184,** 251–268.

95. Teillet, M. A., P. Cochard, and N. M. Le Douarin (1978) Relative roles of the mesenchymal tissues and of the complex neural tube–notochord on the expression of adrenergic metabolism in neural crest cells. *Zoon* **6,** 115–122.

96. Teillet, M. A. and N. M. Le Douarin (1983) Consequences of neural tube and notochord excision on the development of the peripheral nervous system in the chick embryo. *Dev. Biol.* **98,** 192–211.

97. Teitelman, G., H. Baker, T. H. Joh, and D. J. Reis (1979) Appearance of catecholamine-synthesizing enzymes during development of rat sympathetic nervous system: possible role of tissue environment. *Proc. Natl. Acad. Sci. USA* **76,** 509–513.

98. Thiery, J. P., J. L. Duband, and A. Delouvée (1982) Pathways and mechanisms of avian trunk neural crest cell migration and localization. *Dev. Biol.* **93,** 324–343.

99. Vincent, M., J. L. Duband, and J. P. Thiery (1983) A cell surface determinant expressed early on migrating avian neural crest cells. *Dev. Brain Res.* **9**, 235–238.
100. Vincent, M. and J. P. Thiery (1984) A cell surface marker for neural crest and placodal cells: further evolution in peripheral and central nervous system. *Dev. Biol.* **103**, 468–481.
101. Walicke, P. A., R. B. Campenot, and P. H. Patterson (1977) Determination of transmitter function by neuronal activity. *Proc. Natl. Acad. Sci. USA* **77**, 5767–5771.
102. Walicke, P. A. and P. H. Patterson (1981) On the role of Ca^{2+} in the transmitter choice made by cultured sympathetic neurons. *J. Neurosci.* **1**, 343–350.
103. Wolinsky, E. and P. H. Patterson (1983) Tyrosine hydroxylase activity decreases with induction of cholinergic properties in cultured sympathetic neurons. *J. Neurosci.* **3**, 1495–1500.
104. Ziller, C., E. Dupin, P. Brazeau, D. Paulin, and N. M. Le Douarin (1983) Early segregation of a neuronal precursor cell line in the neural crest as revealed by culture in a chemically defined medium. *Cell* **32**, 627–638.

Chapter 2

Development of Peripheral Parasympathetic Neurons and Synapses

Ezio Giacobini

1. Introduction

The present chapter represents an attempt to summarize our current notion of the process of structural, functional, and biochemical differentiation and development of peripheral parasympathetic neurons and synapses. The relative simplicity of the peripheral nervous system (PNS), as compared to the central nervous system (CNS), and a vast body of information derived from classical neuroembryology, largely from the chick embryo, makes autonomic ganglia and their synapses particularly suitable for developmental analysis. The access to quantitative microchemistry, single cell recording, iontophoresis, immunohistochemistry, and monoclonal antibodies has facilitated this analysis and brought about new and detailed information. New findings are available in relation to any stage of development, starting from the neural crest and the first location of precursor cells, to the late period of adult development and aging. For a general review of the subject, refer to articles that have appeared on neurotransmitter biosynthesis during development (5–7,37–39, 45,48,60,74,79,117); on chemical differentiation in vitro (3,11,13, 57,102,103,106); and on synaptogenesis *in situ* (4,26,37–39,42,43, 45,47,48,50,76). In addition, an extensive review on the development of the autonomic nervous system has appeared as a publication of the Ciba Foundation Symposium 1983 (30) and recent chapters on this subject are found in the Handbook of Neurochemistry (48) and in refs. 46 and 47.

2. Early Chemical Differentiation of Autonomic Neurons and Synapses In Vivo

2.1. Cholinergic Synapses and Neurons in Sympathetic Ganglia

The mature autonomic system contains diverse populations of neurons that can be characterized biochemically by means of their neurotransmitter synthesis, metabolism, and release. The factors responsible for generating such a large neuronal pool are unknown. Several mechanisms, including influence of the microenvironment, have been proposed to play a decisive role in the determination of the neurotransmitter phenotypic expression both in vitro (14,101) and in vivo (20,75). Several methodological approaches have been used in order to study the biochemical differentiation of the autonomic nervous system. Because of the large diversity of tissue components, the ideal method would be one that could measure various biochemical changes at a cellular level. Adequate sensitivity and reliable methods for single cell analysis are available (37,40,41); however, in most cases it is difficult to isolate the sample from the primitive embryonic nervous system, particularly during the migration phase. Once the neurons have reached their primary location they can be analyzed as a group or even as single cells using microquantitative methods (37,40,41). At earlier stages, within the neural crest and during migration, biochemical changes can be more easily observed by using nonquantitative histochemical approaches such as immunohistochemistry, fluorescence, monoclonal antibodies, and autoradiography. Because of particular technical problems, little is known about levels of neurotransmitters, enzyme activity, and metabolism, in general, in neural crest components. However, we know that neuronal derivatives of the neural crest express their phenotypic characters early in development and long before synaptogenesis is completed. Noradrenergic characters such as tyrosine hydroxylase (TH), dopamine β-hydroxylase (DBH), and phenylethanolamine-transferase (PNMT) activity have been demonstrated in primordia of rat sympathetic ganglia (*19*) using immunocytochemistry. The appearance of catecholamine (CA) transmitters has been demonstrated by histofluorescence (*19*). This work has shown that TH, DBH, and CA appear simultaneously in the sympathetic ganglia primordia and transiently in a population of neuroblasts in the gut. On the other hand, TH, DBH, and CA have not been detected yet in the neural crest, the migrating crest cells, or any other embryonic structure at those early stages (*19*). In the rat, the appearance of these noradrenergic characters is evident in neuroblasts within the sympathetic primordia and in the gut not earlier than 11.5 d of gestation. The presence of intense CA fluorescence suggests that storage, as well as uptake mechanisms, may also be present at this time.

Using whole avian embryos and quantitative techniques, Ignarro and Shidemen (*65*) have shown that noradrenergic enzyme activities appear sequentially between d 4 and 6. However, whole chick embryos are able to convert [³H]tyrosine to L-Dopa on the first day of incubation (di), to dopamine on d 2, and to norepinephrine (NE) on d 4. As endogenous NE and epinephrine (E) are initially detected on d 3, it is possible that the early embryos obtain these CA from extra-embryonic sources, such as the yolk of fertilized eggs. On the other hand, Kirby and Gilmore (*70*) suggested that chick embryo notochord can synthesize and store CA. In the chicken heart, TH, dopadecarboxylase (DDC), DBH, and PNMT were measured at 1, 2, 4, and 6 di, respectively, whereas L-Dopa and dopamine were first detected at 4 and 6 di (*65*). In primordia of sympathetic ganglia, NE is first detected by flourescence at 3 di (*31*); however, even before the formation of recognizable ganglia, an uptake of [³H]NE can be detected in individual cells near the dorsal aorta (*108*).

Expression of adrenergic phenotype is regulated differently from the noradrenergic phenotype because PNMT is not detectable in developing neuroblasts or adrenal medullary cells when TH or DBH is already present (*11*). Apparently the development of PNMT requires a glucocorticoid stimulus (*104,116*).

In cultures of premigratory head and trunk neural crest cells, activities of both adrenergic (DBH) and cholinergic (cholineacetyltransferase, ChAc) neurons were detected by Kahn et al. (*69*). However, although DBH activity was detected at all ages examined in vitro, ChAc activity was first detected only after 5 d in vitro. Whether the noradrenergic and cholinergic phenotype is already determined prior to migration of the crest cells or whether it is modulated during migration by microenvironmental factors remains to be demonstrated. In avian embryos, Cohen (*20*) showed that the initial appearance of CA histofluorescence is influenced by the interaction with mesenchymal factors and, therefore, ventral crest migration would be necessary for phenotypic expression. On the other hand, the parallel rise in DBH and ChAc activity found in vitro (*69*) suggests that a separate cholinergic population of cells may already exist in the primary crest culture. Preliminary results in our laboratory suggest the presence of acetylcholine (ACh) in early neural crest structures (Giacobini et al., unpublished). It seems, therefore, that the crest neuroblast is not fully predetermined with regard to the initial expression of neurotransmitter, and may be influenced in its "choice" by environmental signals that are still to be characterized (*see* Fig. 1). The magnitude and effect of these environmental signals are not yet clear. It is also unclear whether such environmental signals directly influence and determine the neuronal phenotypic expression or simply select a predetermined population of neurons that already exhibit noradrenergic or cholinergic properties. As discussed in the next section, some crest populations may actually possess cholinergic characteristics at the earliest stages. Finally, it should be pointed out that noradrenergic and cholinergic characteristics such as synthesis, uptake, storage, and so on, are not all expressed at one time and may be regulated differently (*62,85,86*). This concept is illustrated by the findings in cholinergic neurons discussed in the following sections and summarized in Fig. 1.

2.1.1. Characterization of Ganglionic Subpopulations of Parasympathetic Neurons

A relationship remains to the established between subpopulations of neurons defined by morphological and biochemical criteria in vivo and their requirements of specific factors to

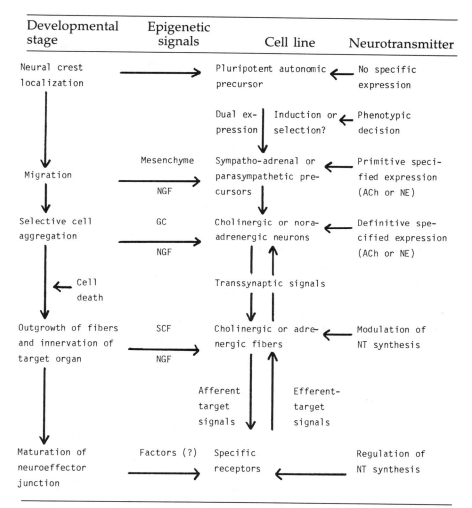

Fig. 1. Schematic representation of neurotransmitter development of autonomic ganglion cells. NT = neurotransmitter; NGF = nerve growth factor; GC = glucocorticoids; SCF = support cell (glial?) factors.

promote differentiation under in vitro conditions. Levi-Montalcini and Hamburger (*81*) could not detect any effects on chick embryo ciliary ganglia when NGF-producing mouse sarcomas were implanted into the embryo. When ciliary ganglia (8-d stage) are cultured in NGF-containing medium, no neurite outgrowth occurs (*57*). Edgar et al. (*28*) attempted to characterize two

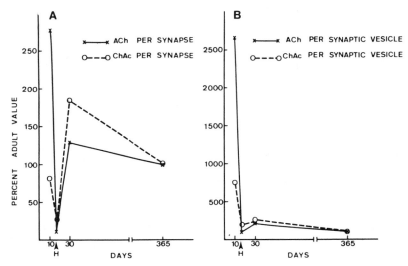

Fig. 2. Amounts of acetylcholine (ACh) and choline acetyltransferase (ChAc) per synapse and per synaptic vesicle. A, the amounts of ACh and ChAc per synapse. B, the amount of ACh and ChAc per synaptic vesicle. The values were derived by dividing the amounts of ACh (82,84) and ChAc (27,87) per whole ganglion by the total number of synapses per segment or the total number of clear synaptic vesicles in synaptic boutons per segment (modified from ref. 64).

major subpopulations of 12-d old neurons of chick paravertebral ganglia that survived in culture in response to either HCM (heart cells conditioned medium) or NGF (nerve growth factor). They showed that HCM- and NGF-responsive neurons differ in their contents of TH and ChAc activities necessary for aminergic and cholinergic expression, respectively. It is difficult to say whether the high ChAc levels found in HCM-responsive neurons represent a selection "in culture" of those cells with high ChAc activity in vivo or whether this is a product of induction by HCM as shown for neonatal rat sympathetic neurons (101). Edgar et al. (28) demonstrated the presence in 12-d in ovo chick sympathetic ganglia of neuroblasts capable of synthesizing ACh in tissue culture, and have suggested that the low levels of ChAC detected in vivo at 12 d in ovo may reflect the presence of cholinergic neuroblasts, rather than activity in presynaptic endings. To examine this hypothesis, Hruschak et al. (64) studies this pattern in vivo in chick sympathetic ganglia. They calculated the amounts of ACh and the activity of ChAC per synapse as functions of age (see Fig.

2A). Both curves are biphasic; high initial levels per synapse are found at 10 d *in ovo*, but the levels are dramatically lower 1 d after hatching. Subsequent development sees an increase again of both these parameters. The early high level of ACh and ChAc per synapse, and their subsequent fall, are consistent with the early presence and subsequent death of cholinergic neuroblasts. The increase in these parameters after 1 d following hatching probably reflects the maturation of synapses, as discussed below.

In electron micrographs, we saw substantial increases in the number of synaptic vesicles per synapse profile with age, and a particularly marked increase in the number of vesicles per synapse profile 1–30 d after hatching. The increase observed in ACh and ChAc per synapse after 1 d following hatching might reflect increases with age in the number of synaptic vesicles per synapse; consequently, we also calculated ACh content and ChAc activity per synaptic vesicle (*see* Fig. 28). Both parameters are extremely high at 10 d *in ovo*, and drop to near adult levels by 1 d after hatching.

The high initial value for the ACh-to-vesicle ratio could reflect the presence during development of a presynaptic, nonvesicular pool of acetylcholine; however, it seems most likely, in view of the data of Edgar et al. (*28*), that the relatively high levels of ACh and ChAc per synaptic vesicle at 10 d *in ovo* reflect the activity of cholinergic neuroblasts within the ganglia. If this hypothesis is correct, then the precipitous drop in both parameters (per synaptic vesicle) between 10 d *in ovo* 1 d after hatching, and their relative stability during subsequent development, suggest the death of the hypothesized cholinergic neuroblasts or the suppression of their cholinergic properties before hatching. In addition, in the same ganglia NE levels and TH and DBH activity show a peak at around 7–9 di (*see* Fig. 3A) that is followed by a decrease and a simultaneous and gradual increase in ACh levels and ChAc activity (*see* Fig. 3B) (*42*). This is a time when cholinergic synapses are very scanty, which also suggests a reduction in the adrenergic population and the appearance of a small cholinergic group of neurons (*42,64*).

Our observations in vivo are consistent with the hypothesis that specific growth factors such as HCM, NGF, or others may selectively promote survival and differentiation of distinct subpopulations of embryonic sympathetic neurons that differ in their storage, synthesis, and release of neurotransmitters. Johnson et al. (*67,68*) have shown that under certain culture conditions, neo-

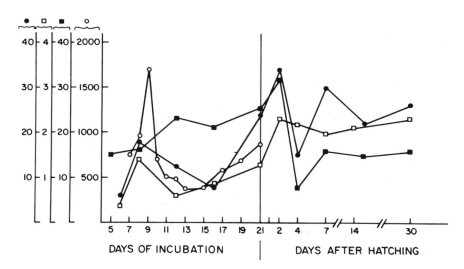

Fig. 3A.   Developmental pattern of NE, DBH, TH, and DDC activity in chicken sympathetic ganglia. $\bigcirc$ = NE, ng/mg protein (66); $\bullet$ = DBH, pmol octopamine/ug, protein/h; $\square$ = TH, pmol CO_2/μg dry weight/h; $\blacksquare$ = DDC, pmol CO_2/μg protein/h. Means and standard errors are not reported (modified from ref. 37).

natal rat superior cervical ganglion neurons paradoxically display certain cholinergic functions, such as hexamethonium-sensitive synaptic contacts and accumulation of ChAc. The specific activity of ChAc falls in cell explants during the first 3–4 d in culture and increases during the next 30 d. The authors conclude that these cultures contain one population of neurons that is initially adrenergic and that over time, under artificial conditions, develops cholinergic mechanisms. Similar conclusions were reached by other studies (106) using microcultures of cells dissociated from developing rat sympathetic ganglia in the presence of cardiac myocytes. This, with in vitro conditions, adrenegic or cholinergic properties can be expressed in different proportions and transitional states in the adrenergic-to-cholinergic direction can be observed by following particular neurons over time. The main point to be demonstrated is whether the notion of multiple function, as expressed in developing neurons, in vitro, can be applied also to adult neurons under normal conditions of development in vivo (44).

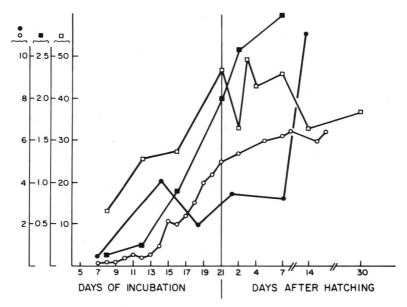

Fig. 3B. Developmental pattern of cholinergic receptors, ChAc, and AChE activity in chicken sympathetic ganglia. ○ = ACh receptor (α-BTX binding), fmol bound/ganglion; □ = AChe, μmol ACh/mg dry weight/h; ■ = ChAc, nmol ACh/ganglion/h; ● = ACh (pmol/ganglion). Means and standard errors are not reported (modified from ref. *43*).

2.1.1.1. Parasympathetic Ganglia

Pre- and postrecognition events in development of autonomic ganglia are summarized in Table 1. The so-called "vagal region" of the neural crest (levels 1–7) provides all cholinergic enteric ganglia (*75*). However, the fate of the ganglioblasts originating from the various regions is not irreversibly fixed (Table 1). The migration pattern of the crest cells can be changed by transplanting the vagal region of the neural crest at the adrenomedullary level, and vice versa (*75*). Under these conditions, adrenergic ganglioblasts migrate into the gut and originate cholinergic ganglion cells, whereas cephalic neural crest cells develop into adrenomedullary cells (*75*). Therefore, the potential for developing into cholinergic or adrenergic cells appears to be widespread in the whole neural crest (Fig. 1). The phenotype expressed by the neuroblasts, as far

Table 1
Pre- and Postrecognition Events in Development of Autonomic Ganglia

Embryonic and postembryonic events		Biochemical events
Premigration	Stable neural crest Precursors	No specialization
Prerecognition	Migration Differentiation Attraction Survival–cell death Development	Differentiation and expression of neurotransmitter characteristics
	RECOGNITION	Consolidation and modulation
Postrecognition	Structural differentiation Function–regulation Maintenance–modulation	Transsynaptic regulation
Maturation and aging	Regeneration–degeneration Cell death–aging	Chemical denervation

as neurotransmitter metabolism is concerned, seems to be regulated by the microenvironment to which the cells are exposed during migration (Table 1 and Fig. 1). The ciliary ganglion of the chicken originates from the neural crest of the midbrain and the anterior hindbrain (*56,80*). Thus, the neurons that give rise to the cholinergic population of the chick ciliary ganglion (*17*) derive from the mesencephalic neural crest. The neural crest cells migrate along the oculomotor nerve to their final position, and by stage 18 (ST 18) (*55*) condense to form the primordium of the ciliary ganglion (*98*). In the chicken, the adult ganglion contains two types of neurons that are cholinergic, that is, small choroid cells and larger ciliary cells that innervate the striated muscle of the iris and the ciliary body (*72,73*). The normal development of the ciliary ganglion of the chick has been extensively studied (*17,50*). This ganglion expresses a cholinergic differentiation from a very early stage, showing measurable levels of ChAc activity (*17*) and ACh (*84*) at 5 di, as well as a high affinity $Na[^+]$-dependent Ch (choline) uptake in the terminals in the iris (*85*). The neurotransmitter identification of "cholinergic" neural crest elements *in situ* is still largely unknown. Several attempts have been made to study the initial stages of differentiation in in vitro systems. Monoclonal antibodies that bind to chick ciliary ganglion neuronal cell bodies and neurites were isolated from 1 to 18-d-old embryos and tested in cell cultures (*1–2*). Five hybrid clones have been produced that secrete antibodies to cell surface components present on ciliary ganglion neurons (*2*). Two of these antibodies also bind to a subpopulation of cranial neural crest cells in vitro (*1–2*). Greenberg and Schrier (*53*) and Cohen (*21*) showed that chick mesencephalic crests in culture show ChAc activity, whereas quail trunk crests differentiate into cells displaying CA fluorescence. Kahn et al. (*69*) described the development of both ACh- and CA-synthesizing enzymes in cultures of quail cranial and trunk neural crest. It should be noted that Greenberg and Schrier (*53*) found only low ChAc activity in freshly dissected neural tubes including neural crest, although Kahn et al. (*69*) could not detect the enzyme activity in any crest culture before 5 d in vitro. The conversion [^{3}H]Ch to [^{3}H]ACh was seen in mesencephilic crests before culture (*110*), and in quail ciliary ganglia at 4 di (*78*). Le Douarin et al. (*78*) demonstrated that if a 4–6-d-old quail ciliary ganglion is transplanted into a 2-d chick at the adrenomedullary level of the neural crest before the migration of the host neural crest has started, the ganglion cells migrate following the same route as normal crest cells. The neurotransmitter syntheis is different according to the different site of their final localization. Therefore, ciliary ganglion cells destined to become cholinergic, when situated in adrenergic ganglia or in the suprarenal gland, start synthesizing catecholamines (*123*). Thus, the differentiation of the neurotransmitter enzymes in autonomic neurons seem to depend on tissue interactions taking place after the cells have left the neural primordia and have grouped into ganglionic structures (Table 1 and Fig. 1). Fauquet et al. (*34*) studied the development of autonomic neuronal precursors in cultures of microsurgically excised quail neural crest grown alone or associated with other embryonic tissues. The results of this study indicate that during ontogeny, cholinergic characteristics appear earlier than adrenergic characteristics in the neuron precursors, and that the early decision in differentiation coincides with the maturation of the somitic mesenchyme. The stimulating effect of the mesenchyme on ACh or CA synthesis could be mediated by soluble factors or by contact mechanisms.

In relation to these findings, three basic questions arise with regard to differentiation of autonomic neurons. First, when and how does heterogeneity of neurotransmitter expression arise? Second, how do neuronal cells recognize their destination and target? Third, is there any precommitment or are all crest cells capable, in principle, of expressing any type of neurotransmitter and of reaching any target? The current view is that cells of the neural crest are pluripotent and perhaps even homogeneous. Thus, differentiation depends on enviroonmental clues. In summary, all levels of the neural crest seem to be potentially able to give rise to neurons that can express either ACh or CA metabolism (Fig. 1). Positive reaction to AChE activity distinguishes neural crest cells from neural tube cells; therefore, AChE histochemistry can be used as a "rational marker" for tracing early stages of migration (76). However, at later stages of development some cells will lose AChE activity and binding properties for monoclonal antibodies (1). According to LeDouarin (76,77) and Smith et al. (110), ACh synthesizing ability represents an early characteristic of neuroblasts, whereas the capacity to synthesize CA is acquired later in embryogenesis and is probably under the influence of interaction with the notochord and somitic mesenchyme (Fig. 1). Synthesis and degradation of ACh are neuronal properties expressed in migrating cells that are not dependent on morphological characteristics. Autonomic ganglion cells are potentially capable of expressing more neurotransmitter phenotypes than those normally expressed in ganglia. This capability, however, can be demonstrated only under particular in vitro conditions (123).

3. Stages of Chemical Differentiation of Autonomic Synapses

In contrast to formation of contacts among nerve cells grown in vitro (44), the establishment of functional synaptic connections *in situ* represents a long-term process consisting of consecutive steps of complex structural and biochemical modifications (42,43). On a chronological scale, synaptogenesis occurs not only during the embryonic phase of life, but continues during the period following birth or hatching (46,64). Recent studies have shown that in the avian PNS the process of synaptic growth may continue throughout adult life, up to 12 mo or later (64,82). In order to better understand the complex process of formation of autonomic ganglionic synapses and neuroeffector junctions, we have arbitrarily separated the series of events preceding recognition of the target cell (organ) from those that follow the establishment of a permanent and functional contact between pre- and postjunctional elements (Table 1). A similar scheme has been proposed by Burnstock (*15*). Given the fact that precursor cells in the neural crest do not seem to show specialization, although still located in the crest (Table 1), the choice of neurotransmitter metabolism remains labile for at least a certain period of embryogenesis. Our scheme recognizes three major periods: (a) the prerecognition events; (b) the recognition process; and (c) the postrecogntion events. These involve three major steps of neurochemical changes (Table 1): (a) transmitter differentiation during migration, depending on interaction with mesenchymal tissues; (b) synaptic recognition and structural differentiation taking place during the process of segregation in ganglia; and (c) the stabilization and modulation processes, probably influenced by the functional afferent and efferent input (Table 1). The influence of the embryonic microenvironment and of cellular interactions seems to be of primary importance for the correct expression and differentiation of neurotransmitter characteristics (Fig. 1). This probably involves a spectrum of both diffusible and membrane-bound factors originating from certain types of nonneuronal cells and exerting dramatic and antagonistic effects on transmitter differentiation.

3.1. Factors Directed Toward Parasympathetic Neurons and Synapses

Approximately 50% of ciliary ganglionic cells undergo neuronal death between d 8 and 14 of embryonic age, at the time when synapses are formed with the iris muscle (*72,73*). One can assume that because only those neurons that are able to establish viable contacts with the targets will survive, intraocular targets might provide them with critical neurotrophic factors at this and subsequent stages of development (*121*). Eight-day chick embryo ciliary ganglion neurons dissociated and seeded in monolayer culture will not survive beyond 24 h with media containing only fetal calf or horse serum, but appear to require special extrinsic factors (*121*). It is interesting that these factors seem to be present in a variety of tissues and preparations, including: (a) skeletal muscle cells (*95–96*); (b) medium preconditioned over heart cell cultures (*58,59*) or extracts of chick heart (*61,91*); (c) extracts of whole chick embryo (*118*); and (a) myotube cultures or even myotube membranes (*119*). Comparative studies of the trophic effects of various extracts were performed (*121*) establishing trophic units (TU) as the activity present in 1 mL of final medium capable of supporting half maximal survival. Bioassays were used to analyze regional distribution of trophic (survival-promoting) activity from different portions of chick embryos (121). The eye of the 12-d chick embryo revealed a high amount of trophic activity. Particular fractions of the eye preparation accounted for almost 90% of the total eye activity with a specific activity of 2400 TU/mg. (121).

Unfortunately, no structural characterization, and only partial purification of such factors, has been achieved so far. The fact that trophic activity directed toward ciliary neurons is also present in tissues other than the eye, such as embryonic carcasses, viscera, skeletal muscle, and heart cells, is puzzling. It seems that ciliary ganglion cells cultured in vitro may gain trophic effects from peripheral tissues other than those they never innervate in vivo. This may indicate that cholinergic neurons located in different parts of the nervous system, including perhaps CNS, may be sensitive to ubiquitously acting cholinergic-type of factors, or that such factors are not selective in their action as far as different cholinergic (PNS or CNS) neurons are concerned.

These studies also indicate that under in vitro conditions, neither nerve target-organ formation nor direct contact with target cells are absolute or necessary conditions for survival and development of ciliary ganglion neurons if appropriate factors are present in the culture medium. This fact emphasizes the difference between in vitro and in vivo requirements for neuronal development (44). Furthermore, it indicates that parasympathetic neurons are not irreversibly committed to cell death in vivo and that their survival may be regulated by several conditions, one of which is probably the availability of the factor(s) supplied by postsynaptic cells.

The outgrowth of axons from parasympathetic ganglia other than the ciliary ganglion, such as the submandibular ganglion, has been shown to depend on the presence of target epithelium (24) and to be stimulated by factors isolated from the submandibulary gland (25).

Stimulation of axonal outgrowth in these ganglia has been postulated to be caused by some proteic factor of epithelial nature. High K^+ concentrations present in the medium seem to stimulate growth of neurons in culture above the maximal levels obtained by eye extracts alone (97). The K^+ effect seems to be mediated by membrane depolarization that might relate to conditions of preganglionic input and synaptic transmission under in vivo conditions (18). At later stages of development (ST 35–40) in the chicken embryo, ciliary ganglion neurons show a decline in their ability to extend neurites when placed in culture (23). This ability to extend processes is not recovered during normal development in vivo; however, if the ganglia are removed and cultured before being dissociated, the neurons may recover their ability to outgrow the neurites rapidly (23). These changes may reflect a conversion of the neurons to a situation more suitable for supporting neurite outgrowth that may enhance regenerative properties of axons following axotomy.

3.1.1. Role of Preganglionic Input on Neuronal Development

Pregaglionic input has been demonstrated to be important for the normal development of autonomic neurons and synapses (6, 42,45,47).

The injection *in ovo* of chlorisondamine, a nicotinic ganglionic blocking drug, caused a long-term hypotrophy in both the ciliary ganglion and the lumbar sympathetic ganglia (17) (See Fig. 4). Choline acetyltransferase activity was reduced in both ganglia, whereas development of the iris was not significantly altered. Injections of α-bungarotoxin delayed development in the striated iris muscle, but had little effect on the ontogeny of ciliary and sympathetic ganglia (*see* Fig. 5) (17). Such experiments demonstrated for the first time that usual development of the iris is dependent on a functional innervation by the ciliary ganglion cells. Although synaptic transmission did not appear to be essential for the survival of developing ciliary neurons, blocking of postsynaptic receptors significantly altered the developmental course in the ganglia after ganglionic receptor blockade, and in the iris after neuromuscular junction receptor blockade. The results of these experiments were confirmed by Wright (122), who in addition provided evidence that the treatment with chlorisondamine did, in fact, reduce cell survival in the ciliary ganglion by a factor of 24%. Thus, receptor interaction during development seems to be a critical factor in mediating afferent influences on cell survival in the ciliary ganglion.

The effects of high K^+ concentrations on the growth and development of axonal growth under in vitro conditions (97), which seems to be mediated by membrane depolarization, might relate to synaptic transmission under in vivo conditions (see previous section).

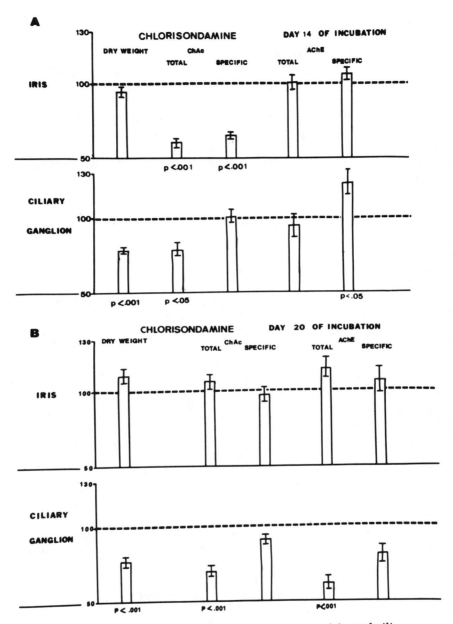

Fig. 4. Effect of chlorisondamine treatment on iris and ciliary ganglion at A, 14 and B, 20 d of incubation. All values are expressed as a percentage of control values. Bars represent mean ±SEM. Enzyme activity is expressed per tissue (total) and per mg dry weight (specific). The significance levels are relative to control values in this and following figures. Number of embryos in each group at 14 di was 8–9 (between 15 and 18 determinations), and at 20 di was 4–5 (between 8 and 10 determinations) (modified from ref. 16).

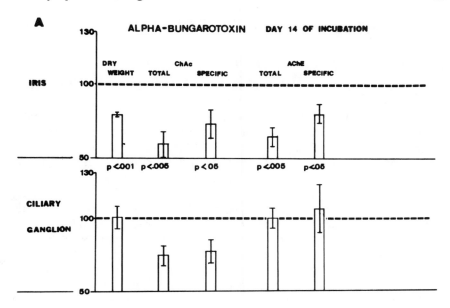

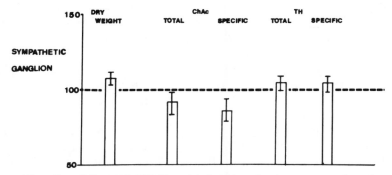

Fig. 5. Effect of ABTX at 14 d of incubation on iris A, ciliary ganglion and B, lumbar sympathetic ganglia. All values are expressed as a percentage of control values. Bars represent mean ±SEM. Number of embryos in each group was five (10 determinations per group for iris and ciliary ganglion, and 10–15 determinations per group for sympathetic ganglia) (modified from ref. *16*).

4. Synaptogenesis in PNS

4.1. Sympathetic Ganglia

The presynaptic cholinergic fibers that make contact with mainly noradrenergic neurons in sympathetic ganglia originate from cell bodies localized in the intermediolateral columns of the spinal cord. The postsynaptic neurons, predominantly noradrenergic, establish contact with a large variety of target structures. The dual representation of neuronal populations in sympathetic ganglia has been discussed previously. Cholineacetyltransferase, the enzyme that synthesizes ACh, is strictly localized to cholinergic fibers and neurons, and presynaptic terminals, whereas AChE can be found both in adrenergic and cholinergic neurons and fibers (*36*). Tyrosine hydroxylase, the rate-limiting enzyme in CA synthesis, is specifically localized to noradrenergic postsynaptic neurons and has been used as a marker for these neuronal populations. The biochemical maturation of cholinergic synapses in the paravertebral ganglia of the chick has been described in detail (*42*). Cholineacetyltransferase and AChE activities are first detected at 5 and 6 di, respectively (*27,32,51*). Acetylcholine is present at 7 di (*85*), and nicotinic ACh receptors are first detected at 12 di (*54*). It should be emphasized that each of these parameters is individually regulated and, therefore, rather than representing a general marker for cholinergic or adrenergic synapses, reflect different aspects and phases of synaptogenesis (Fig. 1). Keeping this limitation in mind, the parameters can be used as relative indexes of maturation of various neural elements. A body of biochemical studies in chick sympathetic ganglia supports the view (*4, 39,40,42,60*) that development of enzyme activities specifically related to neurotransmitter biosynthesis are regulated by: (a) transsynaptic influences provided by preganglionic fibers (*39,40,42*); and (b) interactions with postganglionic target organs (*16*). In spite of a wealth of biochemical data concerning development both in the chick (*42*) and in the rat (*5*), there remained a need for quantitative morphological information to be correlated to the biochemical markers already studied (*64*). Synaptogenesis was studied in the lumbar sympathetic ganglia of chicken by electron microscopic morphometry between 10 di and 1 yr of age (*64*). At 10 di, fewer than 1% of the adult number of synapses are present. The total number of synapses/ganglion increases progressively with age; however, the majority of synapses/ganglion are

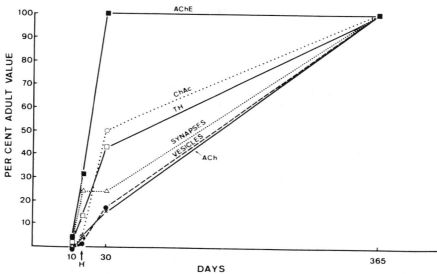

Fig. 6. Comparison of morphological and biochemical indices of synaptic and postsynaptic development in chick sympathetic ganglia. AChE = acetylcholinesterase activity; TH = tyrosine hydroxylase activity; ACh = amount of acetylcholine. The foregoing are relative enzymatic activities of ACh content per whole ganglion. Synapses = total number of clear synaptic vesicles in synaptic boutons per segment (modified from ref. 64).

formed after 30 d following hatching (Fig. 5). The total number of clear synaptic vesicles/ganglion follows a pattern similar to that for the number of synapses (Fig. 5). Both the number of synapses and the number of synaptic vesicles increase roughly in concert with biochemical markers of presynaptic elements.

The studies of Hruschak et al. (64) show for the first time a strong correlation between the developmental appearance of synapses and synaptic vesicles, on the hand, and of preganglionic (ACh levels and ChAc activity) and postganglionic (TH activity) neurotransmitter systems, on the other hand (*see* Fig. 6). An exception is the activity of AChE, which reaches its maximum earlier in development than other morphological or biochemical parameters. There is a particularly strong correlation between the levels of ACh (84), the major neurotransmitter of the preganglionic terminals, and the total number of clear synaptic vesicles (Fig. 6). Both of these parameters differ substantially at 30 d after hatching from the activity of ChAc. This discrepancy may reflect the presence of a nonvesicular pool of ChAc, or indicate developmental changes in the relative activity of ChAc associated with each vesicle. These results are consistent with the concept of transsynaptic regulation of ganglionic development (4,10).

Earlier studies performed in mice during perinatal ontogeny suggested a marked increase in synapse numbers within the sympathetic ganglion (12). The developmental increase in synapse numbers immediately precedes a rise in postsynaptic TH. These observations were the first to suggest that the development of neurotransmitter related enzymes, such as TH, might be dependent on the formation of synaptic contact. Additional evidence for this new concept was the result of decentralization of ganglia in neonatal mice and rats, demonstrating that this operating prevented the normal development increase in TH activity (12). These experiments in sympathetic ganglia (4,17) suggest that presynaptic cholinergic neurons regulate the development of postsynaptic TH or ChAc activity through a transsynaptic process. The regulation of such postsynaptic development is quite complex and involves the interaction of both pre- and postsynaptic elements with a number of developmental signals (*see* Table 2). Various approaches have been used to dissect out and identify the nature of such signals (115). Drugs competing for postsynaptic receptor sites, thus preventing depolarization, also prevent the normal biochemical development of the ganglion in both rodents (8,9) and avians (18,33), suggesting that presynaptic neurons regulate postsynaptic development through cholinergic interaction with nicotinic receptors. In other words, the neurotransmitter itself may act as a "growth factor" and regulator of normal ontogeny. The same mechanism has been observed in development of ciliary ganglia (17,18). A second important aspect of developmental mechanisms is the so-called target regulation, i.e., retrograde transsynaptic factors, originating from targets that may modulate the development of the autonomic neurons in ganglionic structures. Depending on their appearance with respect to innervation, they can be chronologically divided into "early" and "late" signals (47) (Table 2).

In the category "late signals," the physiological activity of the target organ (muscular contraction), as well as neurotransmission, are essential in maintaining synthesis and regulation of membrane components (receptors) and neurotransmitter enzymes (TH, ChAc, and so on) (Table 2). The origin of both signals are postsynaptic; however, their site of action is different, the first

Table 2
Developmental Signals

	Effect	Site of effect	Origin of signal	Direction
E A R L Y	Promotion of growth or survival of neuron	Cell body	Target organ	Retrograde
	Expression of neurotransmitter phenotype	Cell body	Target organ or environment	Retrograde
	Promotion of synaptic contacts	Synapse (muscle)	Noninnervated organ	Retrograde
	Shut off of innervation	Synapse (muscle)	Innervated organ	Retrograde
L A T E	Differentiation of target organ	Target organ (muscle)	Neurotransmission at target organ	Anterograde
	Differentiation of presynaptic organ	Target organ (muscle)	Neurotransmission at target organ	Anterograde
	Differentiation of synaptic contacts	Presynaptic sites (nerve)	Target organ	Retrograde
	Synthesis and regulation of membrane components	Postsynaptic sites (muscle)	Physiological activity of target organ	Local
	Synthesis and regulation of neurotransmitter enzymes	Cell body	Neurotransmission at target organ	Retrograde

acting postsynaptically on the muscle and the second presynaptically on the cell body. Studies in tissue cultures are particularly useful in identifying factors regulating cellular interactions (6,11,13). Drugs such as reserpine, which selectively deplete CA storage, can be useful as experimental tools in manipulating neurotransmitter levels in the embryo. Chick embryos are particularly suitable to pharmacological manipulation because the administration of a drug into a closed system, such as an egg, eliminates maternal drug metabolism. Sparber and Shideman (111,112) were the first to study the effect of reserpine administered before hatching on behavior and brain stem CA. Fairman et al. (*33*) demonstrated that a single 10-μg dose of reserpine administered into the yolk sac produced a significant increase in TH activity that was still present 30 d after hatching. The administration of chlorisondamine, a selective blocker of nicotinic receptors, almost completely abolished the effect of reserpine. These results confirmed and extended the results obtained by other authors on adult rat sympathetic ganglia that first suggested the transsynaptic induction of TH as a long-term regulatory mechanism in the sympathetic system (*92,93*). A distinction between regulation of enzyme synthesis by neuronal activity and NGF was later introduced by Thoenen et al. (*117*).

In conclusion, these studies clearly demonstrated that the synthesis of specific proteins accumulated in the terminals of noradrenergic neurons is regulated by the activity of the preganglionic cholinergic nerves and by NFG both during development and in the adult. The mechanism of action of NGF on the developing synpathetic neuron is still being explored.

4.1.1. Parasympathetic Ganglia

In the chick embryo, cholinergic cells of the enteric plexus of Meissner and Auerback and the ganglion of Remak arise from the vagal (S_1 to S_7) and lumbosacral (posterior to S_{28}) levels of the neural crest (*78*). Cholinergic cells of the ciliary ganglion are derived from the mesencephalic neural crest (*78*). Mesencephalic crest cells seem to possess some basic markers of cholinergic metabolism from the onset of their migratory phase (*110*).

Acetylcholinesterase activity has been detected histochemically in cells dissected from the neural tube and the neural crest at 35, 48, 60, and 74 h of incubation, prior to the appearance of specific morphological characteristics of neuroblasts (*90*). This could indicate either an early differentiation of the cell population that will later form parasympathetic ganglia, such as the ciliary ganglior. (*17*), or the presence of an undifferentiated precursor cell with cholinergic characteristics. This second hypothesis is supported by the finding of measurable ACh levels in different segments of the neural crest (Giacobini, unpublished), via the ability of the mesenchephalic crest to synthesize ACh before explanation (*34*). It has, however, been suggested by several authors, mostly based on results obtained from in vivo experiments, that stable cholinergic differentiation might occur only at later stages, i.e., when cells have reached their final destination and have started to establish connections with the target organ. Two parasympathetic ganglia deriving from neural crest ganglioblasts, such as the ciliary ganglion and the ganglion of Remak, have been examined from the earliest stages of their appearance.

The avian ciliary ganglion shows ChAc activity, AChE activity, and ACh levels at 5 di (17,84). A high-affinity uptake mechanism for Ch is present in the fibers of the ciliary ganglion innervating the primitive iris bud at 4.5–5 di (85). At 5 di, Ch uptake in ciliary ganglion fibers is almost 80% of the 1-yr value. In addition, quail ciliary ganglia have the ability to convert [^{3}H]-Ch to [^{3}H]-ACh at 4 di (0.02 pmol/ganglion) (110). At 6 di, this synthetic ability has increased 15-fold (78). Similarly, the Remak ganglion shows ChAc activity and is able to synthesize ACh at 4.5 di (78). Chick ciliary ganglion neurons grown in dissociated cell culture show a high-affinity uptake mechanism for Ch with an apparent K_m of 0.3 μM (1). Almost all of the neurons can be labeled with [^{3}H]-Ch. This is consistent with the findings in vivo demonstrating that in both adult and developing ciliary ganglion neurons, the high-affinity Ch uptake mechanisms is selectively localized to nerve terminals (85,114). We have described the appearance and development of cholinergic characteristics in ciliary ganglia of the chick starting at 4.5 di up to 9 yr of age (50,83). In these studies the changes in enzyme activity, receptor binding, and neurotransmitter and precursor levels have been closely correlated to electrophysiological and morphological findings in the same structure. Our findings show that in both ciliary ganglion and iris, neurotransmission reaches adult values prior to the establishment of adult levels of ACh and number of cholinergic receptors (AChR), as well as of a fully developed AChE and ChAc activity. In the ciliary ganglion, the highest rate of appearance of cholinergic receptors is prior to the rise of cholinergic enzyme activity, whereas in the iris it is practically simultaneous. This implies that the postsynaptic system interacting with the neurotransmitter (i.e., AChR or AChE) starts developing simultaneously with or even prior to the achievement of full capacity in neurotransmitter synthesis in the cell body and presynaptic terminals.

In earlier publications (42,43), we had proposed a pattern of development that takes into account simultaneous processes of maturation of cell bodies and synapses in autonomic ganglia. For a detailed account of such processes, refer to specialized papers (17,42,43,50).

Briefly, as soon as the migration of the neural crest cells that will constitute the ciliary ganglion is completed at around 4 di (ST 25), the ciliary ganglion cells are biochemically differentiated into cholinergic neurons, and at 5 di, low levels of ACh (1–2 pmol/ganglion) are synthesized. This period is followed by a rapid ninefold increase in ACh in both cell bodies and terminals in the iris at the time when neurotransmission reaches adult levels in the iris. In the ciliary ganglion, however, neurotransmission is fully developed by 7 di when ACh levels are still quite low. It is, therefore, possible that minimal levels of ACh and ChAc activity, such as those found in this first period of ganglionic development, are sufficient to maintain ganglionic function. Levels of ACh and Ch seem to undergo a continuous growth that is initiated in the first week of embryonic development and proceeds into adulthood up to 2 yr of age (82). In the ciliary ganglion, morphological and biochemical observations show that this period corresponds to one of continuous synaptic growth. In a second period, marked by a decrease in ACh and Ch levels, a phase of progressive deterioration of cholinergic metabolism takes place (46,83) (*see* Table 3).

With regard to the development of AChR, these are clearly detectable in both ciliary ganglion and iris before initiation of neurotransmission, although only low presynaptic ChAc and AChE activities are present (16). Under in vitro conditions, neurons obtained from 8-d embryo chick ciliary ganglion show α-bungarotoxin binding sites and AChRs on their membrane, as judged by intophoretic application of ACh (107). Biochemical and histochemical evidence at the EM level suggests that the earliest appearance of cytoplasmic-located AChE may be independent on the establishment of functional synaptic contacts (17,47, 48,87,100). The function of this enzyme at these early stages of development remains obscure. At later stages of development it is possible that retrograde influences may regulate neuronal AChE activity located on postsynaptic membrane. Simultaneous with the maturation of neurotransmission, receptor levels and AChE activity rapidly increase in the postsynaptic elements. When both the number of receptors and neurotransmission have reached adult values in the ganglion (12 di), the ChAc that has accumulated in the cell bodies begins to be transported in large amounts down their axons to the terminals (17). Increased levels of ChAc can then begin to synthesize the amounts of ACh necessary to maintain neurotransmission at the neuromuscular junction. Ex-

Table 3
Cholinergic Parameters of Ciliary Ganglia (CG) and Irises (I) During Aging[a]

		3 mo	1 yr	Age 2 yr	4-5 yr	6-7 yr
ACh[b]	CG	70 ± 4	161 ± 40	384 ± 18	187 ± 25	96 ± 13
	I	317 ± 36	402 ± 42	289 ± 50	198 ± 40	68 ± 11
Ch[b]	CG	529 ± 221	556 ± 65	518 ± 171	221 ± 50	146 ± 58
	I	885 ± 168	1388 ± 310	1012 ± 459	202 ± 31	107 ± 29
Ch/ACh[b]	CG	7.6	3.5	1.4	1.2	1.5
	I	2.8	3.5	3.5	1.0	1.6
Ch uptake[c]	I	136 ± 4	129 ± 15	65 ± 6	45 ± 3	28 ± 5
ChAc[d]	CG	107 ± 3	133 ± 4	91 ± 9	95 ± 5	70 ± 4
	I	70 ± 2	220 ± 31	70 ± 5	116 ± 7	103 ± 7
AChE[d]	CG	287 ± 19	291 ± 21	273 ± 38	362 ± 23	203 ± 26
	I	2459 ± 210	2863 ± 245	2089 ± 384	2962 ± 356	2230 ± 249
ABTX binding[e]	CG	25 ± 10	70 ± 40	50 ± 12	105 ± 30	60 ± 18
	I	100 ± 25	148 ± 40	400 ± 35	215 ± 40	160 ± 32

[a] $n \geq 5$; values are given ±SEM
[b] pmol (from refs. 82 and 84).
[c] pmol/5 min (from ref. 86).
[d] ChAc, nmol ACh synthesized/15 min: AChE, nmol ACh hydrolyzed/15 min (from ref. 86).
[e] ABTX binding, fmol/ganglion or /iris, 3-mo and 1-, 1.3-, 3-, 5-, and 6-yr values are reported (from ref. 88).

periments performed with various cholinergic blockers, including α-bungarotoxin, show that synaptic transmission expressed by normal receptor density and function seem not to be essential for the survival of developing neurons (*18*). However, blockade of postsynaptic receptors influence significantly the development of both ganglia and target organs exerting a general hypotrophic effect on weight and enzymes (see previous section). Thus, we can conclude that synaptic transmission plays a regulatory role in synaptogenesis.

5. Biochemical Development of Neuroeffector Junctions

5.1. Autonomic Ganglia and Their Targets

The peripheral nervous system of the chick offers a suitable model for the study of development of neuroeffector junctions. A particular advantage is offered by the ciliary–ganglion–iris preparation (*49*) that offers the possibility of studying development of neurotransmitter synthesis in homogeneous populations of nerve cells and terminals. The chick iris is easily accessible at early stages of development (*85*). It contains a discrete population of NE, containing nerve terminals and fibers (*29*), and of cholinergic fibers (*17*). Histofluorence for NE of these terminals and fibers is first seen at 13–14 di, and the number of fibers reaches adult levels by 16 di (*71*). However, we found significant levels of NE in the iris at stages prior to the appearance of fluorescence (7 di) (*63*). The levels increase approximately fourfold from 14 di to 3 mo after hatching. The rapid increase to nearly adult values in NE levels and fluorescence suggests a rapid maturation of biosynthetic and storage mechanisms. The significant increases in NE levels are paralleled, but preceded, by increases in ACh levels in cholinergic terminals in the iris (*82,84*). In the posthatching period, the age-related changes in NE levels follow a pattern very similar to that of ACh in the same iris preparation. Thus, differences in NE synthesis and accumulations are marked during the early embryonic period, whereas both ACh and NE neurotransmitters may share a common susceptibility to aging. Neurotransmitter uptake has been considered an early index of synaptogenesis and innervation. In the chick iris, the various characteristics of [^{3}H]-NE uptake do not appear simultaneously during development (62). Na^+-dependence and ouabain sensitivity are both first seen at 10 di, although temperature sensitivity develops only at hatching. Desmethylimipramine (DMI) and cocaine inhibition develop at 10 di, whereas metanephrine inhibition of extraneuronal uptake develops to a significant degree only hatching. [^{3}H]-NE uptake develops at around 10 di, and there is a large increase in V_{max} from 10 to 14 di, which is a period of rapid growth of sympathetic innervation of the iris (62). The appearance of Na^+-dependent and ouabain-, cocaine-, and DMI-sensitive uptake of [^{3}H]-NE that follows Michaelis-Menten kinetics, closely follows the developmental pattern of other noradrenergic characteristics in chick sympathetic ganglia (42), including the curve of NE levels (63). This suggests that biochemical development of this neurotransmitter is closely related to the maturation of its uptake mechanism in the nerve terminal. The high-affinity carrier-mediated Ch uptake represents in the adult nervous tissue the major source of Ch used for the synthesis of ACh in nerve terminals (115). Therefore, it is considered a crucial site of regulation of cholinergic neurotransmitter metabolism in these neurons. Choline uptake has also been reported to constitute an early index of cholinergic innervation in the brain. In conjunction with an electron microscopy study of the innervation of the chick iris. Ch uptake has been characterized as starting at 4.5 di and continuing throughout the adult (115) and aging period (85). Our study clearly demonstrated that Ch uptake constitutes an early and reliable biochemical marker for the presence of cholinergic innervation in target organs. The high-affinity Ch uptake is present in the iris cholinergic fibers prior to the formation of neuromuscular junction and the onset of neurotransmission. It represents, therefore, one of the earliest signs of specialization detectable in the membrane of growing nerve fibers. As shown in the morphological study (85), it is present at the stage in which differentiation of the target organ occurs. The parallel trend of development of the V_{max} for Ch uptake and ACh levels strongly supports the role of Ch uptake as a regulatory mechanism that may be rate limiting for the synthesis of the neurotransmitter. It is interesting to note that V_{max} values for Ch uptake decreases significantly at 5 yr and continue to decrease at 7 yr, although K_m does not change significantly during this period (Table 3) (86). Total ACh and Ch levels follow a similar trend, decreasing progressively from 1 to 7 yr (Table 3). Both NE and ACh decline in aging iris preparations to levels below those found in early embryonic tissue. These observations suggest that the effect of aging on peripheral cholinergic and adrenergic neurons is not generalized, but specifically directed toward the neuronal periphery (terminals), as opposed to cell bodies (46).

5.1.1. Establishment of Synaptic Connections Between Parasympathetic Neurons and With Other Targets In Vitro

Ciliary ganglion neurons from chick embryo can be grown in cell culture following enzymatic and mechanical dissociation (22,89,95,113,119,120). The inclusion of embryo extract in the growth medium was found to be essential. Also, nonneural cells increased the survival of the neurons in culture. Therefore, it appears that a number of factors are involved in the growth and survival of these cells in vitro (95–96). Choline acetyltransferase activity increases 100-fold over the first 2 wk in vitro, mimicking the developmental pattern seen in vivo (52). The cultured neurons also retain some of the electrophysiological properties seen in vivo, such as normal resting potential, action potential amplitude, and an after potential (120). Characteristics, such as the passive membrane properties, differ from those of neurons in ganglia *in situ*. Morphological evidence for chemical synapses between ciliary ganglion neurons was reported, but stimulation of neurons failed to evoke synaptic transmission (105), thus indicating loss of neuronal chemosensitivity in culture lacking proper targets. Contrary to these findings, in a subsequent study, Margiotta and Berg (89) reported that cholinergic synaptic transmission occurs between ciliary neurons. The implication is that ciliary ganglion neurons establish synapses on each other in all culture and that neurotransmitter release occurs either spontaneously or can be evoked by spontaneous action potentials in the presynaptic neurons. It should be noted that synapses between ganglionic neurons has been described in vitro for rat sympathetic neurons (35,99). The formation of synapses between ciliary ganglion neurons in culture represents one of numerous anomalous characteristics of these cells in culture that may reflect the absence of their normal preganglionic innervation and the physiological target (44).

When striated muscle is included in the culture, neuronal chemosensitivity is maintained (95,113,119). Many of the neurons prepared from 8–14-d-old embryos and grown with skeletal myotubes form functional synapses. The membrane remnants of myotubes ruptured by osmotic shock also supported evidence of the responsiveness of the cultured neurons to ACh (*119*). These observations suggest the dependency on the target for survival and the presence of target-derived proteins, in addition to other "trophic" proteins or factors that seem to be important for neuronal differentiation and growth.

6. Synaptic Regression: The Final Act of Synaptic Development?

Several results that are reported in the literature seem to support the view that growth and senescence of neurons form an uninterrupted continuum of events (*46*) (Tables 1 and 3). The phase that has been identified as "synaptic regression" is characterized by negative variations in enzyme activities and neurotransmitter levels. In peripheral cholinergic synapses, significant modifications of precursor uptake seem to constitute one major factor responsible for the decreased synthesis of transmitters (*46*). Further, the turnover rate and release of ACh is slowed down in aging terminals. In both cholinergic and adrenergic synapses the effect of aging seems not to be generalized, but specifically directed toward the periphery (terminals), as opposed to cell bodies. In view of this fact, it is interesting to compare the morphological and biochemical changes occurring in two cholinergic neuroeffector junctions, such as the muscle and plate in the rat diaphragm and the neuromuscular junctions in the chick iris (*see* Table 4). Morphologically, the nerve endings in the iris show pol-

Table 4
Comparison of Biochemical Data of Cholinergic Terminals in Rat Diaphragm and Chicken Iris During Aging

	Rat diaphragm[a]	Chicken irises[b]
ACh amount/end plate	−49%	−83%
ACh release (total)	−62%	Decreased
ACh release (fraction)	−5%	Decreased
Sensitivity to HC-3	Decreased	Decreased
ChAc activity	Unchanged	−47%
AChE activity	Unchanged	−22%
Receptor binding (αBTX)	—	−60%
Choline (total)	—	−92%
Choline uptake (V_{max})	—	−79%

[a]From ref. *109*.
[b]From refs. *46* and *94*.

ymorphic signs of damage with a significant reduction of synaptic vesicles, increased neurofilaments, and mitochondria (*94*). These modifications are symptomatic of severe synaptic degeneration producing a reduced area of synaptic contact between nerve and muscle leading to specific peripheral damage. The comparison between rat diaphragm and chick iris demonstrates some differences between the two structures that might be ascribed to different functional conditions. Both structures, however, show significant changes in cholinergic attributes and metabolism during aging, leading to a strongly reduced function (Table 4).

7. Conclusions

In this chapter we have examined biochemical changes that occur during the period in which synaptic contacts are established between developing autonomic neurons and their targets. These changes have been followed to adult stages. The results that have been reported and discussed here seem to support the view that growth, adulthood, and senescence of autonomic synapses form an uninterrupted continuum of events starting at the post-embryological stages and progressing to a phase of "synaptic regression" leading to strongly reduced function. This latter phase involves biochemical modifications that can now be detected at their earliest stages. This chapter emphasizes the importance of using microquantitative biochemical techniques, as well defined neuronal populations such as autonomic ganglia, in order to characterize phenomena such as chemical differentiation, modifications in neurotransmitter biosynthesis during development, and synaptogenesis in general. It is also important to evaluate these changes in light of physiological events leading to development of functional synapses. Problems such as biochemical heterogeneity, plasticity, and stability of synaptic connections should all be considered in a study of the developing and adult nervous system. The possibility of perturbing such conditions in order to understand and manipulate regeneration processes constitutes the most attractive goal for developmental neurobiology today.

Acknowledgments

The investigations carried out in the author's laboratory were supported by PHS grants NS-11496, NS-11430, and NS-15086,

NSF grant GB-41475 to E. G., and grants from the University of Connecticuit Research Foundation.

References

1. Barald, K. and D. K. Berg (1979) Ciliary ganglion neurons in cell culture: high affinity choline uptake and autoradiographic choline labelling. *Devel. Biol.* **72**, 15–23.
2. Barald, K. (1981) Cell surface specific monoclonal antibodies to chick ciliary ganglion neurons abstract. *Soc. Neurosci.* **7**, 120.
3. Barald, K. (1982) Monoclonal Antibodies to Embroynic Neurons, in *Neuronal Development* (N. C. Spitzer, ed.) Plenum Press, New York, pp. 101–119.
4. Black, I. B. (1974) Growth and Development of Cholinergic and Adrenergic Neurons in a Sympathetic Ganglin: Reciprocal Regulation at the Synapse, in *Dynamics of Degeneration and Growth in Neurons* (K. Fuxe, L. Olson, and Y. Zotterman, eds.) Pergamon Press, Oxford, pp. 455–467.
5. Black, I. B. (1979) Regulation of the Growth and Development of Sympathetic Neurons In Vivo, in *Cellular Neurobiol.* Alan R. Liss, New York.
6. Black, I. B. (1978) Regulation of autonomic development. *Ann. Rev. Neurosci.* **1**, 183–214.
7. Black, I. B. (1982) Stages of neurotransmitter development in autonomic neurons. *Science* **215**, 1198–1204.
8. Black, I. B. and S. C. Geen (1973) Transsynaptic regulation of adrenergic neuron development: inhibition of ganglionic blockade. *Brain Res.* **63**, 291–302.
9. Black, I. B. and S. C. Geen (1974) Inhibition of the biochemical and morphological maturation of adrenergic neurons by nicotinic receptor blockade. *J. Neurchem.* **22**, 301–306.
10. Black, I. B. and C. Mytilineou (1976) Transsynaptic regulation of the development and end organ innervation by sympathetic neurons. *Brain Res.* **101**, 503–521.
11. Black, I. B. and P. H. Patterson (1980) Developmental Regulation of Neurotransmitter Phenotype, in *Current Topics in Developmental Biology* Vol. 15 (A. A. Moscona, A. Monroy, and R. Kevin Hunt eds.) Academic Press, New York.
12. Black, I. B., I. A. Hendry, and L. L. Iversen (1971) Transsynaptic regulation of growth and development of adrenergic neurons in a mouse sympathetic ganglion. *Brain Res.* **34**, 229–240.
13. Black I. B., M. D. Coughlin, and P. Cochard (1979) Factors regulating neuronal differentiation. *Soc. Neurosci. Symp.* **4**, 184–207.

14. Bunge, R., M. Johnson and C. D. Ross (1978) Nature and nurture in development of the autonomic neuron. *Science* **199,**1409–1416.

15. Burnstock, G. (1981) Current Approaches to Development of the Autonomic Nervous System: Clues to Clinical Problems, in *Development of the Autonomic System,* (K. Elliot and G. Lawrenson, eds.) Pitman Books, London, pp. 1–14.

16. Chiappinelli, V. A. and E. Giacobini (1978) Time course of appearance of α-bungarotoxin binding sites during development of chick ciliary ganglion and iris. *Neurochem. Res.* **3,** 465–478.

17. Chiappinelli, V. A., E. Giacobini, G. Pilar, and H. Uchimura (1976) Induction of cholinergic enzymes in chick ciliary ganglion and iris muscle cells during synaptic formation. *J. Physiol.* **257,** 749–766.

18. Chiappinelli, V. A., K. Fairman, and E. Giacobini (1978) Effects of nicotinic antagonists on the development of the chick lumbar sympathetic ganglia, ciliary ganglion, and iris. *Develop. Neurosci.* **1,** 191–202.

19. Cochard, P., M. Goldstein, and I. Black (1978) Ontogenetic appearance and disappearance of tryosine hydroxylase and catecholamines in the rat embryo. *Proc. Natl. Acad. Sci. USA* **75,** 2986–2990.

20. Cohen, A. M. (1972) Factors directing the expression of sympathetic traits in cells of neural crest origin. *J. Exp. Zool.* **1979,** 179–192.

21. Cohen, A. M. (1977) Independent expression of the adrenergic phenotype of neural crest cells in vitro. *Proc. Natl. Acad. Sci. USA* **74(7),** 2899–2903.

22. Collins, F. (1978) Axon initiation by ciliary neurons in culture. *Develop. Biol.* **65,** 50–57.

23. Collins, F. and M. R. Lee (1982) A reversible developmental change in the ability of ciliary ganglion neurons to extend neurites in culture. *J. Neurosci.* **2(4),** 424–430.

24. Coughlin, M. D. (1975) Target organ stimulation of parasymapathetic nerve growth in the developing mouse submandibular gland. *Develop. Biol.* **43,** 140–158.

25. Coughlin, M. D. and M. P. Rathbone (1977) Factors involved in the stimulation of parasympathetic nerve outgrowth. *Develop. Biol.* **61,** 131–139.

26. Dennis, M. J. (1981) Development of the neuromuscular junction: inductive interactions between cells. *Ann. Rev. Neurosci.* **4,** 43–68.

27. Dolezalova, H., E. Giacobini, G. Giacobini, A. Rossi, and G. Toschi (1974) Developmental variations of choline acetyltransferase, dopamine, β-hydroxylase and monoamine oxidase in chicken embryo and chicken sympathetic ganglia. *Brain Res.* **73,** 309–320.

28. Edgar, D., Y-A. Barde, and H. Thoenen (1981) Subpopulations of

cultured chick sympathetic neurons differ in their requirements for survival factors. *Nature* **289**, 294–295.

29. Ehinger, B. (1967) Adrenergic nerves in the avian eye and ciliary ganglion. *Z. Zellforsch. mikrosk. Anat.* **82**, 577–588.

30. Elliot, K. and G. Lawrenson (1981) *Development of the Autonomic Nervous System*, CIBA Foundation Symposium 83, Pitman, Bath, England, pp. 1–389.

31. Enemar, A., B. Falck, and R. Hakanson (1965) Observations on the appearance of norepinephrine in the sympathetic nervous system of the chick embryo. *Develop. Biol.* **11**, 268–282.

32. Fairman, K., E. Giacobini, and V. Chiappinelli. (1976) Developmental variations of tyrosine hydroxylase and acetylcholinesterase in embryonic and posthatching chicken sympathetic ganglia. *Brain Res.* **102**, 301–312.

33. Fairman, K., V. Chiappinelli, E. Giacobini, and L. Yurkewicz (1977) The effect of a single dose of reserpine administered prior to incubation on the development of tyrosine hydroxylase activity in chick sympathetic ganglia. *Brain Res.* **122**, 503–512.

34. Fauquet, M., J. Smith, C. Ziller, and N. M. LeDouarin (1981) Differentiation of autonomic neuron precursors in vitro: cholinergic and adrenergic traits in cultured neural crest cells. *J. Neurosci.* **1(5)**, 478–492.

35. Furshpan, E. J., R. P. MacLeish, P. H. O'Lague, and D. D. Potter (1976) Chemical transmission between rat sympathetic neurons and cardiac myocytes developing in microcultures: evidence for cholinergic, adrenergic, and dual-function neurons. *Proc. Natl. Acad. Sci. (USA)* **73**, 4225–4229.

36. Giacobini, E. (1959) The distribution and localization of cholinesterases in nerve cells. *Acad. Diss. Acta Physiol. Scand.* **45** (Suppl. 156), 1–45.

37. Giacobini, E. (1970) Biochemistry of Synaptic Plasticity Studied in Single Neurons, in *Biochemistry of Simple Nuronal Models*, Vol. 2 (E. Costa and E. Giacobini, eds.) Raven Press, New York, pp. 9–64.

38. Giacobini, E. (1971) Biochemistry of the Developing Autonomic Neuron, in *Advances in Experimental Medicine and Biology*, Vol. 13 (R. Paoletti and A. N. Davison, eds.) New York, Plenum, pp. 145–155.

39. Giacobini, E. (1975a) Neuronal control of neurotransmitters biosynethesis during development. *J. Neurosci. Res.* **1**, 315–331.

40. Giacobini, E. (1975b) The use of microchemical techniques for the identification of new transmitter molecules in neurons. *J. Neurosci. Res.* **1**, 1–18.

41. Giacobini, E. (1977) Validity of Single Neuron Chemical Analysis, in *Biochemistry of Characterized Neurons* (N. N. Osborne, ed.) Pergamon, Oxford, pp. 3–17.

42. Giacobini, E. (1978) Regulation of Neurotransmitter Biosynthesis

During Development in the Peripheral Nervous System, in *Maturation of Neurotransmission* (A. Vernakakis, E. Giacobini, and G. Filogamo, eds.) S. Karger, Basel, pp. 41–64.

43. Giacobini, E. (1979) Synaptogenesis: Chemistry, Structure, or Function? Which Comes First?, in *Neural Growth and Differentiation*, (E. Meisami and M. A. B. Brazier, eds.) Raven Pres, New York, pp. 153–167.

44. Giacobini, E. (1980a) Discrepancies and Differences Between Nerve Cells Growing In Vitro and In Situ: A Discussion, in *Tissue Culture in Neurobiology* (E. Giacobini, A. Vernadakis, and A. Shahar, eds.) Raven Press, New York, pp. 187–204.

45. Giacobini, E. (1980b) Biochemical control of synapse formation in vivo. *Pont. Acad. Scient. Scripta Varia.* **45**, 451–482.

46. Giacobini, E. (1982) Aging of Autonomic Synapse, in *Advances in Cellular Neurobiology*, Vol. 3 (S. Fedoroff and L. Hertz, eds.) Academic Press, New York, pp. 173–214.

47. Giacobini, E. (1983a) Formation and Differentiation of Synaptic Contacts in the Peripheral Nervous System, in *Neural Transmission, Learning, and Memory*, (R. Caputto and C. Ajmone Marsan, eds.) Raven Press, New York, pp. 113–123.

48. Giacobini, E. (1983b) Biochemical Differentiation and Development of Autonomic Neurons and Synapses, in *Handbook of Neurochemistry* (A. Lajtha, ed.) Plenum, New York, pp. 467–488.

49. Giacobini, E. and V. Chiappinelli, (1977) The Ciliary Ganglion: A Model of Cholinergic Synaptogensis, in *Synaptogenesis* (L. Tauc, ed.) Naturalia et Biologia Publ., Paris, pp. 89–116.

50. Giacobini, E. and M. Marchi (1981) Acetylcholine Biosynthesis in Developing Cholinergic Synapses, in *Cholinergic Mechanisms* (G. Pepeu and H. Ladinsky, eds.) Plenum, New York, pp. 1–24.

51. Giacobini, G., P. C. Marchisio, E. Giacobini, and S-H. Koscow (1970) Developmental changes of cholinesterases and monamine oxidase in c hick embryo spinal and sympathetic ganglia. *J. Neurochem.* **17**, 1177–1185.

52. Giacobini, E., G. Pilar, J. Suszkin, and H. Uchimura (1979) Normal distribution and denervation changes of neurotransmitter related enzymes in cholinergic neurons. *J. Physiol.* **286**, 233–253.

53. Greenberg, J. H. and B. K. Schrier (1977) Development of choline acetyltransferase activity in chick cranial neural crest cells in culture. *Develop. Biol.* **61**, 86–93.

54. Greene, L. A. (1976) Binding of α-bungarotoxin to chick sympathetic ganglia: properties of the receptor and its rate of appearance during development. *Brain Res.* **11**, 135–145.

55. Hamburger, V. and H. L. Hamilton (1951) A series of normal stages in the development of the chick embryo. *J. Morphol.* **88**, 49–92.

56. Hammond, W. S. and C. L. Yntema (1958) Origin of ciliary ganglia in the chick. *J. Comp. Neurol.* **110,** 367–385.
57. Hedlund, K-O. and T. Ebendal (1978) Different ganglia from the chick embryo in studies on neuron development in culture. *Zoon* **6,** 217–223.
58. Helfand, S. L., G. A. Smith, and N. K. Wessells (1976) Survival and development in culture of dissociated parasympathetic neurons from ciliary ganglia. *Devel. Biol.* **50,** 541–547.
59. Helfand, S. L., R. J. Riopelle, and N. K. Wessells (1978) Nonequivalence of conditioned medium and nerve growth factor for sympathetic, parasympathetic, and sensory neurons. *Exp. Cell Res.* **113,** 39–45.
60. Hendry, I. A. (1976) Control in the Development of the Vertebrate Sympathetic Nervous System, in *Reviews of Neuroscience* (S. Ehrenpreis and I. J. Kopin, eds.) Raven Press, New York, pp. 149–194.
61. Hill, C. E., I. A. Hendry, and R. E. Bonyhady (1981) Avian parasympathetic neurotrophic factors: age-related increases and lack of regional specificity. *Develop. Biol.* **85,** 258–261.
62. Hoffman, D. W. and E. Giacobini (1980) Characteristics of norepinephrine uptake in developing peripheral nerve terminals. *Brain Res.* **201,** 57–70.
63. Hoffman, D. W., S. K. Salzman, M. Marchi, and E. Giacobini (1980) Norepinephrine levels in peripheral nerve terminals during development and aging. *J. Neurochem.* **34(6),** 1785–1787.
64. Hruschak, K. A., V. I. Friedrich, Jr., and E. Giacobini (1982) Synaptogenesis in chick paravertebral sympathetic ganglia: a morphometric analysis. *Develop. Brain Res.* **4,** 229–240.
65. Ignarro, L. J. and F. E. Shideman (1968) Appearance and concentrations of catecholamines and their biosynthesis in the embryonic and developing chick. *J. Pharmacol. Exp. Therap.* **159(1),** 38–48.
66. Jacobwits, D. M., L. A. Greene and N. B. Thoa (1976) Chomaffin Cells in Culture, in *SIF Cells* (O. Erank, ed.) Elsevier, Amsterdam, pp. 215–222.
67. Johnson, M. I., C. D. Ross, and R. P. Bunge (1980) Morphological and biochemical studies on the development of cholinergic properties in cultured sympathetic neurons. *J. Cell Biol. (II)* **84,** 692–704.
68. Johnson, M. I., C. D. Ross, M. Meyers, E. L. Spitznagel, and R. P. Bunge (1980) Morphological and biochemical studies on the development of cholinergic properties in cultured sympathetic neurons. *J. Cell Biol.(I)* **84,** 680–691.
69. Kahn, C. R., J. T. Coyle, and A. M. Cohen (1980) Head and trunk neural crest in vitro: autonomic neuron differentiation. *Develop. Biol.* **77,** 340–348.

70. Kirby, M. L. and S. A Gilmore (1976) A correlative histofluorescence and light microscopic study of the formation of the sympathetic trunks in chick embryos. *Anat. Rec.* **186,** 437–450.
71. Kirby, M. L., I. M. Diab and T. Mattio (1978) Development of adrenergic innervation of the iris. *Anat. Rec.***55,** 123–130.
72. Landmesser, L. and G. Pilar (1974a) Synapse formation during embryogenesis on ganglion cells lacking a periphery. *J. Physiol.* **241,** 715–736.
73. Landmesser, L. and G. Pilar (1974b) Synaptic transmission and cell death during normal ganglionic development. *J. Physiol.* **241,** 737–747.
74. Lanier, L. P., A. J. Dunn, and C. van Hartesveldt (1976) Development of Neurotransmitters and Their Function in Brain, in *Reviews of Neuroscience*, Vol. 2 (S. Ehrenpreis and I. J. Kopin, eds.) Raven Press, New York, Vol. 2, pp. 195–256.
75. Le Douarin, N. M. (1977) *Cell Interactions in Differentiation.* (M. Karkinen-Jaaskelainen, ed.) Academic Press, London, pp. 171–190.
76. Le Douarin, N. M. (1980) Migration and differentiation of neural crest cells, in *Current Topics in Neurobiology* Vol. 16 (A. A. Moscona, A. Monroy, and R. Kevin Hunt, eds.) Academic Press, New York, pp. 31–85.
77. Le Douarin, N. M. (1982) In vivo and in vitro studies of the development of the autonomic nervous system (abstract) *Int. Soc. Dev. Neurosci.* **52,** 138.
78. Le Douarin, N. M., M. A. Teillet, C. Ziller, and J. Smith (1978) Adrenergic differentiation of the cholinergic ciliary and Remak ganglia in avian embryo after in vivo transplantation. *Proc. Natl. Acad. Sci. USA* **75,** 2030–2034.
79. Le Douarin, N. M., J. Smith, M. A. Teillet, S. C. Le Lievre, and C. Ziller (1980) The neural crest and its developmental analysis in avian embryo chimeras. *TINS* Feb., 39–42.
80. Levi-Montalcini, R. and R. Amprino (1947) Recherches experimentales sur l'oriigine dur ganglion ciliare dans l'embryon de Poulet. *Arch. Biol.* **58,** 265–288.
81. Levi-Montalcini, R. and V. Hamburger (1953) A diffusible agent of mouse sarcoma, producing hyperplasia of sympathetic ganglia and hyperneurotization of viscera in the chick embryo. *J. Exp. Zool.* **123,** 233–288.
82. Marchi, N. and E. Giacobini (1980) Development and aging of cholinergic synapses. II. Continuous growth of acetylcholine and choline levels in autonomic ganglia and iris of the chick. *Devel. Neurosci.* **3,** 39–48.
83. Marchi, M. and E. Giacobini (1981) *Cholinergic Mechanisms* (G. Pepeu and H. Ladinsky, eds.) Plenum, New York, pp. 25–46.
84. Marchi, M., E. Giacobini, and K. Hruschak (1979) Development of

aging of cholinergic synapses. I. Endogenous levels of acetylcholine and choline in developing autonomic ganglia and iris of the chick. *Dev. Neurosci.* **2**, 201–212.

85. Marchi, M., D. W. Hoffman, I. Mussini, and E. Giacobini (1980a) Development and aging of cholinergic synapses. III. Choline uptake in the developing iris of the chick. *Dev. Neurosci.* **3**, 185–198.

86. Marchi, M., D. W. Hoffman, E. Giacobini, and T. Fredrickson (1980b) Age dependent changes in choline uptake of the chick iris. *Brain Res.* **195**, 423–431.

87. Marchi, M., D. W. Hoffman, E. Giacobini, and T. Fredrickson (1980c) Development and aging of cholinergic synapses. IV. Acetylcholinesterase and choline acetyltransferase activities in autonomic ganglia and iris of the chick. *Dev. Neurosci.* **3**, 235–247.

88. Marchi, M., L. Yurkewicz, E. Giacobini, and T. Fredrickson (1981) Development and aging of cholinergic synapses. V. Changes in nicotinic cholinergic receptor binding in ciliary ganglia and irises of the chicken. *Dev. Neurosci.* **4**, 258–266.

89. Margiotta, J. F. and D. K. Berg (1982) Functional synapses are established between ciliary ganglion neurons in dissociated cell culture. *Nature* **296(5853)**, 152–154.

90. Markow, R. M. (1979) Acetylocholinesterase in neural tube cells. *Arkh. Anat. Histol. Embriol.* (Russ.) **76**, 21–27.

91. McLennan, I. S. and I. A. Hendry (1978) Parasympathetic neuronal survival induced by factors from muscle. *Neurosci Lett.* **10**, 269–273.

92. Mueller, R. A., H. Thoenen, and J. Axelrod (1969) Increase in tyrosine hydroxylase activity after reserpine administration. *J. Pharmacol. Exp. Therap.* **169**, 74–79.

93. Mueller, R. A., H. Thoenen, and J. Axelrod (1970) Inhibition of neuronally induced tyrosine hydroxylase by nicotinic receptor blockade. *Eur. J. Pharmacol.* **10**, 51–56.

94. Mussini, I. and E. Giacobini (1982) Axonal and synaptic changes in the iris of aging chicks (abstract). Int. Soc. Dev. Neurosci. 3rd Int. Meeting. **RT10**, 166.

95. Nishi, R. and D. K. Berg (1972) Dissociated ciliary ganglion neurons in vitro: survival and synapse formation. *Proc. Natl. Acad. Sci. USA* **74(11)**, 5171–5175.

96. Nishi, R. and D. K. Berg, (1981a) Two components from eye tissue that differentially stimulate the growth and development of ciliary ganglion neurons in cell culture. *J. Neurosci.* **1(15)**, 505–513.

97. Nishi, R. and D. K. Berg (1981b) Effects of high K$^+$ concentrations on the growth and development of ciliary ganglion neurons in cell culture. *Develop. Biol.* **87**, 301–307.

98. Noden, D. M. (1975) An analysis of the migratory behavior of avian cephalic neural crest cells. *Develop. Biol.* **42**, 106–130.

99. O'Lague, P. H., K. Obata, P. Claude, E. J. Furshpan, and D. D. Potter (1974) Evidence for cholinergic synapses between dissociated rat sympathetic neuron in cell culture. *Proc. Natl. Acad. Sci. USA* **71**, 3602–3606.

100. Olivieri-Sangiacomo, C., A. Del Fá and C. Gangitano (19833) Developmental distributive pattern of acetylcholinesterase in chick embryo ciliary ganglion. *Develop. Brain Res.* **7**, 61–69.

101. Patterson, P. H. (1978) Environmental determination of autonomic neurotransmitter functions. *Ann. Rev. Neurosci.* **1**, 1–17.

102. Patterson, P. H. (1979) Environmental determination of neurotransmitter functions. *Soc. Neurosci. Symp.* **4**, 172–183.

103. Patterson, P. H., D. D. Potter, and E. J. Furshpan (1978) The chemical differentiation of nerve cells. *Sci. Amer.* **239(1)**, 50–59.

104. Phillipson, O. T. and K. E. Moore (1975) Effects of dexamethasone and nerve growth factor on phenylethanolamine *N*-methyltransferase and adrenaline in organ cultures of newborn rat superior cervical ganglion. *J. Neurochem.* **25**, 295–298.

105. Pilar, G. and J. Tuttle (1978) Ciliary ganglion neurons may form ineffective synapses in dissociated cell culture (abstract). *Soc. Neurosci.* **4**, 593.

106. Potter, D. D., S. C. Landis, and E. J. Furshpan (1981) Chemical differentiation of sympathetic neurons, in *Neurosecretion and Brain Peptides* (J. B. Martin, S. Reichlin, and K. L. Bick, eds.) Raven Press, New York, pp. 275–285.

107. Ravdin, P., R. Nishi, and D. Berg (1978) Alpha-bungarotoxin binding sites and acetylcholine receptors in ciliary ganglion neurons in culture (abstract). *Proc. Soc. Neurosci.* 594.

108. Rothman, T. P., M. D. Gershon, and H. Holtzer (1978) The relationship of cell division to the acquisition of adrenergic characteristics by developing sympathetic ganglion cell precursors. *Develop. Biol.* **65**, 322–341.

109. Smith,D. O. Physiological and Structural Changes at the Neuromuscular Junction During Aging, in *The Aging Brain*, Vol. 20 (E. Giacobini, G. Filogamo, G. Giacobini, and A. Vernadakis, eds.) Raven Press, New York, pp. 123–137.

110. Smith, J., M. Fauquet, C. Ziller, and N. M. LeDouarin (1979) Acetylcholine syntheesis by mesencephalic neural crest cells in the process of migration in vivo *Nature* **282(5741)**, 853–855.

111. Sparber, S. B. and F. E. Shideman (1969) Estimation of catecholamines in the brain of embryonic and newly hatched chickens and the effects of reserpine. *Dev. Psychobiol.* **2**, 115–119.

112. Sparber, S. B. and F. E. Shideman (1970) Elevated catecholamines in thirty-day-old chicken brain after depletion during development. *Dev. Psychobiol.* **3**, 123–129.

113. Stevens, W. F., D. W. Slaaf, J. Hooisma, T. Magchielse, and E. Meeter (1978) Neurotransmission and Specificity of Innervation in

Mixed Culture of Embryonic Ciliary Ganglia and Skeletal Muscle Cells, in *Maturation of the Nervous System, Progress in Brain Research*, Vol. 48 (M. A. Corner, et al., eds.) Elsevier/North-Holland, Amsterdam, pp. 21–29.

114. Suszkiw, J. B. and G. Pilar (1976) Selective localization of a high affinity choline uptake system and its role in ACh formation in cholinergic nerve terminals. *J. Neurochem.* **26**, 1133–1138.

115. Sze, P. Y., M. Marchi, A. C. Towle, and E. Giacobini (1983) Increased uptake of [^{3}H]choline by rat superior cervical ganglion: an effect of dexamethasone. *Neuropharmacol.* **22(6)**, 711–716.

116. Teitelman, G., H. Baker, T. H. Joh, and D. J. Reis (1979) Appearance of catecholamine-synthesizing enzymes during development of rat sympathetic nervous system: possible role of tissue environment. *Proc. Natl. Acad. Sci.USA* **76**, 509–513.

117. Thoenen, H., I. A. Hendry, K. Stockel, U. Paravicini, and F. Oesch (1974) Regulation of Enzyme Synthesis by Neuronal Activity and by Nerve Growth Factor, in *Dynamics of Degeneration and Growth in Neurons* (K. Fuxe, L. Olson, and Y. Zolteman, eds.) Pergamon, Oxford, pp. 315–328.

118. Tuttle, J. B. (1977) Dissociated cell culture of chick ciliary ganglion (abstract). *Soc. Neurosci.* **3**, 1695.

119. Tuttle, J. B. (1983) Interaction with membrane remnants of target myotubes maintains transmitter sensitivity of cultured neurons. *Science* **220**, 977–979.

120. Tuttle, J. B., J. B. Suszkiw, and M. Ard (1980) Long-term survival and development of dissociated parasympathetic neurons in culture. *Brain Res.* **183**, 161–180.

121. Varon, S. and R. Adler (1980) Nerve Growth Factors and Control of Nerve Growth, in *Current Topics in Developmental Biology*, Vol. 16 (A. A. Moscona, A. Monroy, and R. Kevin Hunt, eds.) Academic Press, New York, pp. 207–219.

122. Wright, L. (1981) Cell survival in chick embryo ciliary ganglion is reduced by chronic ganglionic blockade. *Develop. Brain Res.* **1**, 283–286.

123. Ziller, J. Smith, M. Fauquet, and N. M. Le Douarin (1979) Environmentally Directed Nerve Cell Differentiation: In Vivo and In Vitro Studies, in *Development and Chemical Specificity of Neurons, Progress in Brain Research*, vol. 51 (M. Cuenod, G. W. Kreutzberg, and F. E. Bloom, eds.) Elsevier/North-Holland, Amsterdam, pp. 59–74.

Chapter 3

Development of the Sympathoadrenal Axis[a]

Theodore A. Slotkin

1. Introduction

The adrenal medulla is an integral part of the peripheral sympathetic system and participates in physiological and metabolic regulation of autonomic effector organs. Activation of sympathetic
pathways by "fight-or-flight" situations typically leads to increased release of adrenal catecholamines in response to centrally
derived activation of the splanchnic nerve that innervates the
chromaffin cells. In this regard, the adrenal medulla behaves essentially as a sympathetic neuronal projection, with the notable
exceptions that the catecholamines are discharged into the
bloodstream instead of a synaptic cleft, and that in most species
the predominant compound released is epinephrine as opposed
to norepinephrine. Adrenal catecholamines, once released into
the circulation, are free to act upon adrenergic receptors throughout the periphery, initiating events that are designed to enhance
the ability of the organism to survive or respond appropriately to
stress situations.

Over the past two decades, it has become increasingly clear
that the adrenal medulla of the fetal and neonatal organism may

<hr>

[a] This review covers work published through 1983.

play an even greater role in physiological mechanisms than it does in the adult, and that adrenal catecholamines may subserve unique functions during critical periods of development. Studies with newborn humans indicate a surge of sympathoadrenal activity associated with vaginal delivery (32), and it is thought that the marked elevation of circulating catecholamines assists the neonate in establishing normal metabolic, cardiovascular, and respiratory functions (29,50); interference with the catecholamine surge is associated with decreased survival potential (32,56). In some species, and most notably in the Sprague-Dawley rat, sympathetic innervation of autonomic end organs is absent or nonfunctional at birth (18,25,42,58), and in these cases the neonate is totally dependent upon adrenomedullary catecholamines for achieving adrenergic responses to stress. For this reason the rat provides a convenient model in which to study the development of the adrenal medulla and sympathoadrenal function, and much of the information concerning the biochemical and morphological processes occurring in this tissue has been evaluated in that species. On the other hand, classical physiological evaluations are more readily performed on large animals, such as the sheep, pig, or cow—species whose sympathetic nervous systems are much more mature at birth (18,25,29,42,50). It is only relatively recently that comparable studies have been feasible in the rat, ·but available evidence suggests that very similar processes delineate the development of sympathoadrenal function in all species, with major differences only in the time frame over which maturation occurs. Accordingly, it is now possible to review sympathoadrenal development from the point of view of the interrelationships among specific cellular events occurring during the maturation process and physiological function of the adrenal medulla as a catecholamine-secreting tissue that mediates vital processes. In keeping with this theme, the current review will concentrate on these specific events with only passing mention of the embryological origin and differentiation of the primitive chromaffin cells. For detailed information on these aspects of sympathoadrenal development, the reader should consult recent reviews (18,20).

The development of the sympathoadrenal system can best be understood with reference to the major components involved in control of catecholamine synthesis, storage, and release (*see* Fig. 1). In the mature animal, nerve impulse activity in the splanchnic nerve elicits release of acetylcholine from nerve terminals juxta-

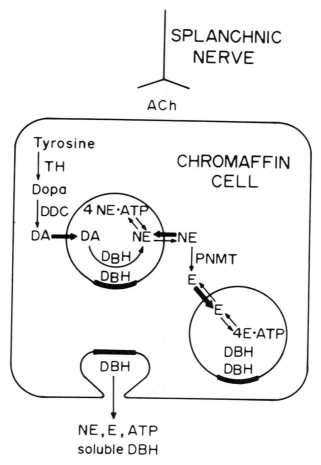

Fig. 1. Schematic diagram of the sympathoadrenal system showing innervation of chromaffin cells by the splanchnic nerve, which re-
leases acetylcholine (ACh). After interacting with nicotinic receptors,
ACh promotes the exocytotic release of norepinephrine (NE), epinephrine (E), adenine nucleotides (notably ATP), and soluble proteins (such
as dopamine β-hydroxylase, DBH) and peptides. ACh also stimulates
the synthesis of catecholamines by induction of tyrosine hydroxylase
(TH), which catalyzes the formation of dopa from tyrosine, and of DBH,
which converts dopamine (DA) into NE. The NE that is formed is either
stored in the granule in a complex with ATP, or returns to the cytoplasm
to be converted to E by the action of phenylethanolamine *N*-
methyltransferase (PNMT); E is then taken back up and stored in the
granule as well.

posed to chromaffin cells. The released transmitter interacts with nicotinic cholinergic receptors on the cell surface to produce an opening of calcium channels in the chromaffin cell membranes; increased intracellular calcium triggers the exocytotic release of the contents of chromaffin granules, including the catecholamines, soluble proteins such as dopamine β-hydroxylase and the chromogranins, adenine nucleotides, and a variety of peptides whose identity and function are only now being explored (*15*). Stimulation of nicotinic receptors also leads to a compensatory induction of tyrosine hydroxylase, the rate-limiting enzyme in catecholamine biosynthesis, as well as increasing the synthesis rate of the chomaffin granules that store and secrete the catecholamines (*15,55*); the granules are also the site where the enzymatic conversion of dopamine to norepinephrine occurs. As will be described below, the development of the granules is a key factor in the maturation of adrenomedullary activity. Mature function of the sympathoadrenal axis thus requires the development of cholinergic receptor systems in chromaffin cells, specialized enzymes, and organelles to permit synthesis, storage, and secretion of catecholamines, and neuronal projections to the cells from the central nervous system; furthermore, target sites for released catecholamines must possess adrenergic receptors linked to cellular function in order for the adrenal medulla to have a significant impact on physiology of the fetus or neonate. The current review will detail how these maturational processes come about.

2. Maturation of the Chromaffin Cell

2.1. Ultrastructural and Biochemical Characteristics

It is now generally accepted that the chromaffin cells arise embryologically from the neural crest, just as sympathetic neurons do (*13,20*). During early migration and differentiation, the cells undergo three distinct stages, distinguished by their histological characteristics: primitive sympathetic cells, pheochromoblasts, and finally, chromaffin cells. Cells showing a chromaffin reaction and containing dense-core granules may be found even as early as 12–15 wk in the human fetus, 20 d of gestation in the rabbit, and during and shortly after the major period of organogenesis (12 d) in the rat, although fully mature-appearing cells often may not be prominent until later (*20*). Thus in the rat, typically electron-dense chromaffin granules are detectable after glutaraldehyde/osmium tetroxide fixation at 12 d of gestation; only norepinephrine-containing granules can be seen initially, and clear-cut differentiation into norepinephrine- and epinephrine-containing subtypes does not appear until the postnatal period (*21,23*).

Biochemical analysis of developing adrenomedullary tissue also indicates an early origin for catecholamine-synthesizing cells. Tyrosine hydroxylase activity is detectable on the first day of incubation in chick embryos, and all enzymes necessary to synthesize catecholamines can be detected by the sixth day (*28*). In the newborn rat, adrenal tyrosine hydroxylase is present in low amounts initially and rises subtantially during the ensuing weeks (*see* Fig. 2) (*2,44*). Additionally, the activity of dopamine β-hydroxylase is quite small at birth; since this latter enzyme is a marker for catecholamine storage granules, it is likely that the development of chromaffin granules (i. e., storage capabilities) is a major factor that determines the developmental profile of adrenomedullary catecholamine levels. In support of this hypothesis, the developmental profiles of dopamine β-hydroxylase activity and of tritiated epinephrine-uptake capacity (a measure of granule function) closely parallel those of the catecholamines themselves (Fig. 2) (*2,51,52*). Furthermore, electron microscopy indicates that the granules appear just before catecholamines can be detected in the rat adrenal and shortly thereafter become more electron-dense than would normally be expected (*23*). Analytical ultracentrifugation of chromaffin granules isolated from newborn rats indicates that early in development the granules are "overstuffed" with catecholamines relative to those seen in mature animals; thus, a greater proportion of granules in the neonate sediment in $1.6M$ sucrose and the amine transport and storage properties of these granules are correspondingly abnormal (*51–53*). These findings all support the view that the maturation of storage capabilities of chromaffin granules is a major limiting factor in the development of adrenal catecholamine stores. Similar conclusions have been reached in nonadrenal catecholamine-containing tissues, such as sympathetic neurons (*6,58*) and central noradrenergic pathways (*31*). In each case, the appearance of storage vesicles, assessed either through biochemical or morphological techniques, appears to be one of the last steps in develop-

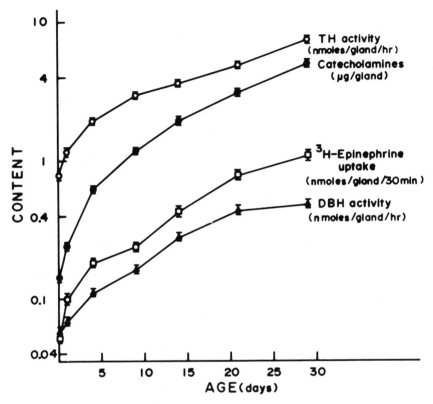

Fig. 2.   Development of key biochemical factors in the neonatal rat adrenal medulla: tyrosine hydroxylase activity (TH), dopamine β-hydroxylase activity (DBH), uptake capacity of the granules for epinephrine, and catecholamine content of the tissue (1,12,51–53).

ment of the tissue and is directly associated with the maturational rise in catecholamine content and the onset of synaptic function.

2.2. Factors Controlling Chromaffin Cell Development

The pivotal role played by storage granules in the development of adrenal catecholamines raises the question of how the synthesis and turnover of these organelles are regulated during ontogeny. Clearly, a major signal determining the number and properties of the granules is the degree of splanchnic nerve input to the developing adrenal medulla. In the mature animal, neurally induced catecholamine secretion is accompained by transsynaptic induction of tyrosine hydroxylase, the rate-limiting enzyme in catecholamine biosynthesis, and of dopamine β-hydroxylase, the enzymatic marker for storage granules (15). Neural signals "turn on" the synthesis of catecholamines, as well as the formation of new granules. Does neural input, then, control the maturational increases in catecholamine storage capabilities? A substantial body of evidence indicates that it does. In the neonatal rat, where the adrenal is relatively granule-deficient, the splanchnic nerve is not functionally connected to the adrenal and is incapable of evoking stimulation of granule synthesis (12,51,52). Correspondingly, granular turnover is low, enabling those granules present to become fully saturated with catecholamines; this explains why the organelles appear to be more dense on sucrose gradients. Splanchnic nerve function becomes established at about the end of the first postnatal week (as shall be discussed in a later section), and is accompanied by a period of intense sympathetic nerve discharge. Accordingly, granule synthesis and turnover become markedly elevated at that time, leading to large increments in dopamine β-hydroxylase activity, and a proliferation of new, immature granules that are typically lower in density and relatively catecholamine-poor (51,52). Interference with this later stage of neural control of development can lead to permanent deficiencies in the synthesis and storage capabilities of adrenal catecholamine granules; sectioning the splanchnic nerve at 3 wk of age leads to arrest of adrenomedullary development (44).

If neural input via the splanchnic nerve is so critical to later developmental stages of the adrenal medulla, what factors might control early development? Understandably, we can rule out neural input during these stages, as major development of adrenal catecholamine synthesis and storage occurs prior to onset of splanchnic nerve control of the tissue. Indeed, there is now substantial evidence that premature exposure of developing chromaffin cells to neural signals may actually interfere with normal maturation. Administration of agents that cause precocious onset of splanchnic neuronal control of adrenomedullary function (thyroid hormone, opiates) invariably result in deficits in ontogenetic increases in catecholamines, tyrosine hydroxylase, and dopamine β-hydroxylase (3,5,34,36,57). Thus, there appear to be two developmental phases for chromaffin cells—one in which factors other than neurogenic input regulate adrenomedullary maturation and a subsequent stage in which neuronal input is a predominant signal.

A variety of studies have explored whether the signals controlling chromaffin cell development during these earlier stages are intrinsic (genetic) and/or environmental in origin. Catecholamine-containing cells that have been excised during fetal development will continue to differentiate for a substantial period in vitro on much the same timetable as they would have done in vivo (*14,45*). This finding supports at least some intrinsic mechanism for control of phenotypic development of catecholaminergic cells that may provide major developmental control of the tissue prior to exposure to transsynaptic signals. The characteristics of developing chromaffin tissue during this preneurogenic period are subject to modification by a variety of humoral factors, most notably hormones. The adrenal medulla is, of course, uniquely situated to receive exposure to relatively high concentrations of cortical steroids that have long been known to influence adrenomedullary function. In the mature animal, adrenocortical atrophy induced by hypophysectomy causes a corresponding reduction in adrenomedullary catecholamine biosynthetic and storage capabilities, with the enzyme responsible for epinephrine biosynthesis (phenylethanolamine N-methyltransferase) exhibiting the greatest sensitivity to steroids (*62,63*). In the developing rat, the differentiation of chromaffin cells into epinephrine- and norepinephrine-specific types also appears to depend on the influence of glucocorticoid hormones (*17*); this accounts for the earlier appearance of noradrenergic cells than adrenergic cells. Later in development, exogenous glucocorticoid administration leads to deficits in catecholamine biosynthesis and storage (*38*), perhaps secondarily to the steroid-induced dramatic reduction in ACTH secretion by the pituitary (*24*).

A second set of hormonal signals regulating chromaffin cell development is provided by the thyroid. Ordinarily, normal levels of circulating thyroid hormones are thought to be required for development of neuronal tissues (including catecholaminergic neurons) to proceed (*4,27*). In the adrenal medulla, however, a reciprocal relationship appears to exist between thyroid hormones and chromaffin cell maturation. Exposure of the neonate to excess thyroid hormone levels produces a slowing of adrenomedullary development, with deficits in tyrosine hydroxylase and dopamine β-hydroxylase, as well as catecholamines, persisting well beyond the point at which thyroid hormone administration is terminated

(*36*); eventually, neonatal hyperthyroidism results in reductions in adrenomedullary volume and in the number of chromaffin cells (*33*). Conversely, perinatal hypothyroidism, induced by maternal and neonatal administration of propylthiouracil, exerts a promoting effect on adrenal development, with a premature rise in enzyme activity and amine levels (*39*). It is unclear at this time whether these effects involve a direct participation of thyroid hormones in chromaffin cell development or whether they are secondary to the profound influence of thyroid status on the timing of onset of splanchnic nerve control of the adrenal medulla (see below). In either case, the influence on adrenomedullary maturation could represent a functional, homeostatic adjustment; because thyroid hormone excess leads to supersensitivity of adrenergic effector sites to catecholamines, a downward adjustment of adrenomedullary activity would thus tend to obviate any abnormality in net adrenergic function.

3. Role of Innervation In Onset of Adrenomedullary Function

3.1. Development of Innervation by the Splanchnic Nerve

In the mature animal, adrenomedullary activity is controlled largely by the central nervous system through stimulation of the splanchnic nerve. The release of catecholamines, rate of catecholamine biosynthesis, and formation of chromaffin granules are all regulated "transsynaptically"; that is, by acetylcholine released from the splanchnic nerve terminals and acting across the cholinergic synapse on the nicotinic receptors of the chromaffin cell (*15*). The importance of such neuronal input in the mature animal can be illustrated by sectioning the splanchnic nerve or by blocking adrenomedullary nicotinic receptors with pharmacological agents such as chlorisondamine (*12,18,43*); either procedure can eliminate the catecholamine secretory response to stressors as well as the induction of tyrosine hydroxylase or dopamine β-hydroxylase.

Because the adrenal medulla goes through substantial development prior to the ingrowth of splanchnic nerve terminals, there is an extended period in which the central nervous system is incapable of regulating these processes. In species that mature relatively early (such as the sheep), central input to the adrenal develops before birth (*19,29*), whereas in many other species functional innervation is not apparent until after delivery (*18,58*). The processes involved in the onset of neural activity have been delineated best for the rat, a species in which neural maturation occurs almost entirely postnatally. Splanchnic control of adrenomedullary function is totally absent in the neonatal rat, appears toward the end of the first week postnatally, and becomes fully mature by 10 d of age (*54,58*). This is best demonstrated by the absence of a catecholamine secretory response to insulin-induced hypoglycemia, a stimulus that ordinarily produces reflexively mediated catecholamine release (*see* Fig. 3) (*5,52*). In order for this reflex to take place, four components must be present: (a) the central nervous system must sense the hypoglycemia and send stimuli down the efferent sympathetic pathway, (b) splanchnic nerve connections to the adrenal medulla must be functional, (c) the adrenal medulla must contain nicotinic receptors, and (d) the receptors

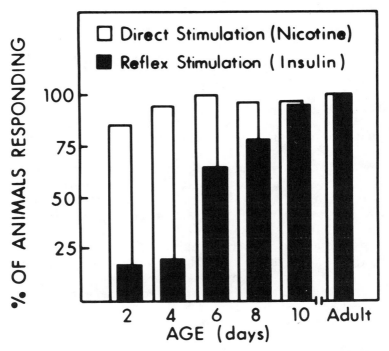

Fig. 3. Development of catecholamine secretory responses in vivo to direct stimulation with nicotine and to reflex stimulation by insulin in the neonatal rat (*5,46,52*).

must be coupled to the secretory response. Studies with excised neonatal adrenals incubated in vitro with depolarizing concentrations of potassium indicate that neonatal chromaffin cells contain all the machinery necessary to permit exocytotic release of granule contents (52); high potassium evokes profound release of catecholamines that is completely prevented by removing calcium from the incubation medium, thus fulfilling the major criteria for exocytosis (*see* Fig. 4). Similarly, administration of a test dose of nicotine in vivo (Fig. 3) readily causes a catecholamine secretory

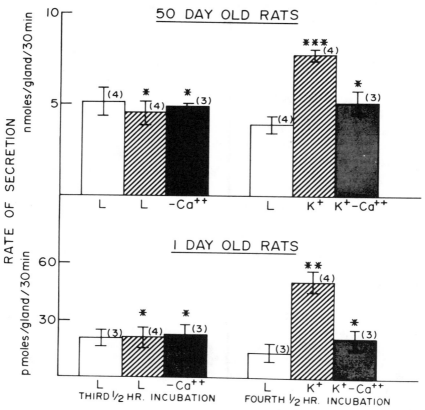

Fig. 4.   Demonstration of exocytosis in excised adrenals of adult (top) and neonatal rats (bottom) incubated in Locke's solution (L) with high potassium with or without calcuim in the medium (52). Panels on the left indicate spontaneous release during control periods with normal Locke's or with Locke's without calcium. Panels on the right show the ability of high levels of potassium to cause secretion and the requirement of calcium for the secretory effect.

response (5,8,46), indicating that the receptors and receptor-mediated responses are present in the adrenal and are linked to the exocytotic release mechanisms even at 2 d of age, well before the onset of neurally evoked response capabilities. Nerve recordings in neonatal rats also indicate that the central nervous system is fully responsive to stimuli that evoke sympathoadrenal reflexes; both hypoglycemia and asphyxia in 2-d-olds can elicit activation of preganglionic sympathetic efferents equivalent to that seen in more mature animals (60). It is therefore likely that the neonatal adrenal is unresponsive to neurogenic stimuli because of deficiencies in the function of splanchnic connections to the tissue; this hypothesis is supported by: (a) ultrastructural studies showing that few nerve terminals are present in the adrenal at birth, but that marked proliferation occurs during the ensuing week (41), and (b) measurements of choline acetyltransferase activity, a marker for cholinergic nerve terminals, that indicate very low values at birth, but a sixfold increase in the next few days (33).

In the mature rat, control of catecholamine biosynthesis proceeds through the same neurogenic pathway as does secretion; exposure of the adrenal to the acetylcholine released by the splanchnic nerve produces induction of both tyrosine hydroxylase and dopamine β-hydroxylase (15). Again, this mechanism does not operate early in postnatal life in the rat (12) since no increase in enzyme activity is seen with reflex stimulation (e.g., reserpine) given to 2- or 4-d-old rats (*see* Fig. 5). As is true for catecholamine secretion, the onset of neurogenic control of catecholamine biosynthesis appears toward the end of the first week postnatally and becomes fully mature shortly thereafter (12,54,58). Direct stimulation of the tissue with nicotine produces tyrosine hydroxylase induction in most animals even at 2 d of age (46), once again confirming that the adrenal possesses functional receptors and would be capable of responding (if stimulated) well before the actual appearance of the neurogenic response. The absence of neurogenic control of catecholamine biosynthesis in the neonate has a significant impact on drug effects on the neonatal adrenal. For example, there is a prolongation of the time required to restore amine levels after depletion of catecholamines by administration of reserpine or opiates to neonates (2,5,12,34), a situation similar to that seen in mature rats in which the adrenals have been denervated (3,43).

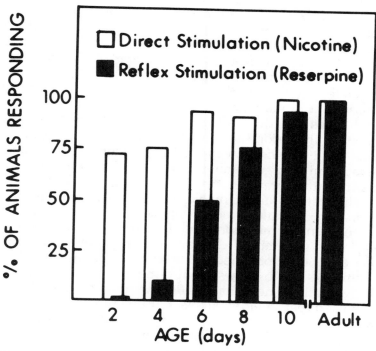

Fig. 5. Development of induction capabilities of tyrosine hydroxylase in response to direct, in vivo stimulation with nicotine or reflex stimulation caused by a catecholamine-depleting drug (reserpine) in the neonatal rat (8,12,46).

Just as the development of chromaffin cells is under control of hormonal factors, the onset of functional innervation of the adrenal by the splanchnic nerve appears to be subject to endocrine coordination. In this case, a predominating role is played by thyroid hormones. Administration of a single dose of triiodothyronine to a 1-d-old rat will produce functional splanchnic innervation of the adrenal medulla within 24 h (36,37); this is in keeping with the ability of thyroid hormones to promote synaptogenesis at nicotinic cholinergic sites (59). Studies with animals rendered hypothyroid with propylthiouracil also confirm that normal circulating levels of thyroid hormones participate in establishing the timetable at which functional innervation ordinarily occurs; hypothyroidism produces a delay in the onset of splanchnic control of adrenomedullary secretion of catecholamines in response to hypoglycemia (26). Thus, hormonal status can influence adrenomedullary development through two distinct mechanisms: direct effects on chromaffin cell development and indirect effects through influencing the onset of neural input to the adrenal.

Immediately following the appearance of splanchnic nerve transmission to the adrenal medulla, the sympathoadrenal system undergoes a period of transient hyperactivity that is most prominent in the second to third postnatal week in the rat. Evidence for this otherwise unexpected phenomenon is provided by the elevation in storage granule turnover and the presence of numerous immature granules that have abnormally low densities and unusual catecholamine uptake and storage properties (51,52). The immature granules resemble those seen during extensive sympathoadrenal stimulation in adult animals (55). Corroborative evidence for a generalized elevation of sympathetic tone during this phase has been obtained by direct nerve recordings of sympathetic trunks and by assessment of neurotransmitter turnover rates in sympathetic neurons (49,60). The significance of transient sympathetic hyperactivity is unclear at this time, but it does appear to be related to the onset of control of sympathetic outflow by afferent inputs such as baroreceptors (9). As discussed in a later section, hyperactivity may aid in overcoming the initial subsensitivity of autonomic effector organs to adrenergic stimuli and almost certainly aids in achieving the final set-point of end-organ reactivity to catecholamines.

The close coordination of hormonal and neurogenic factors regulating the ontogeny of nerve input and chromaffin cell development indicates why treatments that cause precocious appearance of sympathetic innervation should not be viewed as beneficial, or even innocuous. Because factors such as development of catecholamine synthesis and storage, sympathetic impulse activity and synaptic transmission, and end-organ sensitivity are all interrelated, any factor that upsets the timing of these events is likely to result in a net deleterious result as the animal matures. Thus, neonatal hyper- or hypothyroidism both have long-term negative effects on sympathetic and sympathoadrenal function that may appear only as the animal reaches adulthood (26,36,37,39). Similar deficits occur with other agents that influence the development of sympathetic tissues, their receptors, or their nerve supplies (2,5,8,34,46,57).

3.2. Adrenomedullary Function Prior to Neurogenic Control

Despite the absence of functional neuronal connections to the immature adrenal medulla, the tissue is, in fact, not passive to stressful stimuli. Instead, catecholamine release may proceed by a nonneurogenic mechanism that is not present in mature animals. A unique secretory mechanism in fetal and neonatal calves was first described by Comline and Silver (*18*), who showed that asphyxia produced release of adrenal catecholamines well before the onset of splanchnic control of the tissue. In the neonatal rat, similar nonneurogenic secretory responses have been described for a number of stressors (*7,12,16,54,56*). The differences between neurogenic and nonneurogenic release mechanisms can be illustrated by the acute effects of hypoxia (*see* Table 1). In mature animals, stress-induced catecholamine secretion is reflexively mediated and can therefore be blocked by chlorisondamine, a nicotinic antagonist that prevents splanchnic nerve signals from reaching the chromaffin cells of the adrenal medulla. In neonatal rats, hypoxia produces a catecholamine secretory response, but chlorisondamine cannot block the effect. In the neonates, then, catecholamine secretion is occurring but is not caused by neuronal stimulation of the adrenal.

The nonneurogenic mechanism for catecholamine release in the rat disappears concurrently with the appearance of neurogenic control of the adrenal (*12,16,56*); after the end of the first postnatal week, when neurogenic control has appeared, the secretory response to hypoxia is successfully blocked by the nicotinic antagonist (Table 1). The question remains, however, whether the loss

Table 1
Secretion of Adrenal Catecholamines Caused by Hypoxia
in the Neonatal Rat[a]

Age, d	Exposure	Degree of secretion, percent	
		Saline	Chlorisondamine
1	5% oxygen, 75 min	22 ± 5^b	26 ± 6^b
8	7% oxygen, 120 min	38 ± 4^b	4 ± 5

[a]Differentiation of neurogenic (chlorisondamine-sensitive) and nonneurogenic (chlorisondamine-insensitive) components) (*56*).
[b]Significant depletion. Data represent ±SEM.

of nonneurogenic response capabilities is causally related to the development of splanchnic nerve function. New studies from our laboratory support the view that the onset of nerve stimuli does participate in terminating the phase in which nonneurogenic responses are possible. Acceleration of the onset of splanchnic nerve function caused either by repeated maternal stress during late gestation, or by neonatal hyperthyroidism, is invariably accompanied by the loss of nonneurogenic catecholamine secretory responses. Surgical denervation of the adrenals at 3 d of age, a point just before the onset of splanchnic nerve function, leads to the preservation of the nonneurogenic secretory mechanism (manuscript in preparation). Furthermore, prolonged denervation of mature adrenal medullae may lead to a partial restoration of at least some nonneurogenic response capabilities. For example, administration of morphine to adult rats causes a secretion of adrenal catecholamines that is entirely reflexive in character, i.e., secretion is blocked completely by acute denervation of the adrenals (1,3,64). Upon chronic denervation, however, a small but significant secretory response to morphine reappears and is not neurally dependent. Although chronic interference with nerve input to mature adrenals has been found to restore at least some nonneurogenic response components to a variety of pharmacologic agents, in no case has a full response capability been found that is comparable to that seen in the neonate. It is thus likely that the onset of neuronal input plays some role in the loss of nonneurogenic responses potential, but that other factors contribute to this phenomenon that are not dependent solely upon nerve input.

The existence of a nonneurogenic mechanism for catecholamine secretion raises the question of processes other than nerve input that might control the amount of catecholamine released. In this regard, it is important to note that opiate receptors are found associated with chromaffin cells and that these cells may also be particularly rich in opiate peptides (61). Recently, it has been noted that the changes that take place in chronically denervated mature rat adrenals include a dramatic rise in the level of opiate peptides (47). The coincidental reappearance of at least some of the components of nonneurogenic responses in denervated adult adrenals and increased levels of enkephalin-like peptides raises the possibility that adrenomedullary opioids could be involved in modulating the nonneurogenic catecholamine secretion seen in immature (non-innervated) adrenals. Alternatively, high levels of endorphins released by the pituitary could act as hormones affecting adrenomedullary release. In either case, the hypothesis that endogenous opioids are involved in neonatal adrenal function in some fashion is supported by recent studies in which administration of naloxone, an opiate antagonist, was found to potentiate, and methadone to inhibit, the nonneurogenic catecholamine release in the neonatal rat (16). In contrast to the situation in neonates, neither naloxone nor methadone given to adults had a significant effect on neurogenic secretion. Potentiation by naloxone of neonatal catecholamine release suggests strongly that endogenous opiates are serving in vivo to modulate the nonneurogenic process.

It should be noted that the catecholamine secretion seen in the absence of nerve input in the immature adrenal differs in at least one other major way from the neurogenic secretion seen in mature animals. In the absence of nerve input, no induction of catecholamine biosynthetic enzymes can occur to help replenish depleted stores of amine (12); consequently, after neonatal secretion of catecholamine stores, restitution to normal levels may take a prolonged period because of the failure to induce tyrosine hydroxylase or dopamine β-hydroxylase. Apparently, no nonneurogenic mechanism exists that enables the adrenal to adjust catecholamine biosynthetic capabilities to match physiologic demand for the released amines. Consequently, the nonneurogenic secretory mechanism should probably be viewed as a mechanism for dealing with acute demands on the fetus and neonate.

4. The Functional Significance of Adrenomedullary Cathecholamine Release in the Fetus and Neonate

It is clear that the adrenal medulla possesses the functional capability to secrete catecholamines early in development, prior to the onset of neurogenic control of the tissue. The question remains as to whether the catecholamines thus released are of functional significance to the fetus or neonate. Ordinarily, the release of adrenal catecholamines is associated with either extreme stress or with specific responses, particularly to metabolic situations that require adrenergic stimulation. In the fetus, most of the metabolic needs are met by the mother and involve placental transfer of oxygen and essential nutrients. The fetus must respond, however, to challenges posed by various perinatal stressors that could compromise placental blood flow or delivery of these factors. In particular, the stress of birth is associated with an astonishing increase in plasma catecholamine levels, much of which is derived from adrenal sources (29,32,50); a number of events in the perinatal period would definitely require catecholaminergic stimulation to aid in neonatal survival (29,50). First, the absorption of lung fluid and secretion of surfactant are necessary components in the respiratory adjustment of the newborn to extrauterine life; both these factors are subject to β-adrenergic control and are thus potentially dependent upon sympathoadrenal activity. Second, the birth process is potentially associated with hypoxia and/or hypercapnia, situations that also usually require sympathetic stimulation in order for appropriate circulatory and metabolic adjustments to be made. Last, the neonate can no longer depend on placental delivery of nutrients and must therefore regulate his own metabolic balance; again, the homeostatic mechanisms participating in these processes are potentially sympathoadrenal in origin.

For obvious reasons, the majority of studies on the physiological role of catecholamines in the fetus and neonate have involved studies in fairly large animals that tend to have a relatively mature sympathetic nervous system at birth (18,19,25,29,50). The difficulty, then, is in separating adrenergic effects that depend on adrenomedullary activity from those that involve sympathetic neuronal innervation of autonomic effectors. Thus, although hypoglycemia and hypoxia are capable of causing adrenal catecholamine release from the fetal and neonatal sheep adrenal, it is clear that major components of the net physiological response to these situations involve catecholamines derived instead from sympathetic neurons. Recent experiments with chemically sympathectomized or adrenomedullectomized sheep have confirmed that major contributions to circulating catecholamines during fetal/neonatal stress involve neuronal elements as well as the adrenal (30).

In order to provide a definitive role for adrenal catecholamines, we must consider a species that lacks functional innervation of sympathetic target tissues at birth, namely the rat (54,58). As already described, the rat adrenal does not receive neurogenic input at birth, but is capable of direct secretory responses to some stressors. Notably, hypoglycemia is not capable of evoking any adrenomedullary secretion whatsoever until the ingrowth of the adrenal nerve supply has occurred (5,8,52). That the neonatal rat lacks the ability to produce a sympathetic response to hypoglycemia indicates toleration of potential fluctuations in blood glucose without the need for sympathoadrenal adjustments; to some extent, this is offset by the nearly continuous feeding of the neonatal rat by the dam. It should be noted that in larger species that usually do not nurse continuously, adrenal responses to hypoglycemia or to feeding are quite prominent, but again require the presence of an intact nerve supply to the adrenal (that has generally developed by birth in those species) (18,19,29,30,50).

In contrast to the lack of adrenal reactivity to hypoglycemia in neonatal rats, adrenal catecholamine release is prominent in response to hypoxia, a stress to which the neonate is almost certainly exposed during the birth process and immediately thereafter (56). As already discussed, this release is nonneurogenic, i.e., it does not require the presence of functional nerve supplies to the adrenal; again, this is mechanism that has also been described in larger species, but is primarily replaced by neurogenic secretion by birth in those animals. Exposure of neonatal rats to 5% oxygen for up to 90 min produces a small degree of mortality (*see* Table 2). The importance of the adrenomedullary catecholamine release that accompanies this stress can be demon-

TABLE 2
Hypoxia-Induced Mortality: Effects of Surgical or
Drug Treatments in Neonatal Rats

Pretreatment	Mortality, percent
None	29
Sham surgery	35
Adrenalectomy	94[a]
Adrenalectomy and corticosterone	100[a]
Bretylium	20
Phenoxybenzamine (systemic)	100[a]
Phenoxybenzamine (central)	0
Atenolol	39
ICI-118551	92[a]

[a]Significant increase in mortality from group labeled "none." Exposure lasted 90 min at 5% oxygen (56).

strated by the marked increase in mortality that occurs if the animals are bilaterally adrenalectomized 24 h prior to exposure to hypoxia; mortality in adrenalectomized animals is nearly total when compared to sham-operated littermates. Massive doses of corticosterone (17 mg/kg) provide no protection from the hypoxia-induced mortality in adrenalectomized pups, and measurement of circulating corticosterone by radioimmunoassay confirms that survival during neonatal hypoxia is not dependent upon the adrenocortical component of adrenal activity. In this particular case, we can additionally rule out the potential participation of sympathetic neurons, which are nonfunctional at this stage of development (58); interference with sympathetic neuronal release of catecholamines by bretylium, which does not affect adrenomedullary catecholamine release, does not evoke an increase in susceptibility to hypoxia-induced mortality. Accordingly, the neonatal rat model demonstrates conclusively that it is the secretion of catecholamines specifically from the adrenal medulla that provides the protective signal enabling survival during hypoxia.

If catecholamines derived from the neonatal adrenal medulla are the key to the physiological adjustments during hypoxic stress, then blockade of adrenergic receptors should exacerbate the effects of exposure to low oxygen. Administration of phenoxybenzamine, an α-adrenergic antagonist, to neonatal rats was found to produce an increase in hypoxia-induced mortality equivalent to that seen with adrenalectomy (Table 2) (56); phenoxybenzamine introduced directly into the central nervous system by intracisternal injection did not cause any significant increase in mortality, suggesting that the effect of systemic phenoxybenzamine resulted from actions in the periphery. Cardiac-specific β_1-receptor blockade with atenolol did not cause any such promotion of the lethal effects of hypoxia, but β_2-blockade with ICI-118551 did. Thus, the protective effect of catecholamines appears to involve actions outside the central nervous system involving specifically the α-adrenergic and β_2-adrenergic receptors that are present at neonatal effector sites. The β_2-sites are known to be involved in neonatal respiratory adjustments (30), but the α-receptors are not: the adverse effects of phenoxybenzamine are found even when total anoxia (100% nitrogen atmosphere) is substituted for hypoxia; if the catecholamines were acting on α-receptors involved in respiratory function, then their protective effect would have been absent in anoxia, a situation in which respiratory stimulation would not aid survival. It is thus apparent that potential participation of the amines in circulatory and metabolic regulation should receive attention in future work.

A final major factor that should be resolved is the reason for the impressive degree of catecholamine release associated with the stress of birth (*32*) and for the period of intensive sympathetic hyperactivity that accompanies the onset of central control of sympathetic tone (*9,59,60*). In terms of the necessity for a high degree of adrenergic effect during the birth process, excessive catecholamine release is consistent with homeostatic requirements for catecholamine action. Although neonatal tissues possess functional adrenergic receptors even before the development of autonomic innervation of effector sites, end-organ function tends to be substantially subsensitive to catecholamines relative to the reactivity seen at maturity (*35,48*). Consequently, a greater catecholamine release may be compatible simply with maintenance of typical adrenergic effects. More intriguing possibilities for the reasons for high sympathoadrenal activity in the neonate include the likelihood that bursts of activity help to adjust the final sensitivity of adrenergic effector sites during later developmental phases. For example, β-adrenergic sensitivity of the heart to circulating catecholamines is quite low at birth in the rat and actually goes through its greatest increases when sympathoadrenal activity is high (*48*). Recent studies of sympathetic ganglion development also indicate that circulating epinephrine from the adrenal medulla may play a major role in maturation of elements of sympathetic nerve projections throughout the organism (*40*); adrenalectomy was observed to retard ganglionic development, an effect that was reversed by epinephrine administration, but not by adrenal steroids. Thus, the participation of adrenal catecholamines in the fetus and neonate may extend well beyond the classical short-term physiological roles usually assigned to these substances and could include trophic effects on elements involved in development of neuroeffector transmission.

5. Conclusions

Control of adrenomedullary development exhibits two distinct phases. Initially, when splanchnic nerve input is absent, chromaffin cells develop largely according to intrinsic genetic information and are subject to modification by hormonal and other metabolic inputs. These include thyroid and steroid factors. After the establishment of nerve connections, further development of the tissue occurs under transsynaptic control, involving stimulation of chromaffin cells by the acetylcholine released from splanchnic nerve terminals. Prominent indices of both phases of chromaffin cell maturation can be provided by examination of cellular constituents that are specific to catecholaminergic cells, notably tyrosine hydroxylase activity and the number and properties of the chromaffin granules. Indeed, developmental changes in the granules appear to play a major role in determining the functional maturation of adrenomedullary cells.

The release of catecholamines from the developing adrenal medulla also goes through two phases. The mature mechanism by which catecholamine secretion and biosynthesis are normally regulated, namely transsynaptic neurogenic input from the splanchnic nerve, is present at birth in many large animal species, but does not appear until the end of the first week of postnatal life in the rat. Prior to that time, some (but not all) stressors can elicit catecholamine secretion by a nonneurogenic mechanism that operates in the absence of central control of the adrenal medulla. The nonneurogenic mechanism appears to be modulated in part by endogenous opiate mechanisms and disappears concurrently with the onset of functional neuronal connections to the adrenal. The presence of adrenomedullary catecholamine release early in life enables critical survival functions to be expressed without the requirement for central nervous system input. Additionally, circulating catecholamines derived from the adrenal may play key roles in the adjustment of autonomic end-organ sensitivity to sympathetic effectors, as well as in development of sympathetic nerve elements themselves.

Final consideration must be given to the historical role of the adrenal medulla as a model for studying other branches of the sympathetic system. Results obtained for development of function of sympathetic neurons and synapses indicate maturational profiles resembling that of the adrenal. Thus, the onset of sympathetic control of cardiac function in the rat displays much the same time course and drug sensitivities as in the adrenal (9,54,58). Cardiac β-adrenergic receptors are present and functional at birth and are coupled to physiological and biochemical responses of the myocardium, just as the nicotinic receptors in the adrenal are present and functional (6,8,46). As is true of the adrenal, reflex response capabilities in the cardiac pathway do not appear until the end of the first postnatal week and the onset of neurogenic responses corresponds to the development of sympathetic ganglionic synapses and sympathetic nerve terminals in the myocardium (5,6); again, this resembles the case seen in the adrenal medulla. Agents that elicit precocious synaptogenesis or retard synaptic development produce the same effects in the sympathoadrenal and cardiac-sympathetic axes (5,6,8,11,22, 36,37,57), and the preponderant role of thyroid hormones in determining the time course of development has been noted in both (26,36,37,39,59). Finally, nonneurogenic effects have been observed in the cardiac–sympathetic pathway (10) just as in the sympathoadrenal axis. These results indicate that the sympathoadrenal system is likely to prove useful in further studies of the mechanisms underlying and controlling neuronal development and neonatal physiology.

Acknowledgments

This research is supported by USPHS HD-09713.

References

1. Anderson, T. R. and T. A. Slotkin (1975) Effects of morphine on the rat adrenal medulla. *Biochem. Pharmacol.* **24,** 671–679.
2. Anderson, T. R. and T. A. Slotkin (1975) Maturation of the adrenal medulla IV. Effects of morphine. *Biochem. Pharmacol* **24,**2469–2474.
3. Anderson, T. R. and T. A. Slotkin (1976) The role of neural input in the effects of morphine on the rat adrenal medulla. *Biochem. Pharmacol.* **25,** 1071–1074.
4. Balazs, R., W. A. Cocks, J. T. Eayrs, and S. Kovacs (1971) Biochemical Effects of Thyroid Hormones on the Developing Brain, in *Conference on Hormones in Development* (M. Hamburgh and E. J. W. Barrington, eds.) Appleton-Century-Crofts, New York, pp. 357–379.
5. Bareis, D. L. and T. A. Slotkin (1978) Responses of heart ornithine decarboxylase and adrenal catecholamines to methadone and sympathetic stimulants in developing and adult rats. *J. Pharmacol. Exp. Ther.* **205,** 164–174
6. Bareis, D. L. and T. A. Slotkin (1980) Maturation of sympathetic neurotransmission in the rat heart. I. Ontogeny of the synaptic vesicle uptake mechanism and correlation with development of

synaptic function. Effects of neonatal methadone administration on development of synaptic vesicles. *J. Pharmacol. Exp. Ther.* **212,** 120–125.

7. Bartolome, J., M. Bartolome, F. Seidler, T. R. Anderson, and T. A. Slotkin (1979) Effects of early postnatal guanethidine administration on adrenal medulla and brain of developing rats. *Biochem. Pharmacol.* **25,** 2378–2390.

8. Bartolome, J., C. Lau, and T. A. Slotkin (1977) Ornithine decarboxylase in developing rat heart and brain: Role of sympathetic development for responses to autonomic stimulants and the effects of reserpine on maturation. *J. Pharmacol. Exp. Ther.* **202,** 510–518.

9. Bartolome, J., E. Mills, C. Lau, and T. A. Slotkin (1980) Maturation of sympathetic neurotransmission in the rat heart. V. Development of baroreceptor control of sympathetic tone. *J. Pharmacol. Exp. Ther.* **215,** 569–600.

10. Bartolome, J., S. M. Schanberg, and T. A. Slotkin (1981) Premature development of cardiac sympathetic neurotransmission in the fetal alcohol syndrome. *Life Sci.* **28,** 571–576.

11. Bartolome, J., F. J. Seidler, T. R. Anderson, and T. A. Slotkin (1976) Effects of prenatal reserpine administration on development of the rat adrenal medulla and central nervous system. *J. Pharmacol. Exp. Ther.* **179,** 293–302.

12. Bartolome, J. and T. A. Slotkin (1976) Effects of postnatal reserpine administration on sympathoadrenal development in the rat. *Biochem. Pharamacol.* **25,** 1513–1519.

13. Black, I. B. (1978) Regulation of autonomic development. *Ann. Rev. Neuroscience* **1,** 281–214,

14. Black, I. B. (1982) Stages of neurotransmitter development in autonomic neurons. *Science* **215,** 1198–1204,

15. Carmichael, S. W. *The Adrenal Medulla,* vol. 1. Eden Press, Westmount.

16. Chantry, C. J., F. J. Seidler, and T. A. Slotkin (1982) Non-neurogenic mechanism for release of catecholamines from the adrenal medulla of neonatal rats: Possible modulation by opiate receptors. *Neuroscience* **7,** 673–678.

17. Ciaranello, R. D., D. Jacobowitz, and J. Axelrod (1973) Effect of dexamethasone on phenylethanolamine N-methyltransferase in chromaffin tissue of the neonatal rat. *J. Neurochem.* **20,** 799–805.

18. Comline, R. S. and M. Silver (1966) The development of the adrenal medulla of the fetal and new-born calf. *J. Physiol.* (London) **183,** 305–340,

19. Comline, R. S., I. A. Silver, and M. Silver (1965) Factors responsible for the stimulation of the adrenal medulla during asphyxia in the fetal lamb. *J. Physiol.* (London) **178,** 211–239,

20. Coupland, R. E. (1980) The Development and Fate of Catecholamine Secreting Endocrine Cells, in *Biogenic Amines in Development* (Parvez, H. and Parvez, S., eds.) Amsterdam, Elsevier/North Holland, pp. 3–28.
21. Daikoku, S., O. Takashi, A. Takahashi, and M. Sako (1969) Studies on the functional development of rat adrenal medullary cells. Tokushima *J. Exp. Med.* **16**, 153–162.
22. Dailey, J. W. (1978) Effects of maternally administered resperpine on the development of the cold stress response and its possible relation to adrenergic nervous system function. *Res. Comm. Chem. Path. Pharmacol.* **19**, 389–402
23. Elfvin, L.-G. (1967) The development of the secretory granules in the rat adrenal medulla. *J. Ultrastruct. Res.* **17**, 45–62.
24. Ganong, W. F. (1963) The Central Nervous System and the Synthesis and Release of Adrenocorticotropic Hormone, in: *Advances in Neuroendocrinology* (A. V. Nalbandov, ed.) Univ. of Illinois Press, Urbana, pp. 92–157.
25. Gootman, P. M., N. M. Buckley, and N. Gootman (1979) Postnatal maturation of neural control of the circulation. *Rev. Perinatal Med.* **3**, 1–72.
26. Gripois, D. and A. Diarra (1983) Adrenal epinephrine depletion after insulin-induced hypoglycaemia in young hypothyroid rats. *Mol. Cell. Endocrinol.* **30**, 241–245.
27. Hamburgh, M., L. A. Mendoza, J. F. Burkhart, and F. Weil (1971) Thyroid-Dependent Processes in the Developing Nervous System, in *Conference on Hormones in Development* (M. Hamburgh and E. J. W. Barrington, eds.) Appleton-Century Crofts, New York, pp. 403–415.
28. Ignarro, L. J. and F. E. Shideman (1968) Appearance and concentrations of catecholamines and their biosynthesis in the embryonic and developing chick. *J. Pharmacol. Exp. Ther.* **159**, 38–48.
29. Jones, C. T. (1980) Circulating Catecholamines in the Fetus, Their Origin, Actions, and Significance, in *Biogenic Amines in Development* (H. Parvez and S. Parvez, eds.) Elsevier/North Holland, Amsterdam, pp. 63–86.
30. Jones, C. T. (1983) Adrenal medullary activity in fetal sheep. *Prog. Neuro-Pyschopharmacol. Biol. Psychiat.* Suppl., 41.
31. Kirksey, D. F., F. J. Seidler, and T. A. Slotkin (1978) Ontogeny of (-)^{3}H-norepinephrine uptake into synaptic storage vesicles of rat brain. *Brain Res.* **150**, 367–375.
32. Lagercrantz, H. and P. Bistoletti (1973) Catecholamine release in the newborn infant at birth. *Pediat. Res.* **11**, 889–893.
33. Lau, C. US Environmental Protection Agency, personal communication.
34. Lau, C., M. Bartolome, and T. A. Slotkin (1977) Development of

central and peripheral catecholaminergic systems in rats addicted perinatally to methadone. *Neuropharmacol.* **16**, 473–478.

35. Lau, C., S. P. Burke, and T. A. Slotkin (1982) Maturation of sympathetic neurotransmission in the rat heart. IX. Development of transsynaptic regulation of cardiac adrenergic sensitivity. *J. Pharmacol. Exp. Ther.* **223**,

36. Lau, C. and T. A. Slotkin (1979) Accelerated development of rat sympathetic neurotransmission caused by neonatal triiodothyronine administration. *J. Pharmacol. Exp. Ther.* **208**, 485–490.

37. Lau, C. and T. A. Slotkin (1980) Maturation of sympathetic neurotransmission in the rat heart. II. Enhanced development of presynaptic and postsynaptic components of noradrenergic synapses as a result of neonatal hyperthyroidism. *J. Pharmacol. Exp. Ther.* **212**, 126–130.

38. Lau, C. and T. A. Slotkin (1981) Maturation of sympathetic neurotransmission in the rat heart. VII. Suppression of sympathetic responses by dexamethasone. *J. Pharmacol. Exp. Ther.* **216**, 6–11.

39. Lau, C. and T. A. Slotkin (1982) Maturation of sympathetic neurotransmission in the rat heart. VIII. Slowed development of noradrenergic synapses resulting from hypothyroidism. *J. Pharmacol. Exp. Ther.* **220**, 629–636.

40. Markey, K. A. and P. Y. Sze (1981) Influnece of adrenal epinephrine on postnatal development of tyrosine hydroxylase activity in the superior cervical ganglion. *Devl. Neurosci.* **4**, 267–272.

41. Mikhail, Y. and Z. Mahran (1965) Innervation of the cortical and medullary portion of the adrenal gland of the rat during postnatal life. *Anat. Rec.* **152**, 431–437.

42. Pappano, A.J. (1977) Ontogenetic development of autonomic neuroeffector transmission and transmitter reactivity in embryonic and fetal hearts. *Pharmacol. Rev.* **29**, 3–33.

43. Patrick, R. L. and N. Kirshner (1971) Acetylcholine-induced stimulation of catecholamine recovery in denervated rat adrenals after reserpine-induced depletion. *Mol. Pharmacol.* **7**, 389–396.

44. Patrick, R. L. and N. Kirshner (1972) Developmental changes in rat adrenal tyrosine hydroxylase, dopamine beta-hydroxylase, and catecholamine levels: Effect of denervation. *Devl. Biol.* **29**, 204–213.

45. Patterson, P. H. (1978) Environmental determination of autonomic neurotransmitter functions. *Ann. Rev. Neurosci.* **1**, 1–17.

46. Rosenthal, R. N. and T. A. Slotkin (1977) Development of nicotinic responses in the rat adrenal medulla and long-term effects of neonatal nicotine administration. *Brit. J. Pharmacol.* **60**, 59–64.

47. Schultzberg, M., J. M. Lundberg, T. Hokfelt, L. Terenius, J. Brandt, R. P. Edle, and M. Goldstein, (1978) Enkephalin-like immunoreactivity in gland cells and nerve terminals of the adrenal medulla. *Neuroscience* **3**, 1169–1186,

48. Seidler, F. J. and T. A. Slotkin (1979) Presynaptic and postsynaptic contributions to ontogeny of sympathetic control of heart rate in the preweanling rat. *Brit. J. Pharmacol.* **65**, 531–534.

49. Seidler, F. J. and T. A. Slotkin (1981) Development of central control of norepinephrine turnover and release in the rat heart: Responses to tyramine, 2-deoxyglucose, and hydralazine. *Neuroscience* **6**, 2081–2086,

50. Silver, M. and A. V. Edwards (1980) The Development of the Sympathoadrenal System with an Assessment of the Role of the Adrenal Medulla in the Fetus and Neonate, in: *Biogenic Amines in Development* (H. Parvez and S. Parvez, eds.) Elsevier/North Holland, Amsterdam, pp. 147–212.

51. Slotkin, T. A. (1973) Maturation of the adrenal medulla. I. Uptake and storage of amines in isolated storage vesicles of the rat. *Biochem. Pharmacol.* **22**, 2023–2032.

52. Slotkin, T. A. (1973) Maturation of the adrenal medulla. II. Content and properties of catecholamine storage vesicles of the rat. *Biochem. Pharmacol.* **22**, 2033–2044.

53. Slotkin, T. A. (1975) Maturation of the adrenal medulla. III. Practical and theoretical considerations of age-dependent alterations in kinetics of incorporation of catecholamines and noncatecholamines. *Biochem. Pharmacol.* **24**, 89–97.

54. Slotkin, T. A., C. J. Chantry, and J. Bartolome (1982) Development of central control of adrenal catecholamine biosynthesis and release. *Adv. Biosci.* **36**, 95–102.

55. Slotkin, T. A. and N. Kirshner (1973) Recovery of rat adrenal amine stores after insulin administration. *Mol. Pharmacol.* **9**, 105–116.

56. Slotkin, T. A. and F. J. Seidler (1983) Role of adrenomedullary catecholamine release in survival during neonatal hypoxia *Prog. Neuropsychopharmacol. Biol. Psychiat.*, Suppl. pp. 41–42.

57. Slotkin, T. A., F. J. Seidler, and W. L. Whitmore (1980) Precocious development of sympathoadrenal function in rats whose mothers received methadone. *Life Sci.* **26**, 1657–1663.

58. Slotkin, T. A., P. G. Smith, C. Lau, and D. L. Bareis (1980) Functional Aspects of Development of Catecholamine Biosynthesis and Release in the Sympathetic Nervous System, in *Biogenic Amines in Development* (H. Parvez and S. Parvez, eds.) Elsevier/North Holland, Amsterdam, pp. 29–48.

59. Smith, P. G., E. Mills, and T. A. Slotkin (1981) Maturation of sympathetic neurotransmission in the efferent pathway to the rat heart: Ultrastructural analysis of ganglionic synaptogenesis in euthyroid and hyperthyroid neonates. *Neuroscience* **6**, 911–918,

60. Smith, P. G., T. A. Slotkin, and E. Mills (1982) Development of sympathetic ganglionic transmission in the neonatal rat: Pre- and postganglionic nerve response to asphyxia and 2-deoxyglucose. *Neuroscience* **7**, 501–507.

61. Viveros, O. H., E. J. Diliberto, E. Hazum, and K. Chang (1979) Opiate-like materials in the adrenal medulla: Evidence for storage and secretion with catecholamines. *Mol. Pharmacol.* **16,** 1101–1108.

62. Wurtman, R. J. and J. Axelrod (1966) Control of enzymatic synthesis of adrenaline in the adrenal medulla by adrenal cortical steroids. *J. Biol. Chem.* **241,** 2301–2305.

63. Wurtman, R. J., E. P. Noble, and J. Axelrod (1967) Inhibition of enzymatic synthesis of epinephrine by low doses of glucocorticoids. *Endocrinology* **80,** 825–828.

64. Yoshizaki, T. (1973) Effect of histamine, bradykinin, and morphine on adrenaline release from rat adrenal gland. *Jap. J. Pharmacol.* **23,** 695–699.

Chapter 4

Endocrine Control of Synaptic Development in the Sympathetic Nervous System

The Cardiac–Sympathetic Axis[a]

Theodore A. Slotkin

1. Introduction

One of the most important functions of autonomic innervation is regulation of the cardiovascular system. As reviewed elsewhere in this book, sympathetic and sympathoadrenal factors play key roles in the maintenance and support of the circulation in the fetus and neonate, both in the resting state and during stress situations. Consequently, the timing of onset of synaptic function in the developing sympathetic nervous system is of key importance during the critical perinatal period, and the identification of the endogenous mechanisms that mediate synaptic development is vital.

It has long been known that endocrine secretions profoundly influence developing nervous tissues, as is evident from the

[a]This review covers work published through 1983.

97

devastating neurological sequelae of diseases such as cretinism. Based on morphological and behavioral information in the central nervous system, a comprehensive picture has emerged for the participation of thyroid and adrenal steroid hormones in the migration, differentiation, and general development of nerve cells and synapses (3,5,6,13–15,21–24,30,41,42,45,46,49–52,65). More recently, it has become apparent that similar endocrine factors may serve as primary determinants of synaptic development in the peripheral sympathetic nervous system, setting the time course of functional maturation of sympathetic pathways, as well as regulating the sensitivity of sympathetic target tissues to neuronal stimulation (25,33,34–40,47,62,64,66). This paper will review endocrine participation in development of the cardiac–sympathetic axis, a pathway that has been the most intensively studied for the effects of a variety of hormonal factors.

2. Development of Synaptic Function in the Cardiac–Sympathetic Axis

In some species, and most notably in the Sprague-Dawley rat, functional sympathetic connections to the heart are incomplete or absent in the neonate (7,9,10,27,44,61); thus, these species have often been favored for studies of synaptic development, since the processes regulating the onset of activity can be studied without the necessity of intrauterine techniques. In order for the cardiac–sympathetic pathway to operate, four developmental processes must be completed (*see* Fig. 1): (a) cardiac β-adrenoceptors and receptor-linked response systems must be present and functional in the myocardium; (b) the postganglionic noradrenergic synapses in the myocardium must be capable of synthesizing, storing, and secreting norepinephrine; (c) ganglionic transmission needs to be established to permit cholinergic signals to reach the postganglionic neuron; and (d) the central nervous system must develop the capability of processing afferent information about cardiovascular status (such as baroreceptor input) and making the appropriate adjustments in sympathetic output. As shown in Fig. 1, a variety of standard pharmacological tests can be applied to establish the onset of function at each of these sites. First, the presence and development of myocardial β-receptors can be evaluated through radioligand binding tech-

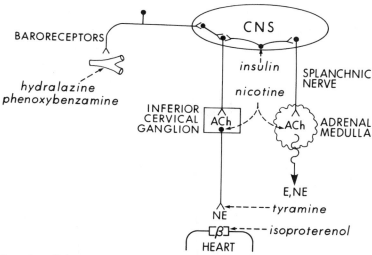

Fig. 1. Schematic representation of the cardiac–sympathetic and sympathoadrenal axes, showing sites of stimulation of substances used for testing onset of sympathetic function. Abbreviations: CNS, central nervous system; ACh, acetylcholine; NE, norepinephrine; E, epinephrine.

niques, and their linkage to functional responses in the heart can be assessed through the use of β-agonists, such as isoproterenol. The presence of the releasable pool of norepinephrine in sympathetic nerve terminals can be demonstrated through the use of tyramine, a drug that acts indirectly by displacement of presynaptic norepinephrine. The ability of the postganglionic neuron to respond to nicotinic cholinergic stimulation can be tested through administration of nicotine, which acts directly on the ganglionic nicotinic receptors. Evoking sympathetic stimulation through actions in the central nervous system, such as that seen with insulin-induced hypoglycemia, enables the function of the entire efferent sympathetic pathway to be examined (from the central nervous system to the end organ). Finally, the ability of the central nervous system to process afferent sensory information about cardiovascular status and adjust sympathetic tone accordingly can be tested with pharmacological agents that cause reductions in blood pressure (i.e., hydralazine, phenoxybenzamine), thus calling forth baroreceptor-mediated increases in demand for sympathetic activity.

2.1. Development of Noradrenergic Receptors and Responses in the Myocardium

It is useful to examine first the development of β-adrenoceptors in the myocardium; unless these are present and linked to functional responses of the heart, the status of sympathetic nerve activity and synaptic maturation would be of little consequence, since the end organ would be incapable of responding to neuronal stimulation. Even in the rat, a species in which sympathetic neuronal development is largely postnatal, cardiac β-receptors and sympathetic effects of directly administered catecholamines have been demonstrated to be present at or before birth *(1,2,10,31,37,43,53)*. Assessment of radioligand-specific binding indicates that the concentration of receptors in the neonatal myocardium, although lower than in the adult, is sufficient to support substantial degrees of sympathetic response, both in terms of cardioacceleration, as well as biochemical changes associated with sympathetic activation, such as stimulation of ornithine decarboxylase or adenylate cyclase activity. Indeed, even in the early periods of organogenesis during fetal life, tachycardia and enzyme activation can be readily demonstrated. These results indicate that the myocardium is capable of responding to sympathetic input very early in life; however, attempts to elicit sympathetic response through neuronal stimulation (as opposed to administration of directly acting β-agonists) invariably indicate the absence of sympathetic function *(7,9,10,40)*. There is also failure to elicit transsynaptic changes in receptor number or response capabilities during this period, indicating an absence of tonic sympathetic input to the myocardial receptors *(34)*. Consequently, sympathetic efferent neuronal elements must be immature at birth, and their development should provide the key to understanding what determines the age of onset of sympathetic pathway function.

2.2. Development of Postganglionic Synaptic Transmission

Neonatal rats do not develop the ability to respond to centrally induced stimulation of the cardiac–sympathetic axis until the end of the first postnatal week *(61)*. To understand why the overall function of this pathway is initially immature, a biochemical tool

DEVELOPMENT OF CARDIAC RESPONSES TO REFLEX SYMPATHETIC STIMULATION

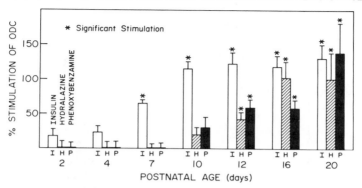

Fig. 2. Cardiac biochemical responses to reflex stimulation in the neonatal rat. Insulin (I) acts through direct effects in the central nervous system consequent to hypoglycemia, whereas hydralazine (H) and phenoxybenzamine (P) require action via baroreceptor-mediated reflexes. Asterisks denote significant stimulation of cardiac ornithine decarboxylase activity (ODC) (data from refs. *7,9,37,40,11,12*).

that has been utilized is the activity of ornithine decarboxylase (*48,57*). In adult rats insulin-induced hypoglycemia elicits sympathetically mediated cardiac ornithine decarboxylase stimulation; this effect can be blocked by pretreatment with β-adrenocepter-blocking, ganglionic-blocking, or catecholamine-depleting agents, indicating an absolute dependence on an intact sympathetic pathway (*7*). The effect on the heart does not appear to depend on release of adrenal catecholamines, but rather involves the cardiac–sympathetic nerve supply exclusively. It is thus of critical importance that the neonatal rat heart does not respond initially to reflex sympathetic stimulation, shown by the absence of ornithine decarboxylase increases after insulin administration (*see* Fig. 2). Although direct β-adrenergic stimulation of the neonatal heart with isoproterenol does increase ornithine decarboxylase (*10*), stimulation of postganglionic neurons with nicotine is ineffective at this time (*7,10*); thus it is most likely that the neonate does not possess an operative postganglionic neuron (See Fig. 3).

Why is the postganglionic neuron nonfunctional in the neonate? Evidence has accumulated that suggests the development of cardiac sympathetic neurotransmission is directly related to maturation of sympathetic nerve terminals in the myocardium, and more specifically, to ontogeny of neurotransmitter synaptic

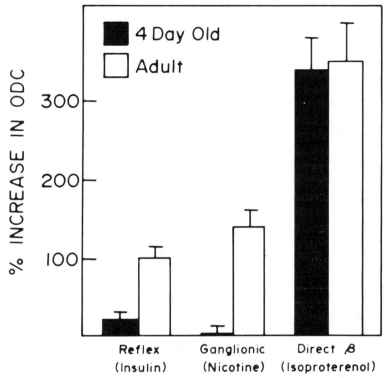

Fig. 3.   Cardiac biochemical responses to stimulants acting at different loci in the efferent sympathetic pathway (data from refs. *1* and *10*).

vesicles. The cardiac ornithine decarboxylase response to insulin administration appears in most rats at about 1 wk of age, the point at which chronotropic responses to tyramine (whose action depends upon the presence of displaceable norepinephrine in noradrenergic nerve terminals) also approach maturity (9). Atwood and Kirshner (4) have studied outgrowth of cardiac noradrenergic nerve terminals, using tritiated norepinephrine accumulation by atrial slices, as an index of presynaptic development. Their results indicate that the greatest nerve terminal proliferation occurs between 4 and 7 d of age, the period immediately preceding the onset of cardiac–sympathetic responses. Subsequent kinetic analysis and subcellular fractionation of the label incorporated into the slices suggested that a major factor responsible for the maturational increase in accumulation is ontogenetic change in transmitter-storage capabilities. Using a recently devel-

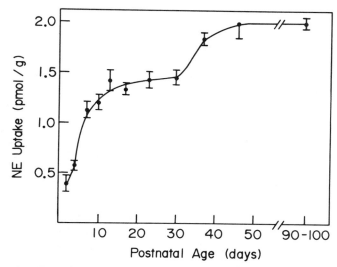

Fig. 4. Development of cardiac sympathetic synaptic vesicles, as indicated by ability of isolated vesicle preparations to take up exogenous radiolabeled norepinephrine (NE) (data from ref. 9).

oped technique to evaluate norepinephrine uptake into isolated vesicle preparations (8), Bareis and Slotkin (9) examined the development of norepinephrine storage. The most rapid phase of vesicle development was found to occur in the first week of postnatal life, evidenced by a four- to fivefold increase in vesicular norepinephrine uptake (*see* Fig. 4).

Ultrastructural examination of developing sympathetic rat heart neurons corroborates the conclusion that this period is associated with maturational increases in noradenergic synapses and with changes in vesicle numbers and properties (61). In the 2-d-old right atrium, there are very few neuronal varicosities in apposition with cardiac muscle cells. Many nerves that are juxtaposed to muscle display typical characteristics of immature neurons: accumulations of smooth endoplasmic reticulum and only a few dense-core vesicles that are intermediate in size between the small and large vesicle populations found in mature terminals. Small granular vesicles are absent in the neonate. By 8 d of age, however, numbers of terminals and vesicles have increased substantially, and the vesicle population has begun to shift in characteristics toward the bimodal distribution of diameters typical of mature neurons. At the morphological level, then, evidence also supports the view that the development of specific elements of the postganglionic neurons, and especially of synaptic vesicles, is an important component in determining the age of onset of competence of the cardiac–sympathetic axis.

But are these ontogenetic changes in the number and properties of synaptic vesicles actually related to overall function of the cardiac–sympathetic nerve pathway? An excellent correlation exists between development of the vesicles and the magnitude of chronotropic responses to tyramine or reflex cardiac ornithine decarboxylase stimulation after insulin (9,61). These data are consistent with the hypothesis that the appearance of functional noradrenergic synaptic vesicles is a rate-limiting process in postnatal development of sympathetic control of the heart. That conclusion is further supported by the direct relationship between drug- and hormone-induced alterations of synaptic vesicle development that invariably occur when the age at which cardiac responses to sympathetic nerve stimulation has been shifted (7,9,37–39,61). The effects of such hormonal manipulations on vesicle development and synaptic function will be detailed in a later section.

2.3. Development of Ganglionic Synaptic Transmission

In addition to synaptic function at postganglionic sites, some attention should be given to development of ganglionic transmission, although less information is available concerning maturation of these synapses in the cardiac–sympathetic pathway. Glucose deprivation induced by administration of 2-deoxyglucose has the same functional effect as insulin-induced hypoglycemia, i.e., it calls forth reflex activation of sympathetic pathways, including the cardiac–sympathetic axis. Direct recording of preganglionic sympathetic nerve-trunk activity after 2-deoxyglucose administration to neonatal rats indicates that the central nervous system is capable of eliciting activation of preganglionic nerve activity (63). Despite this fact, norepinephrine is not released from postganglionic nerve terminals in the myocardium at 2 or 4 d of age, and again, the response does not appear until the end of the first postnatal week (54). This result suggests that neurotransmission across the ganglion is incompetent in the neonate. Indeed, direct quantitative ultrastructural evaluation of ganglionic synaptogenesis in the inferior cervical ganglion (which supplies innervation to the right atrium) indicates low synapse counts at 2 d of age and a fivefold increase in synapses during the period in which function is developing (62). Thus, it is likely that the onset of activity of the efferent sympathetic pathway to the heart is limited by ganglionic synaptogenesis, as well as by synaptic development at the postganglionic sites in the myocardium; both processes appear to occur over the same time period, i.e., in the first postnatal week. As we shall see later, the participation of endocrine factors in sympathetic development involves both ganglionic and postganglionic sites.

2.4. Development of Baroreceptor Regulation of Sympathetic Tone

The successful completion of the structural and biochemical development of elements in the efferent sympathetic pathway to the heart, and the attendant onset of functional capabilities of ganglionic and postganglionic synapses, do not terminate the period in which ontogenetic changes in sympathetic activity occur. Measurements of turnover of myocardial norepinephrine (54) indicate a surge in activity of postganglionic neurons during the third postnatal week (see Fig. 5), and direct electrophysiological measurements of preganglionic and postganglionic tonic nerve activity in other cervical–sympathetic pathways indicate that the third week is associated with a profound elevation in overall sympathetic outflow from the central nervous system (63). What is responsible for this surge of activity that occurs well after the efferent pathway to the heart has become functional? Recent evidence suggests that the development of central processing of baroreceptor input is the final step in maturation of sympathetic capabilities and may initiate the period of hyperactivity (11,12,19,20).

Even after the first postnatal week, the central nervous system is not capable of successfully processing baroreceptor input and adjusting sympathetic tone in response to changes in blood pressure (11,12). Acute lowering of blood pressure through a direct-acting vasodilator (hydralazine) or through α-adrenergic blockade (phenoxybenzamine) fails to elicit either cardioacceleration or stimulation of cardiac ornithine decarboxylase activity in neonatal rats until the end of the second postnatal week, and a fully mature response to hypotension does not occur until nearly another week has elapsed (Fig. 2). Similarly, hydralazine is unable to cause release of cardiac norepinephrine from sympathetic nerve terminals even at a stage in which release in response to tyramine

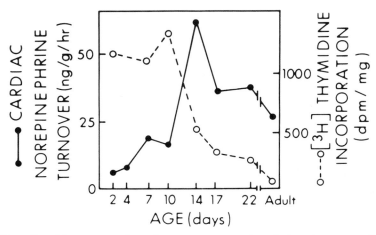

Fig. 5. Turnover of endogenous norepinephrine in sympathetic nerve terminals of the developing rat heart (solid circles and lines) and comparison with DNA synthesis in the myocardium (open circles and broken lines). Turnover was measured after inhibition of norepinephrine synthesis by α-methyl-*p*-tyrosine. DNA synthesis was assessed after administration of radiolabeled thymidine and isolation of trichloroacetic acid precipitable material (data from ref. 54 and from additional work in the author's laboratory).

or 2-deoxyglucose is readily detectable (54). These results are consistent with the view that the final stage in development of cardiac–sympathetic function is the appearance of the ability of afferent sensory information from the baroreceptors to regulate sympathetic outflow.

What are the functional correlates of the completion of this last element in cardiac–sympathetic neurotransmission? First, the onset of function of baroreceptor reflexes initiates the tonic control of heart rate by sympathetic fibers. Prior to that period, although responses to some acute stressors are capable of being elicited, there is no apparent ongoing participation of sympathetic activity in establishing the resting heart rate; thus, administration of chlorisondamine, a ganglionic blocking agent, does not lower the heart rate in the neonatal rat at 1 wk of age (when efferent transmission is functional), but does so only after baroreceptor control is established (53). Second, the initiation of tonic central control of sympathetic outflow is accompanied by a surge of sympathetic activity that subsides during the fourth postnatal week (54,63); this surge may raise sympathetic tone to twice that seen even in adult rats (*see* Fig. 5). The functional significance of the surge has not been definitively identified. However, recent evidence suggests that transient sympathetic hyperactivity may play a role in the final adjustments of end-organ sensitivity to sympathetic stimulation (53). In addition, since the cardiac β-adrenoceptors are subsensitive, a higher degree of sympathetic outflow may be required merely to express significant adrenergic effects. A more intriguing possibility is that the elevation in sympathetic tone serves to regulate the pattern of macromolecular synthesis in cardiac tissue, terminating the phase in which cell division occurs and initiating further cardiac growth by hypertrophy. In the mid-1970s, Claycomb (17) showed that isoproterenol, a β-agonist, could shut off the synthesis of DNA in developing heart tissue in vivo and proposed that adrenergic activity could serve as an endogenous maturational signal. Comparison of the time course of the surge in sympathetic tone (evaluated by the norepinephrine turnover rate) with the synthesis of cardiac DNA (tritiated thymidine incorporation) indicates a precise correlation (Fig. 5). Again, as shall be discussed below, endocrine participation in the onset of this final stage of cardiac–sympathetic development is prominent and has consequences for long-term regulation of sympathetic activity, as well as for maturation of heart tissue.

3. Role of Thyroid Hormones in Development of Sympathetic Function

It is well established that thyroid hormone plays a vital role in general cellular growth and, more particularly, in development of the nervous system. By far the most widely studied effects on immature nervous tissue have been those on the central nervous system. Neonatal hyperthyroidism produces a shift of neonatal cerebellar cells from the "proliferation" phase to the "differentiation" phase (30), thus leading to an initial acceleration of maturation, but an eventual deficit in the number of nerve cells and correspondingly abnormal physiological and behavioral performance. One of the earliest effects of excess thyroid hormone on developing nervous tissue is to evoke biochemical changes consistent with precocious cellular development and synaptogenesis as well as premature increases in polyamine biosynthetic capabilities, and RNA, DNA, and protein synthesis (3,5,42); each of these parameters then also declines prematurely, producing ultimate deficiencies in brain growth.

In contrast to the wealth of information concerning thyroid status and central nervous system maturation, much less is known about hormonal influence on synaptic development and function in the peripheral sympathetic system; indeed, until recently, few studies have appeared on this subject. The belated interest in the sympathetic aspects of thyroid hormone regulation of development is unfortunate in two regards: first, the importance of sympathetic function to the neonate is profound, and the critical role played by thyroid status in establishing sympathetic competence is thus vital to the survival of the neonate and second, the sympathetic system has provided a convenient model in which to establish how specific subcellular alterations (morphological and biochemical) caused by abnormal thyroid status are translated into deficits in synaptic performance. In the latter regard, it is frequently difficult, if not impossible, to identify specific alterations in performance associated with subcellular abnormalities of a particular set of synapses in the central nervous system; but in the peripheral sympathetic system, where the synaptic connections, transmitter, and receptor types and end-organ activity are all clearly defined, the translation of biochemical and morphological defects into deficits in physiological function have been readily established. Thus, the examination of the effects of thyroid status on synaptic maturation in the cardiac–sympathetic axis is of interest not only because of the inherent physiological importance of this pathway, but also as a model system for studying how subcellular events influence development of neuronal function.

3.1. Hyperthyroidism

3.1.1. Actions at Postganglionic Noradrenergic Synapses

In 1979, Lau and Slotkin (40) first reported that neonatal hyperthyroidism in the rat, induced by postnatal administration of triiodothyronine (T3), was capable of accelerating the ontogeny of function of the cardiac–sympathetic axis. Using insulin-induced reflex stimulation of cardiac ornithine decarboxylase as a marker for activity of the efferent sympathetic pathway, they demonstrated that a single injection of T3 given to 1-d-old rats initiated a series of events that enabled successful neurotransmission to occur within 24 h (i.e., at 2 d of age), compared to the normal development of transmission at 6 d of age (*see* Fig. 6, top). But does this, in fact, constitute a demonstration of accelerated neuronal development? Thyroid hormones are known to enhance the response of sympathetic end organs to adrenergic stimulation by increasing the number and activity of β-adrenoceptors (33); the apparent early onset of cardiac–sympathetic competence could conceivably simply represent end-organ supersensitivity to norepinephrine. Supersensitivity is clearly present in the neonatal hyperthyroid state (37); measurement of dihydroalprenolol binding to specific β-receptor sites indicates a twofold increase in the number of receptors within the 24-h period after T3 administration, and continued T3 treatment produces a sustained increment in receptor binding (*see* Fig. 6, middle). When thyroid hormone treatment is discontinued, there is a regression in the number of receptors to normal, and then subnormal, values. Actions at T3 on end-organ reactivity to sympathetic stimulation thus clearly participate in the apparent early onset of neurotransmission in the cardiac–sympathetic axis.

The first indication that hyperthyroidism also affects neuronal components of sympathetic activity was provided by the observation that hormonally induced enhancement of the heart rate response to tyramine, which acts presynaptically at the postganglionic nerve terminal, precedes the appearance of supersensitivity to isoproterenol, which acts directly at the receptor site (37). This situation could occur only if an increase had occurred in the pool of displaceable norepinephrine in the nerve terminal—a phenomenon usually associated with progressive maturation of sympathetic efferent synapses (9). As described above, an excellent correlation can be obtained between synaptic vesicle development and onset of efferent sympathetic function: indeed, such a relationship appears to be universal in that synaptic vesicles are the last presynaptic element to mature in brain and adrenal medulla, as well as in cardiac nerve terminals (9,32,55,56). It is thus of critical importance that T3 administration to 1-d-old rats has been shown to cause an immediate increase (within 24 h) in the uptake of norepinephrine by isolated synaptic vesicle preparations (37), a biochemical index of synaptic vesicle development (*see* Fig. 6, bottom). This is truly a developmental effect and not simply a typical thyroid hormone action; administration of T3 to mature animals does not affect the vesicular system in any fashion (37).

These biochemical findings have been confirmed by morphological techniques as well (61,62). In both euthyroid and hyperthyroid neonates, dense-core vesicles can be readily identified and quantified in nerve bundles supplying the right atrium (*see*

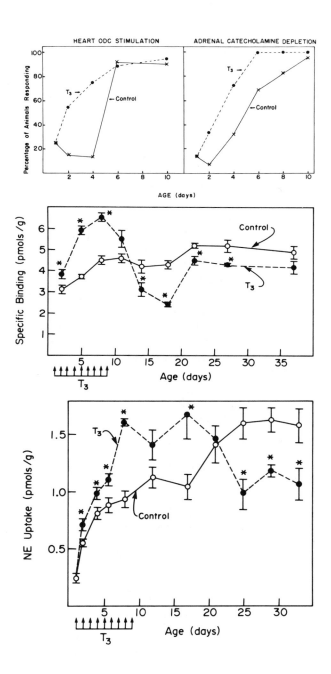

Fig. 7), and in nerve specializations (terminals) juxtaposed to cardiac muscle cells (*see* Fig. 8). Administration of T3 to 1-d-old rats produces an immediate and significant increase in the number of neurotransmitter storage vesicles undergoing axonal transport, as evidenced by an elevation in the number of vesicles per unit area of axoplasm in nerve bundles (Table 1); this occurs without any change in the number of axons present. Correspondingly, the number of vesicles per unit area of nerve terminal also nearly doubles in this period, confirming that the vesicles are reaching their site of physiological function more rapidly in the hyperthyroid neonates. At the same time, the type of vesicle present in the nerve terminals appears to undergo a transition from the single population of intermediate-diameter vesicle typical of immature synapses, to the bimodal or trimodal distribution found in mature cardiac nerve terminals (*see* Fig. 9). Findings utilizing physiological, biochemical, and morphological techniques thus all indicate that hyperthyroidism produces an acceleration of onset of cardiac–sympathetic activity through actions at both presynaptic and postsynaptic loci of noradrenergic synapses; the net initial effect is to promote the appearance of postganglionic function through precocious maturation of vesicle number and type.

3.1.2. Actions at Ganglionic Synapses

The promotion of synaptic development and function provided by neonatal hyperthyroidism is not restricted to the postganglionic noradrenergic synapse. Ultrastructural examination of the inferior cervical ganglion reveals an acceleration of syn-

---

Fig. 6.   Effects of neonatal hyperthyroidism on development of sympathetic function. In the top panel, the ability of cardiac–sympathetic and sympathoadrenal axes to respond to central nervous system stimulation (evoked by insulin-induced hypoglycemia) is compared in control rats and in rats rendered hyperthyroid by administration of triiodothyronine (T3). The developmental curves for proportions of animals responding to insulin stress are significantly different in hyperthyroid rats, as assessed by Fisher's exact test. In the middle panel, the effects of hyperthyroidism on development of β-adrenoceptor binding sites for dihydroalprenolol are compared; asterisks denote significant differences by Student's *t*-test. In the bottom panel, effects on presynaptic noradrenergic nerve terminals are evaluated by ability of synaptic vesicles to take up norepinephrine (NE) (data from refs. 37 and 49).

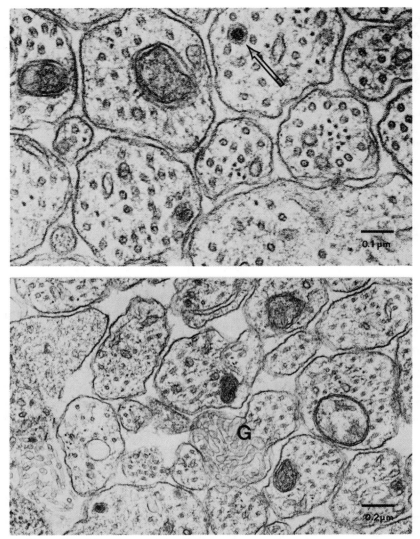

Fig. 7. Electron micrographs of axon bundles in right atria of 2-d-old rat treated 24 h previously with T3 (top), compared with a control littermate (bottom). Animals were pretreated with 5-hydroxydopamine prior to fixation to enhance visualization of dense core vesicles (arrow). Growth cones are also readily found in control tissue (labeled "G" in bottom panel) (work conducted in the author's laboratory by Dr. Peter G. Smith).

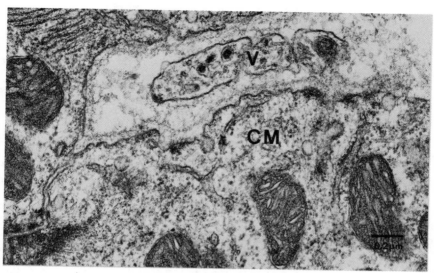

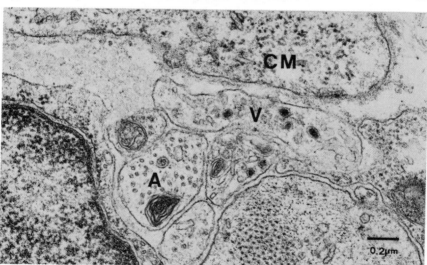

Fig. 8. Electron micrographs of nerve varicosities (V) in 2-d-old rat treated with T3 24 h previously (top), compared with a control littermate (bottom). Other readily identified elements are cardiac muscle cells (CM) and axonal profiles (A). Dense core vesicles are readily visualized within the varicosities (work conducted in the author's laboratory by Dr. Peter G. Smith).

Table 1
Effects of Neonatal Hyperthyroidism on the Cardiac–Sympathetic Axis:
Quantitative Ultrastructural Findings[a]

Site evaluated	Measurement made	Control	T3
Postganglionic axon bundles	Dense core vesicles per square micron of axoplasm	0.33 ± 0.04	0.47 ± 0.07[b]
	Growth cones per square micron of axoplasm	0.40 ± 0.06	0.15 ± 0.03[b]
	Axons per square micron of axoplasm	7.2 ± 0.4	8.1 ± 0.7
Postganglionic nerve terminals	Dense core vesicles per square micron of axoplasm	5.7 ± 1.2	10.4 ± 1.2[b]
Inferior cervical ganglion	Synapses per ganglion cell nucleus	0.05 ± 0.03	0.25 ± 0.05[b]
Value for 8-d-old control: 0.37 ± 0.10, N.S. vs 2 d T3			

[a]One-d-old rats were made hyperthyroid by administration of triiodothyronine (T3) and were evaluated 24 h later. Data for axon bundles and nerve terminals were obtained in the author's laboratory by Dr. Peter G. Smith, and those for ganglia were taken from ref. 62.
[b]Asterisks denote significant differences by Student's *t*-test.

apse formation and maturation in neonatal rats given T3 (62). Indeed, the relative density of mature-appearing synapses 24 h after thyroid hormone administration was found to be elevated fourfold relative to the normal values for that age (Table 1); the increased density was equivalent to that seen ordinarily at the end of the first postnatal week (the time at which neurotransmission develops in control animals), indicating that neurotransmission at this site is probably also accelerated just as it is at the postganglionic synapse. The fact that insulin-induced hypoglycemia is capable of eliciting end-organ stimulation shortly after T3 administration (37,40) is further proof that development of ganglionic transmission has been enhanced by the hormonal manipulation; since this reflex is centrally derived, ganglionic synapses must be operating in order for the heart to be stimulated. Thus, there is conclusive evidence for a general promotion by thyroid hormone of synaptogenesis and synaptic function in both ganglionic and postganglionic sites in the developing cardiac–sympathetic axis. These results resemble those seen in

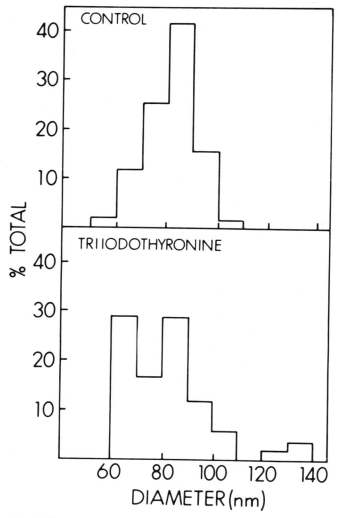

Fig. 9. Effects of neonatal hyperthyroidism on diameters of dense-core vesicles found in cardiac–sympathetic nerve terminals. Distributions are significantly different by chi-square analysis (data obtained in the author's laboratory by Dr. Peter G. Smith).

the central nervous system, and further provide the first direct evidence that the morphological and biochemical changes initiated by thyroid hormone administration are translated into definite alterations in synaptic transmission.

3.1.3. Actions on Baroreceptor Control Mechanisms

Evidence has also accumulated that suggests thyroid hormones
can accelerate the final step in maturation of the cardiac–sympathetic axis, i.e., the appearance of baroreceptor control of sympathetic activity. Using the ornithine decarboxylase enzyme marker for activation of the cardiac–sympathetic axis,
Bartolome et al. (11) demonstrated that daily T3 administration to
neonatal rats produced early development of the capability to respond to challenge with hydralazine, a direct arteriolar vasodilator that activates the cardiac–sympthetic axis through baroreceptor-mediated reflexes (12). Whereas the response to
hydralazine appeared during the third postnatal week in controls,
T3 treatment led to significant reflex responses as early as 8 d of
postnatal age (*see* Fig. 10, left). Again, the neuronal origin of the
enhanced stimulation could be demonstrated by assessing the
hydralazine-induced increase in transmitter turnover (i.e., release
of norepinephrine) from cardiac nerve terminals; release in
T3-pretreated neonates challenged with hydralazine was three
times that in control neonates (*see* Fig. 10, right), indicating an increase in the reflex stimulation of sympathetic efferents during
hypotension.

Thus, thyroid hormone administration appears to accelerate
the onset of function of the cardiac–sympathetic axis of all four
levels of synaptic activity: (a) at the end organ, where β-receptors

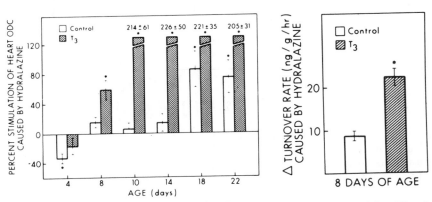

Fig. 10. Effects of neonatal hyperthyroidism, induced by T3 administration, on onset of baroreceptor reflexes, as evaluated by ability of
a vasodilator (hydralazine) to evoke stimulation of cardiac ODC (left)
and to cause norepinephrine release (increased turnover) from sympathetic nerve terminals in the myocardium (right) (data from ref. *11*).

are induced; (b) at the noradrenergic nerve terminal, where pre-cocious arrival of synaptic vesicles initiates premature onset of nerve function; (c) at the sympathetic ganglion, where a massive increase in the number of synapses can occur readily after hor-mone exposure; and (d) in the central nervous system, where the final step in development of control of sympathetic activity oc-curs; namely the onset of processing of baroreceptor sensory in-formation that initiates tonic regulation of sympathetic function.

3.1.4. Long-Term Deficits Induced by Neonatal Hyperthyroidism

It is particularly striking that neonatal hyperthyroidism produces the precocious appearance of all aspects of cardiac–sympathetic neurotransmission (and probably of neurotransmission in other sympathetic sites as well). Should this be viewed as a beneficial effect, permitting the neonate to respond to his environment or to demands of stress situations earlier than would ordinarily be the case? Recent studies indicate that the early elicitation of nerve function is in fact neither beneficial, nor even innocuous. Exami-nation of the pattern of receptor development in the heart reveals the unusual finding that the early phase, in which receptor bind-ing is enhanced, is followed later in maturation by a deficit in the number of receptors (Fig. 6, middle), and a consequent deficiency in the capability of the heart to respond to β-adrenergic stimula-tion (37). Similarly, the initial enhancement in the number of syn-aptic vesicles in the myocardium, as evaluated through the norepinephrine uptake marker, is succeeded by an eventual deficiency in vesicular capacity (Fig. 6, bottom), signaling a corre-sponding abnormality in the presynaptic nerve terminals. Ultrastructural examination of sinoatrial nerve bundles in hyper-thyroid neonates indicates why such defects should appear at later stages of development: along with the promotion of vesicle transport and synaptogenesis in the efferent sympathetic pathway, there is a sharp reduction in the number of growth cones present (Table 1), indicating a premature termination of the developmental phase in which axonal proliferation occurs (61). What is of particular interest is that this deleterious effect is read-ily detectable in the very same rapid timeframe over which the promotion of synaptic maturation occurs, i.e., within 24 h after T3 administration. Thus, following the initial expression of en-hanced maturation of synapses by T3, there will be an eventual deficit in the number and function of nerve terminals as a consequence of the compression of the time available for proliferative events. Again, this resembles the case seen in the central nervous system in which thyroid hormones induce a shift away from early maturational events toward later ones, and thus produce precocious cellular maturation, but with eventual deficits in cell number.

3.1.5. Hyperthyroidism and Cardiac Muscle Development

In view of the profound effects of hyperthyroidism on the timing of onset of sympathetic function, it is not suprising that growth of the heart itself, which is subject to sympathetic influence, is markedly affected by neonatal treatment with thyroid hormones. As was true for neuronal development, the net effects of T3 administration on heart growth are biphasic, with initial enhancement followed by eventual deficits as the animal reaches adulthood (37). It is difficult, however, to separate the influence of thyroid hormones on sympathetic components regulating cardiac development from the direct effects of these agents on cardiac tissue. Hyperthyroidism is known to cause cardiac hypertrophy through mechanisms that are independent of sympathetic input to the heart (18). However, there is some evidence supporting the view that at least a portion of the developmental effects of hyperthyroidism on cardiac tissue result from actions on cardiac sympathetic neurotransmission. First, the activity of cardiac ornithine decarboxylase, which is influenced partly through sympathetic tone, is first enhanced and then suppressed during neonatal T3 treatment (36,40); this is distinctly unlike the direct effects of T3 on ornithine decarboxylase, since continuous thyroid hormone administration to adult rats usually maintains the enzyme activity in an elevated state (35). Second, the pattern of nucleic acid synthesis in developing hearts of hyperthyroid rats is different from that seen in adults, where direct effects of the hormone can be expected to predominate. In immature rats, RNA synthesis actually falls to subnormal values, even while neonatal T3 administration is continuing (36), an observation that is inconsistent with direct effects of thyroid hormones, but is entirely compatible with the postulated role of sympathetic input in cardiac development. It is thus possible, or even likely, that initial enhancement and subsequent deficits of cardiac tissue development in the hyperthyroid state reflect a combination of direct effects of thyroid hormones on the myocardium, along with influence of premature exposure of cardiac tissue to norepinephrine released from sympathetic nerves, occurring secondarily to promotion of synaptic development by hyperthyroidism.

3.2. Hypothyroidism

As shown in the previous section, neonatal hyperthyroidism causes premature onset of cardiac sympathetic function, resulting from accelerated synaptic development at central, ganglionic, and postganglionic sites, along with end-organ supersensitivity to adrenergic stimulation. These effects raise the question of whether normal levels of thyroid hormones play an obligatory role in sympathetic nerve development. Recent studies, utilizing propylthiouracil-induced perinatal hypothyroidism in the rat, have examined whether functional maturation of each of these sites is, in fact, regulated by endogenous thyroid hormones (39). When propylthiouracil treatment was begun 5 d before birth and terminated during the second postnatal week, neonatal rats remained severely hypothyroid through 3–4 wk of age, as evidenced by direct measurement of serum thyroxine levels (*see* Fig. 11, top left). Examination of β-receptor binding of dihydroalprenolol confirmed that development of end-organ sensitivity to adrenergic stimulation requires endogenous thyroid hormones: hypothyroidism totally arrested the maturational increases in the number of binding sites, and recovery toward normal binding did not occur until thyroid hormone levels were allowed to rise after discontinuing propylthiouracil treatment (Fig. 11, top right). Similarly, these animals showed extremely low reactivity to isoproterenol, a direct β-agonist; tachycardia in response to even massive doses of isoproterenol indicated a 40% reduction in maximally achievable heart rate compared to that of controls (Fig. 11, bottom left). Both the effects on receptor binding and heart rate control could be prevented by thyroid hormone replacement therapy during the postnatal period (39). Thus, the participation of endogenous thyroid hormones in development of end-organ receptors, and hence of end-organ sensitivity, is absolute. It should be noted that endocrine participation in receptor and sensitivity development is thus much more important in the neonate than is transsynaptic control of these factors; neither sympathectomy nor repeated administration of sympathetic stimulants in neonates can up- or down-regulate receptor numbers or

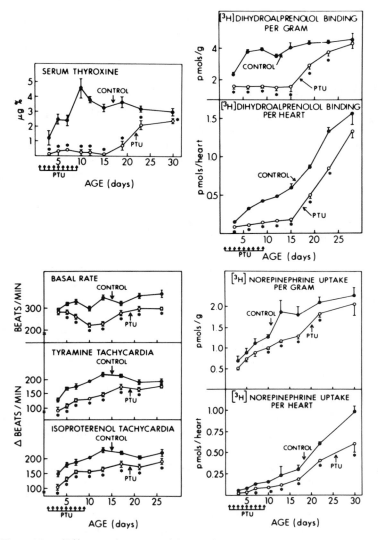

Fig. 11. Effects of neonatal hypothyroidism, induced by perinatal propylthiouracil (PTU) administration, on development of the cardiac–sympathetic axis. Effectiveness of the treatment is shown by the profound effect on serum thyroxine (upper left). Hypothyroidism arrests development of β-adrenoceptor binding sites (upper right), resulting in deficits in physiological responses to sympathomimetics (lower left). Presynaptic nerve terminal development is also inhibited, as shown by retardation in ontogeny of synaptic vesicle-uptake mechanisms (lower right) (data from ref. *39*).

receptor-mediated cardiac responses (*40*), whereas hyper- or hypothyroidism can readily do so (*37,39*). Not until the third postnatal week do neuronal factors such as impulse activity or transmitter availability appear to contribute to receptor regulation, and hence thyroid hormones may provide the most important endogenous signal for ontogeny of receptors early in neonatal life.

Thyroid status also appears to be critical to the normal maturation of noradrenergic nerve terminals in the myocardium (*39*). The development of the capacity of isolated synaptic vesicles to take up and store norepinephrine is severely impaired during neonatal hypothyroidism, and on an absolute basis, does not appear to recover even long after discontinuing propylthiouracil treatment, and thyroid hormone levels have largely returned to normal (Fig. 11, bottom right). Again, thyroid hormone replacement therapy successfully prevents the short- and long-term consequences of propylthiouracil treatment (*39*). The persistence of synaptic deficits in the neonatal hypothyroid rat is consistent with the concept of a "critical period" for the participation of thyroid hormones in nerve development—a principle that originally derives from studies in the central nervous system. Abnormal thyroid status outside the critical period produces short-term, reversible alterations in neuronal structure and function, whereas effects exerted during the critical period are essentially irreversible. In any case, the retardation of development of presynaptic components of cardiac synapses is of clear-cut functional significance; the heart rate response to tyramine (Fig. 11, bottom left), the cardiac norepinephrine content, and the amount of tyramine-releasable norepinephrine (*see* Table 2) are all reduced during perinatal hypothyroidism.

Although no direct studies of the effects of hypothyroidism on development of cardiac sympathetic ganglia have appeared, a number of indirect studies suggest that maturation of ganglionic synapses is also coordinated by endogenous thyroid hormones (*28,29,39,59,61*). First, as described in the previous section, elevation of thyroid hormone levels does enhance the rate of synaptic development in the inferior cervical ganglion, which supplies innervation to the right atrium (*62*). Second, innervation of the adrenal medulla, a ganglion-like tissue that receives typical cholinergic–nicotinic innervation by the splanchnic nerve, is impeded by neonatal hypothyroidism (*29,39*). The development of these synapses occupies the same timeframe as do those in ganglia supplying cardiac innervation, and shows the same reac-

Table 2
Effects of Neonatal Hypothyroidism on Cardiac Norepinephrine Levels,
Turnover, and Release[a]

Age, d	Content, ng		Turnover, ng/hr		% Released by tyramine		% Released by hydralazine	
	CON	PTU[b]	CON	PTU[b]	CON	PTU[b]	CON	PTU[b]
10	28	13	3	1	30	12	16	0
17	98	32	8	5	35	16	23	0
34	236	152	—	—	41	44	69	0
41	324	159	31	18	41	38	68	56

[a]Rats were rendered hypothyroid by maternal administration of propylthiouracil (PTU) begun 5 d before birth continued through 12 d postnatally, concurrently with administration to the pups begun at birth and continued through 12 d; controls (CON) received vehicle on the same schedule. Turnover was assessed by measuring norepinephrine depletion 4 h after inhibiting transmitter synthesis with α-methyl-*p*-tyrosine; release by tyramine or hydralazine was determined after acute challenge with either agent in synthesis-inhibited animals. Data were obtained in the author's laboratory by Ruta J. Slepetis.
[b]PTU-treated animals are different from CON in every parameter, as determined by two-way analysis of variance (factor 1 = age, factor 2 = treatment).

tivity to drugs and hormones administered perinatally (59,61). It is equally obvious, however, that future work should concentrate directly on demonstrating a role for endogenous thyroid hormones in development of ganglionic transmission in the cardiac–sympathetic axis.

New data available for the effects of propylthiouracil administration on development of cardiac norepinephrine turnover and release during hypotension make it clear that maturation of central control of sympathetic reflexes and of sympathetic tone are also regulated by thyroid hormone status. The ability of hydralazine-induced hypotension to evoke transmitter release from cardiac sympathetic terminals ordinarily develops during the third postnatal week consequent to maturation of central processing of baroreceptor sensory input (12,19,20,54). In hypothyroid neonates, hydralazine is unable to cause release of cardiac norepinephrine until the animals reach young adulthood, well beyond the sixth postnatal week (Table 2). Since it is the onset of baroreceptor control of sympathetic efferent tone that normally initiates both the surge of sympathetic activity in the third postnatal week and the subsequent tonic control of heart rate by sympathetic efferents, the deficiency in development of these central pathways in hypothyroid rats produces: a long-term reduction in sympathetic outflow (as shown by a reduction in the spontaneous turnover rate of norepinephrine in sympathetic nerve terminals, Table 2); and a persistent reduction in the basal heart rate (Fig. 11, bottom left).

Finally, because sympathetic tone participates in regulation of development of cardiac muscle, growth of the heart is deficient in the hypothyroid state (39). As was true of hyperthyroidism, the question always remains about which components of myocardial maturation are directly affected by thyroid hormones and which are influenced by sympathetic components whose functional ontogeny is under thyroid influence. It is interesting to note that the perturbation of cardiac ornithine decarboxylase activity, which is under both sympathetic and thyroid control, is similar in neonatal hypothyroidism (39,60) and in animals who have undergone a reduction in sympathetic tone consequent to central catecholaminergic lesions (19,20). In both cases, cardiac norepinephrine turnover is suppressed, ornithine decarboxylase is low, and cardiac growth is deficient. Since ornithine decarboxylase is the enzyme that initiates synthesis of the polyamines, which in turn control synthesis of nucleic acids and proteins in developing cells, disturbances in the maturational pattern of activity of this enzyme have been used to elucidate the impact of abnormal endocrine or neuronal status on cardiac tissue development (48,57). The available data would indicate that the influence on cardiac development of neonatal hypothyroidism and of centrally induced sympathetic hypoactivity are at least similar, suggesting that neuronal components could play a partial role in the contribution of endogenous thyroid hormones to growth and development of cardiac tissue.

Studies with perinatal hypothyroidism thus indicate that endogenous thyroid hormones do participate in the development of the cardiac–sympathetic axis at all levels: they are required for normal maturation of β-adrenoceptors in the myocardium and for development of physiological sensitivity to adrenergic stimulation; they are needed for ontogeny of essential components of presynaptic noradrenergic nerve terminals; they probably participate in ganglionic synaptogenesis; and they are absolutely necessary for the onset of baroreceptor control of sympathetic tone and reactivity. Finally, thyroid hormones participate in maturation of cardiac tissue through direct actions on the myocardium and indirectly through their influence on sympathetic input to the heart.

4. Role of Glucocorticoids in Development of Sympathetic Function

The studies described in the previous section demonstrate that thyroid hormones act to control sympathetic synaptic development at virtually every site in the cardiac–sympathetic axis, and that the roles of thyroid hormones are similar in the central and peripheral nervous system. Although steroids also are thought to participate in synaptic ontogeny, effects of these hormones appear to differ at various neuronal sites. In the brain, glucocorticoid administration slows cellular development, suppresses maturation of catecholamine transmitter levels, and delays behavioral ontogeny (3,15,41). However, in developing peripheral sympathetic ganglia, histofluorescent intensity of biogenic amine-containing cells is markedly increased by glucocorticoids (25), despite deficits produced in the key enzymes of catecholamine biosynthesis (tyrosine hydroxylase and dopamine β-hydroxylase). In the adrenal medulla, catecholamine levels and amine biosynthetic enzymes are suppressed by neonatal glucocorticoid treatment (16,64), a finding that may reflect adrenal atrophy secondary to termination of adrenocorticotropin production caused by exogenous steroids (26).

Thus, in contrast to the case seen with thyroid hormones, there is no clear picture of any broadly applicable principle by which the effects of glucocorticoids on synaptic development can be understood. Despite the biochemical evidence for dramatic, if unpredictable, effects on neuronal maturation, few studies have established whether such changes produce functional alterations in synaptic transmission or in end-organ responses to neuronal input. Indeed, the only system in which this has been extensively evaluated is the cardiac–sympathetic axis of the neonatal rat (38,58). Repeated administration of dexamethasone, a synthetic glucocorticoid steroid with vitually no mineralocorticoid activity, does produce deficiencies in cardiac responsiveness to isoproterenol, indicating a reduction in net sympathetic effect (*see* Fig. 12, left). However, there are no corresponding changes in the number of β-receptors (*see* Table 3), indicating that subsensitivity to adrenergic stimulation results from postreceptor changes in reactivity, possibly from uncoupling of the receptors from cardiac responses. This is an effect unique to the immature rat: No such uncoupling occurs in adults treated with dexamethasone.

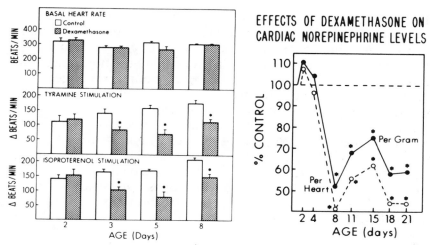

Fig. 12. Effects of neonatal hyperglucocorticism, induced by dexamethasone administration, on development of the cardiac–sympathetic axis. Dexamethasone lowers response capabilities to administration of tyramine or isoproterenol (left), and produces an eventual deficit in cardiac norepinephrine levels (right) (data from refs. *38* and *58*).

Because β-receptors are less effective in influencing heart rate in glucocorticoid-treated neonates, any stimulation of the sympathetic pathway will result in a lower net response. Accordingly, tyramine, which acts on presynaptic noradrenergic nerve terminals, also produces an apparently lower heart rate change in the

Table 3

Effects of Neonatal Dexamethasone Treatment on β-Adrenoceptor Binding Sites and Synaptic Vesicular Uptake of Norepinephrine[a]

Age, d	Specific binding of dihydro-alprenolol, pmol/g		Vesicular uptake of norepinephrine, pmol/g	
	CON	DEX	CON	DEX
2	2.6 ± 0.3	2.5 ± 0.2	0.40 ± 0.05	0.32 ± 0.04
4–5	3.9 ± 0.3	3.6 ± 0.2	0.67 ± 0.05	0.60 ± 0.80
8	3.4 ± 0.1	3.7 ± 0.2	—	—

[a]Animals were given dexamethasone (DEX) daily, beginning at 1 d of age. Controls (CON) received vehicle on the same schedule (data from ref. *38*). None of the differences is statistically significant.

affected animals (Fig. 12, left). However, this does not reflect any actions of dexamethasone on development of the nerve terminals themselves, since there is no corresponding reduction in biochemical markers for presynaptic development (norepinephrine uptake into isolated vesicles, Table 3). Although prolonged treatment with dexamethasone eventually causes a reduction in cardiac norepinephrine content in the developing rat heart (Fig. 12, right), this effect does not occur until well after the onset of gross morphological retardation (arrested weight gain, stunted cardiac growth, neonatal mortality), and the deficits in heart rate responses are no greater during the phase in which transmitter levels are subnormal. Thus, steroid hormones appear to participate in the development of synaptic function in the cardiac–sympathetic axis only indirectly, in that they affect the coupling of β-receptors to cardiac response units without apparent direct influence on the process of synaptic transmission. The role of glucocorticoids is thus probably less important than that of thyroid hormones in establishing synaptic function, but may be relatively more important during earlier phases of embryonic and fetal development, in which they are thought to participate in the differentiation of early sympathetic cells (*13,14,45*).

Dexamethasone does, however, have a profound and immediate effect on biochemical and morphological development of cardiac tissue. Within 24 h of administering dexamethasone to neonatal rats, cardiac ornithine decarboxylase activity is suppressed and polyamine levels are depleted (*58*). In view of the role of these factors in macromolecular synthesis in developing tissues, it is not suprising that growth of the heart is arrested or slowed immediately thereafter. However, unlike the case of thyroid hormones, it is clear that these effects are direct actions of glucocorticoids on heart tissue, rather than neurally mediated: all the biochemical alterations in the myocardium appear in advance of any effect of dexamethasone on the cardiac–sympathetic axis, or on cardiac reactivity to sympathetic stimulation. Indeed, cardiac norepinephrine levels do not become subnormal until nearly a week after cardiac biochemistry is perturbed and several days after cardiac weight is low. The effects of dexamethasone on cardiac development and reactivity to sympathetic stimulation thus superficially resemble those seen in hypothyroidism, but in fact involve totally different mechanisms that, in the case of glucocorticoids, are independent of neuronal participation.

5. Conclusions

The cardiac–sympathetic axis develops in three well-defined phases that occupy separate time periods. In the rat, the fetus and neonate possess a functional cardiac β-receptor complement that is coupled to physiological and biochemical responses in the myocardium. However, the neonate is not capable of cardiac sympathetic responses to postganglionic, ganglionic, or centrally elicited stimulation. During the first postnatal week, noradrenergic nerve terminals in the myocardium and ganglionic synapses undergo sufficient maturational change to enable nerve signals to pass from the central nervous system to the heart, but the sympathetic axis does not tonically control heart rate at this time. During the latter part of the second to third postnatal week, tonic control does appear consequent to ontogeny of central processing of baroreceptor sensory input that regulates sympathetic outflow, and this phase is accompanied by transient sympathetic hyperactivity that assists in setting the final sensitivity of cardiac β-receptors, as well as terminating the phase cellular replication in the myocardium.

Thyroid hormones appear to be the prime regulators of each of these phase of sympathetic development. Hyperthyroidism: (a) promotes the synthesis of β-receptors and enhances cardiac responses to β-adrenergic stimuli; (b) promotes the maturation of noradrenergic synapses in the myocardium and the transport of synaptic vesicles into the nerve terminals; (c) promotes the development of the releasable pool of norepinephrine in postganglionic nerve endings; (d) promotes the proliferation of ganglionic synapses; and (e) promotes the onset of baroreceptor control of sympathetic tone. Thyroid hormone deficiencies during development have the opposite effects, resulting in slowing of development at each of these sites, indicating that endogenous thyroid hormones are critical to sympathetic nerve development. Indeed, during early maturation of sympathetic pathways, thyroid status may represent the most important factor determining the timing of onset of synaptic transmission. The effects of thyroid status also extend to the development of the myocardium itself, probably as a result of direct influence of thyroid hormones on cardiac tissue, along with indirect effects mediated by the hormonal actions on sympathetic maturation and activity. Because of the defined nature of synapses, transmitters, and end-organ responses in the cardiac–sympathetic axis, this pathway can serve as a model system for examining how biochemical and morphological processes that occur during normal and abnormal development are translated into defects in synaptic performance.

It is of particular interest that the accelerated development of sympathetic synaptic components and the resultant premature onset of sympathetic function seen in neonatal hyperthyroidism, or the slowing of development in hypothyroid rats, both result ultimately in deficits in growth and synaptic performance later in life. This indicates that perturbation of the timing of onset of synaptic function cannot be viewed simply in terms of the importance of neurotransmission as isolated from other developmental processes. The participation of nerve activity in macromolecular synthesis in immature cardiac tissue can serve as one example of how early development of neuronal components can lead to deleterious effects as the animal matures. Similarly, effects of other hormonal agents, such as glucocorticoids, that affect synaptic transmission, can do so through indirect actions on sympathetic pathways or general metabolic damage and secondary growth-inhibitory effects. Neuronal maturation thus must be viewed as just one component of the overall development of the organism as a whole, since interactions among metabolic, hormonal, and neuronal factors will all influence the short- and long-term physiological capabilities of the neonate.

Acknowledgments

Supported by USPHS HD-09713.

References

1. Adolf, E. F. (1965) Capacities for regulation of heart rate in fetal infant and adult rats. *Am. J. Physiol.* **209**, 1095–1105.
2. Adolf, E. F. (1971) Ontogeny of heart-rate controls in hamster, rat, and guinea pig. *Am. J. Physiol.* **220**, 1886–1902.
3. Anderson, T. R. and S. M. Schanberg (1975) Effect of thyroxine and cortisol on brain ornithine decarboxylase activity and swimming behavior in developing rat. *Biochem. Pharmacol.* **24**, 495–501.

4. Atwood, G. F. and N. Kirshner (1976) Postnatal development of catecholamine uptake and storage of the newborn rat heart. *Dev. Biol.* **49**, 532–538.

5. Balazs, R., W. A. Cocks, J. T. Eayrs, and S. Kovacs (1971) Biochemical Effects of Thyroid Hormones on the Developing Brain, in *Conference on Hormones in Development* (M. Hamburgh and E. J. W. Barrington, eds.) Appleton-Century-Crofts, New York, pp. 357–379.

6. Balazs, R., P. D. Lewis, and A. J. Patel (1975) Effects of Metabolic Factors on Brain Development, in *Growth and Development of the Brain* (M. A. B. Brazier, ed.) Raven Press, New York, pp. 83–115.

7. Bareis, D. L. and T. A. Slotkin (1978) Responses of heart ornithine decarboxylase and adrenal catecholamines to methadone and sympathetic stimulants in developing and adult rats. *J. Pharmacol. Exp. Ther.* **205**, 164–174.

8. Bareis, D. L. and T. A. Slotkin (1979) Synaptic vesicles isolated from rat heart: *l*-Norepinephrine uptake properties. *J. Neurochem.* **32**, 345–351,

9. Bareis, D. L. and T. A. Slotkin (1980) Maturation of sympathetic neurotransmission in the rat heart. I. Ontogeny of the synaptic vesicle uptake mechanism and correlation with development of synaptic function. Effects of neonatal methadone administration on development of synaptic vesicles. *J. Pharmacol. Exp. Ther.* **212**, 120–125.

10. Bartolome, J., C. Lau, and T. A. Slotkin (1977) Ornithine decarboxylase in developing rat heart and brain: Role of sympathetic development for responses to autonomic stimulants and the effects of reserpine on maturation. *J. Pharmacol. Exp. Ther.* **202**, 510–518.

11. Bartolome, J., C. Lau, and T. A. Slotkin (1982) Neonatal hyperthyroidism causes premature development of baroreceptor-mediated cardiac sympathetic reflexes. *Dev. Neurosci.* **5**, 208–215.

12. Bartolome, J., E. Mills, C. Lau, and T. A. Slotkin (1980) Maturation of sympathetic neurotransmission in the rat heart. V. Development of baroreceptor control of sympathetic tone. *J. Pharmacol. Exp. Ther.* **215**, 596–600.

13. Black, I. B. (1978) Regulation of autonomic development. *Ann. Rev. Neurosci.* **1**, 283–214.

14. Black, I. B. (1982) Stages of neurotransmitter development in autonomic neuron. *Science* **215**, 1198–1204.

15. Bohn, M. C. and J. M. Lauder (1980) Cerebellar granule cell genesis in the hydrocortisone-treated rat. *Dev. Neurosci.* **3**, 81–89.

16. Ciaranello, R. D., D. Jacobowitz, and J. Axelrod (1973) Effect of dexamethasone on phenylethanolamine *N*-methyltransferase in chromaffin tissue of the neonatal rat. *J. Neurochem.* **20**, 779–805.

17. Claycomb, W. C. (1976) Biochemical aspects of cardiac muscle cell differentiation: Possible control of deoxyribonucleic acid synthesis and cell differentiation by adrenergic innervation and cyclic adenosine 3′: 5′monophosphate. *J. Biol. Chem.* **251**, 6082–6089.

18. Cohen, J. (1974) Role of endocrine factors in the pathogenesis of cardiac hypertrophy. *Circ. Res.* **34**, II-49–II-57.

19. Deskin, R., E. Mills, W. L. Whitmore, F. J. Seidler, and T. A. Slotkin (1980) Maturation of sympathetic neurotransmission in the rat heart. VI. The effect of central catecholaminergic lesions. *J. Pharmacol. Exp. Ther.* **215**, 342–347.

20. Deskin, R. and T. A. Slotkin (1981) Central catecholaminergic lesions in the developing rat: Effects on cardiac noradrenaline levels, turnover, and release. *J. Auton. Pharmacol.* **1**, 205–210.

21. Dussault, J. H. (1975) Development of the Hypothalamic–Pituitary–Thyroid Axis in the Neonatal Rat, in *Perinatal Thyroid Physiology and Disease* (D. A. Fisher and G. N. Burrow, eds.) Raven Press, New York, pp. 73–78.

22. Eayrs, J. T. (1961) Age as a factor determining the severity and reversibility of the effects of thyroid deprivation in the rat. *J. Endocrinol.* **22**, 409–419.

23. Eayrs, J. T. (1971) Thyroid and Developing Brain: Anatomical and Behavioral Effects, in *Conference on Hormones in Development* (M. Hamburgh and E. J. W. Barrington, eds.) Appleton-Century-Crofts, New York, pp. 345–355.

24. Eayrs, J. T. and S. H. Taylor (1951) The effect of thyroid deficiency induced by methyl thiouracil on the maturation of the central nervous system. *J. Anat.* **85**, 350–358.

25. Eranko, L. and O. Eranko (1972) Effect of hydrocortisone on histochemically demonstrable catecholamines in the sympathetic ganglia and extra-adrenal chromaffin tissue of the rat. *Acta Physiol. Scand.* **84**, 125–133.

26. Ganong. W. F. (1963) The Central Nervous System and the Synthesis and Release of Adrenocorticotropic Hormone, in *Advances in Neuroendocrinology* (A. V. Nalbandov, ed.) University of Illinois Press, Urbana, pp. 92–157.

27. Gootman, P. M., N. M. Buckley, and N. Gootman (1979) Postnatal maturation of neural control of the circulation. *Rev. Perinatal Med.* **3**, 1–72.

28. Gresik, E. W. (1976) Preliminary observations on the effects of chronic hypothyroidism on the development of the superior cervical ganglion of the rat. *Brain Res.* **110**, 619–622.

29. Gripois, D. and A. Diarra (1983) Adrenal epinephrine depletion after insulin-induced hypoglycaemia in young hypothyroid rats. *Mol. Cell. Endocrinol.* **30**, 241–245.

30. Hamburgh, M., L. A. Mendoza, J. F. Burkhart, and F. Weil (1971) Thyroid-Dependent Processes in the Developing Nervous System, in *Conference on Hormones in Development* (M. Hamburgh and

E. J. W. Barrington, eds.) Appleton-Century Crofts, New York, pp. 403–415.

31. Harden, T. K., B. B. Wolfe, J. R. Sporn, J. P. Perkins, and P. B. Molinoff (1977) Ontogeny of beta-adrenergic receptors in rat cerebral cortex. *Brain Res.* **125**, 99–108.

32. Kirksey, D. F., F. J. Seidler, and T. A. Slotkin (1978) Ontogeny of (−) [³H]-norepinephrine uptake into synaptic storage vesicles of rat brain. *Brain Res.* **150**, 367–375.

33. Kunos, G., C. Brass, W. H. Kan, and L. Mucci (1978) Thyroid control of rat and rabbit cardiac adrenoceptors. *Fed. Proc.* **37**, 684.

34. Lau, C., S. P. Burke, and T. A. Slotkin (1982) Maturation of sympathetic neuotransmission in the rat heart. IX. Development of transsynaptic regulation of cardiac adrenergic sensitivity. *J. Pharmacol. Exp. Ther.* **223**, 675–680.

35. Lau, C. and T. A. Slotkin, (1979) Regulation of rat heart ornithine decarboxylase: Change in affinity for ornithine evoked by neuronal, hormonal, and ontogenetic stimuli. *Mol. Pharmacol.* **16**, 504–512.

36. Lau, C. and T. A. Slotkin, (1980) Regulation of ornithine decarboxylase activity in the developing heart of euthyroid or hyperthyroid rats. *Mol. Pharmacol.* **18**, 247–252,

37. Lau, C. and T. A. Slotkin (1980) Maturation of sympathetic neurotransmission in the rat heart. II. Enhanced development of presynaptic and postsynaptic components of noradrenergic synapses as a result of neonatal hyperthyroidism. *J. Pharmacol. Exp. Ther.* **212**, 126–130.

38. Lau, C. and T. A. Slotkin (1981) Maturation of sympathetic neurotransmission in the rat heart. VII. Suppression of sympathetic responses by dexamethasone. *J. Pharmacol. Exp. Ther.* **216**, 6–11,

39. Lau, C. and T. A. Slotkin (1982) Maturation of sympathetic neurotransmission in the rat heart. VIII. Slowed development of noradrenergic synapses resulting from hypothyroidism. *J. Pharmacol. Exp. Ther.* **220**, 629–636.

40. Lau, C. and T. A. Slotkin, (1979) Accelerated development of rat sympathetic neurotransmission caused by neonatal triiodothyronine administration. *J. Pharmacol. Exp. Ther.* **208**, 485–490.

41. Lengvari, I., B. J. Branch, and A. N. Taylor (1980) Effects of perinatal thyroxine and/or corticosterone treatment on the ontogenesis of hypothalamic and mesencephalic norepinephrine and dopamine content. *Dev. Neurosci.* **3**, 59–65.

42. Nicholson, J. L. and J. Altman (1972) Effects of early hypo- and hyperthyroidism on the development of rat cerebellar cortex. I. Cell proliferation and differentiation. *Brain Res.* (Osaka) **44**, 13–23.

43. Nukari-Siltovuori, A. (1977) Postnatal development of adrenergic and cholinergic sensitivity in the isolated rat atria. *Experientia* **33**, 1611–1612.

44. Pappano, A. J. (1977) Ontogenetic development of autonomic neuroeffector transmission and transmitter reactivity in embryonic and fetal hearts. *Pharmacol. Rev.* **29**, 3–33.
45. Patterson, P. H. (1978) Environmental determination of autonomic neurotransmitter functions. *Ann. Rev. Neurosci.* **1**, 1–17.
46. Rastogi, R. B. and R. L. Singhal (1976) Influence of neonatal and adult hyperthyroidism on behavior and biosynthetic capacity for norepinephrine, dopamine, and 5-hydroxytryptamine in rat brain. *J. Pharmacol. Exp. Ther.* **198**, 609–618.
47. Robecchi, A., S. Di Vittorio, and G. Einaudi (1960) The cardiovascular function during prolonged corticosteroid therapy. *Acta Rheum. Scand.* **6**, 241–250,
48. Russell, D. H. (1980) Ornithine decarboxylase as a biological and pharmacological tool. *Pharmacology* **20**, 117–129.
49. Schapiro, S. (1968) Some physiological, biochemical, and behavioral consequences of neonatal hormone administration: Cortisol and thyroxine. *Gen. Comp. Endocrinol.* **10**, 214–228.
50. Schwark, W. S. and R. R. Keesey (1975) Thyroid hormone control of serotonin in developing rat brain. *Res. Commun. Chem. Path. Pharmacol.* **10**, 37–50.
51. Schwark, W. S. and R. R. Keesey (1976) Influence of thyroid hormone on norepinephrine metabolism in rat brain during maturation. *Res. Commun. Chem. Path. Pharmacol.* **13**, 673–678.
52. Schwark, W. S. and R. R. Keesey (1976) Cretinism: Influence on rate-limiting enzymes of amine synthesis in rat brain. *Life Sci.* **19**, 1699–1704.
53. Seidler, F. J. and T. A. Slotkin (1979) Presynaptic and postsynaptic contributions to ontogeny of sympathetic control of heart rate in the preweanling rat. *Brit. J. Pharmacol.* **65**, 531–534.
54. Seidler, F. J. and T. A. Slotkin (1981) Development of central control of norepinephrine turnover and release in the rat heart: Responses to tyramine, 2-deoxyglucose, and hydralazine. *Neurosci.* **6**, 2081–2086.
55. Slotkin, T. A. (1973) Maturation of the adrenal medulla. I. Uptake and storage of amines in isolated storage vesicles of the rat. *Biochem. Pharmacol.* **22**, 2023–2032.
56. Slotkin, T. A. (1973) Maturation of the adrenal medulla. II. Content and properties of catecholamine storage vesicles of the rat. *Biochem. Pharmacol.* **22**, 2033–2044.
57. Slotkin, T. A. (1979) Ornithine decarboxylase as a tool in developmental neurobiology. *Life Sci.* **24**, 1623–1630.
58. Slotkin, T. A., G. Barnes, C. Lau, F. J. Seidler, P. Trepanier, S. J. Weigel, and W. L. Whitmore (1982) Development of polyamine and biogenic amine systems in brains and hearts of neonatal rats given dexamethasone: Role of biochemical alterations in cellular maturation for producing deficits in ontogeny of neurotransmitter

levels, uptake, storage, and turnover. *J. Pharmacol. Exp. Ther.* **221,** 686–693.

59. Slotkin, T. A., C. J. Chantry, and J. Bartolome (1982) Development of central control of adrenal catecholamine biosynthesis and release. *Adv. Biosci.* **36,** 95–102.

60. Slotkin, T. A., A. Johnson, W. L. Whitmore, and R. J. Slepetis (1984) Ornithine decarboxylase and polyamines in developing rat brain and heart: Effects of perinatal hypothyroidism. *Intl. J. Dev. Neurosci.* **2,** 155–161.

61. Slotkin, T. A., P. G. Smith, C. Lau, and D. L. Bareis (1980) Functional Aspects of Development of Catecholamine Biosynthesis and Release in the Sympathetic Nervous System, in *Biogenic Amines in Development*, (H. Parvez and S. Parvez, eds.) Elsevier/North Holland, Amsterdam, pp. 29–48.

62. Smith, P. G., E. Mills, and T. A. Slotkin (1981) Maturation of sympathetic neurotransmission in the efferent pathway to the rat heart: Ultrascturctural analysis of ganglionic synaptogenesis in euthyroid and hyperthyroid neonates. *Neurosci.* **6,** 911–918.

63. Smith, P. G., T. A. Slotkin, and E. Mills (1982) Development of sympathetic ganglionic transmission in the neonatal rat: Pre- and postganglionic nerve response to asphyxia and 2-deoxyglucose. *Neurosci.* **7,** 501–507.

64. Sorimachi, M. (1977) Impaired maturation of presynaptic cholinergic nerve terminals in the superior cervical ganglia after administration of guanethidine and dexamethasone. *Japan. J. Pharmacol.* **27,** 629–634.

65. Turner, B. B., R. J. Katz, and B. J. Carroll (1979) Neonatal corticosteroid permanently alters brain activity of epinephrine-synthesizing enzyme in stressed rats. *Brain Res.* **166,** 426–403.

66. Wildenthal, K. (1974) Studies of fetal mouse hearts in organ culture: Influence of prolonged exposure to triiodothyronine on cardiac responsiveness to isoproterenol, glucagon, theophylline, acetylcholine, and dibutyryl cyclic 3′, 5′-adenosine monophosphate. *J. Pharamacol. Exp. Ther.* **190,** 272–279.

Chapter 5

Development of ANS Innervation to the Avian Heart

Margaret L. Kirby and Donald E. Stewart

1. Introduction

The development of the autonomic nervous system in avians has been studied and reviewed extensively since late in the nineteenth century. Recent advances in many techniques have made our understanding of autonomic development much more accurate. The advent of cellular marking using quail-chick chimeras and refinements in microsurgical techniques, have been fundamental in mapping the fate of autonomic ganglion cell precursors derived from the neural crest. Electron microscopic evaluation has clarified the migration and differentiation of autonomic neurons. The histofluorescence method for catecholamines has greatly enhanced our evaluation of the location of adrenergic neurons. Many other histochemical and biochemical methods have been developed that should lead eventually to the complete evaluation of other transmitter types, as well as the state of maturity of autonomic neurons.

Several studies have been done in chick embryos to show when the autonomic nerves to the heart are first capable of controlling heart rate and contractility. Pappano reviewed these studies in 1977 (*42*) and several studies have been added subsequently.

Autonomic nerves may influence differentiating tissues even before they assume their adult roles (*10*). For this reason it is important to know the exact sequence of morphological development of the autonomic nerves to the heart in relation to onset of adult-type function of these nerves. Many investigators have used Romanoff's description (*45*) of developing cardiac autonomic innervation, which is based on studies from the first third of the twentieth century. Most of these descriptions have been found recently to be incorrect with respect to temporal aspects of cardiac autonomic innervation.

The purpose of this chapter is to describe the morphological, histochemical, and biochemical development of innervation to the chick heart. The chapter will correlate development with the known onset of functional control of the avian heart only briefly.

2. Autonomic Innervation of the Adult Avian Heart

The autonomic nervous system in avians is divided into sympathetic and parasympathetic systems (which may appropriately be referred to as thoracolumbar and craniosacral, as in mammals).

Parasympathetic preganglionic nerves are distributed with four cranial nerves and nerves of the sacral plexus. The preganglionic neurons for the cervical, thoracic, and abdominal viscera are located in the dorsal motor nucleus of the vagus. The vagus nerve exits the brainstem and, in the chick, has two sensory ganglia. The superior ganglion of the vagus is combined with the superior ganglion of the glossopharyngeal nerve close to the brainstem, and the inferior ganglion of vagus is located at the thoracic inlet (*13*). Branches of this inferior ganglion form the cranial cardiac nerve. Other cardiac branches arise from the recurrent nerve. These branches provide preganglionic innervation to the bulbar plexus located at the base of the arterial pole of the heart. Caudal cardiac branches arise from the vagus at the level of the pulmonary veins. These branches provide preganglionic innervation to the ganglia in the atrial plexus.

The postganglionic parasympathetic cardiac neurons are located in ganglia in the bulbar and atrial plexuses (*45*). The ultrastructure of these cells has never been described in adult chickens, although the cells have been studied during development, as discussed below. Postganglionic fibers leave the atrial and bulbar plexuses to join the cardiac vasculature, and ramify as subepicardial and subendocardial plexuses over the surface of the entire heart (1).

The sympathetic preganglionic cell column is called the nucleus of Terni and is the avian homolog of the intermediolateral cell column of the mammal. The nucleus of Terni is located in the gray matter of the spinal cord, dorsally, and dorsolaterally to the central canal. This column extends from the last cervical spinal segment to the second lumbar segment. Axons from this nucleus leave the spinal cord via the ventral root. The postganglionic sympathetic nerves are located in paravertebral (sympathetic trunks) or prevertebral ganglia. Innervation to the heart originates only from the sympathetic trunks.

The sympathetic innervation of the heart in the adult chicken is from bilateral sympathetic cardiac nerves that arise from the first thoracic ganglia of the sympathetic trunks (3). The nerves travel laterally to the lungs, just medial to the first rib. They pierce the saccopleural membrane and travel ventrally to the lungs. More caudally, they join the cranial venae cavae, eventually joining the vagi. The vagi and sympathetic cardiac nerves enter the pericardial sac and travel along the great vessels to reach the heart (3). The terminal fibers of both autonomic divisions branch extensively in the myocardium. The narrowing end fibers terminate in varicosities arranged in rows along the sarcolemma of the muscle fibers. These disappear without showing any particular morphological specializations.

3. Preganglionic Autonomic Neurons

3.1. Parasympathetic

According to Windle and Austin (55), the dorsal motor nucleus (DMN) can be recognized first at 43–44 h of incubation in the chick. The roots leave the brain at 60 h. In the early embryo, the DMN extends caudally to cervical levels of the spinal cord. However, by the time of hatching, the nucleus does not extend further caudally than the hypoglossal nuclei. The cells in the DMN are all generated from the neural epithelium of the ventral part of the basal plate in the first 4 d of incubation (35). Generation of cells in the nucleus follows a caudo-rostral gradient, with those located farthest caudally generated first (2–3 d), and the rostral cells being generated slightly later (3–4 d). The primary dorsal migration of the DMN to a position comparable to that of the nucleus of Terni is complete by d 7 of incubation (35).

Cell death in the nucleus begins on d 6. Between d 8 and 10, cell death occurs predominantly in the caudal third of the nucleus. Cell death in the rostral third of the nucleus begins on d 10, and 54% of the original neuron population is lost by maturity (56). Wright proposes that the asynchronous pattern of cell death in the DMN may be related to the timing of functional innervation at the target sites. Although the preganglionic cardiac parasympathetic nerves have not been identified in the chick, they are located at the rostral end of the dorsal motor nucleus in the pigeon (12). Since functional cardiac innervation is thought to begin after d 10, this would support Wright's hypothesis that cell death in the rostral part of the nucleus is related to the onset of functional parasympathetic innervation of the heart.

3.2. Sympathetic

In the avian central nervous system, the preganglionic sympathetic cell column (nucleus of Terni) is located dorsolaterally to the central canal. This is in contrast to the preganglionic sympathetic neurons of the mammalian spinal cord that are located in the intermediolateral cell column. The cells of the nucleus of Terni migrate from the ventral motor column at 3.5–4 d, and by 8 d have reached their destination (35,51). Synapses with sympathetic ganglion cells are first seen at 7–8 d, and continue throughout the embryonic period. Caserta and Ross (9) have reported that synapses are formed in the nucleus of Terni beginning on d 10. Since peripheral synapses precede synapse formation on cells in the nucleus of Terni, the latter are not prerequisites for preganglionic axonal growth or formation of synapses with the sympathetic trunks.

The factors that influence and regulate mammalian autonomic neuron development have been extensively studied by Black and coworkers (for review, *see* ref. 7). Although it is uncertain whether the conclusions derived from their work with mammalian tissues are applicable to the avian class, the questions asked regarding autonomic neuron development are identical. A host of factors, including anterograde and retrograde transsynaptic regulation in ganglia, NGF, target organs, and various hormonal mechanisms, have been shown to affect the biochemical and morphological development of autonomic neurons (7). Although pertinent, these issues are beyond the scope of this chapter and will not be reviewed here.

4. Postganglionic Autonomic Neurons

4.1. Neural Crest

All of the postganglionic autonomic neurons are derived from the neural crest. The neural crest is thickened ectoderm around the edge of the embryonic neural plate. As the neural plate closes to form the neural tube, the crest cells are released from their position in the neural fold (52). Although the neural fold consists of three different types of cells, only the cells that migrate are called neural crest cells. Once the neural crest cells are free of the neural fold, they may migrate actively or be translocated passively to their final destination (41).

The neural crest extends from the diencephalon to the caudalmost extent of the neural tube. It is divided into two regions. Cranial neural crest extends from the diencephalon to the level of the fifth somite. The remainder of the neural crest, beginning at the level of somite 6 and extending caudally, is called trunk neural crest (32,40). The trunk neural crest forms all of the cells of the peripheral nervous system at the spinal level, the adrenal medulla, and melanocytes. The cranial neural crest forms many mesenchymal derivatives *in addition* to peripheral nervous elements and melanocytes (19,39).

Trunk neural crest cells migrate ventrolaterally (after neural tube closure) either between the somites or between the neural tube and the somites. Cephalic neural crest cells disperse between the surface ectoderm and the paraxial mesoderm. Most of these cells become located beside and beneath the pharynx and forebrain.

Ectomesenchymal cells are in abundance in the cranial neural crest, but comprise only a small proportion of the trunk neural crest. The ectomesenchymal cells of the cephalic neural crest participate in the formation of periocular tissues and skeletal and connective tissue derivatives of all the branchial arches, including the outflow region of the heart.

4.2. Cardiac Ganglia

Precursors for parasympathetic postganglionic neurons in the heart (cardiac ganglia) arise from the neural crest bilaterally located over somite 1–3. This level of neural crest provides prospective parasympathetic postganglionic neurons, satellite cells, and Schwann cells to the cardiac ganglia. Migration of this region of neural crest begins just after closure of the neural tube, at about stage 10 (30–35 h). Migration proceeds rapidly, and by stage 11 enough cells have migrated ventrolaterally to populate all of the cardiac ganglia (*30*).

In addition to seeding all of the cardiac ganglia with neural cells, the same region of neural crest provides ectomesenchymal cells that are necessary for proper development of the arterial outflow region of the heart (*25*).

Neural crest migration has not been followed from the neural fold over somites 1 and 2 into the ganglia in the conotruncal region of the heart. However, the development of the cardiac ganglia has been studied ultrastructurally from the third day of incubation to hatching (*31*). Cardiac branches of the vagus arise from the inferior (thoracic) vagal ganglia and the recurrent branches of the vagi between 3.5 and 4 d. These branches grow along the fourth and sixth aortic arches toward the base of the heart. These cardiac branches of the vagus are accompanied by small cells that cluster around the conotruncal region. The cells have the characteristic features of migrating neuroblasts from the neural crest, as described by Tennyson (*50*). By d 5 of development, some of these cells have become bipolar neuroblasts, and have processes that can be followed for some distance from the cell body. Supporting and Schwann cells can also be distinguished. Between 5 and 10 d of incubation, the ganglia contain globular neuroblasts with ovoid, eccentrically located nuclei. Axodendritic synapses are the first recognizable synapses in the ganglia. These appear first on d 8. Mitotic figures, as well as degenerating cells, can be seen in the ganglia during this period. From d 11 to hatching, the cardiac ganglia mature. The neurons become large and spherical. The nucleus becomes spherical and occupies a smaller proportion of the cell volume because of the large accumulation of cytoplasm. Axosomatic synapses are occasionally apparent. Satellite cells become much more numerous and have dense ovoid nuclei. The cardiac ganglia at this phase resemble dorsal root ganglia. The neurons themselves are morphologically similar to chick ciliary ganglion cells (*50*).

Rickenbacher and Müller (44) have used a whole-organ technique for acetylcholinesterase demonstration that shows the development and location of the cardiac ganglia and subepicardial plexuses in the chick embryonic heart. Cholinesterase-positive nerves can be identified in the heart on the seventh and eighth day of incubation. Using the cholinesterase technique, the ganglia in both atrial and bulbar plexuses can be visualized as they continue to grow in number and size until d 15. One third of the surface of the base of the heart is covered by ganglia, and the line of demarcation is somewhat abrupt. A considerable subepicardial plexus develops by d 12 and nerve fibers can be traced to the apex of the heart.

The appearance of acetylcholine in the developing chick heart has not been studied, but the uptake of [^{3}H]choline has. Presumably the rate of synthesis of acetylcholine in the chick heart is related directly to the activity of the high-affinity choline uptake system, as it is in isolated rat atria (54). High-affinity [^{3}H]choline uptake in isolated chick atrium can be measured first on d 7 of incubation, and increases rapidly to reach a peak at d 10–12 of development, then drops to a plateau by d 17 of incubation. This level of uptake is maintained through wk 4 of posthatching (*see* Fig. 1).

Cardiac ganglion cell terminals contact the myocardium in the *en passant* form of synapses. Because the development of postganglionic terminals in cardiac muscle has not been studied at the ultrastructural level, it is difficult to determine when morphological contact of the parasympathetic terminals is made with the myocardium or conducting system. It has been established that the atropine-sensitive cholinergic neuroeffector transmission begins on d 12 of incubation, although pretreatment with physostigmine allows transmission to begin on d 10. This is indirect evidence that the synapses *en passant* are present and functional by d 10 (42).

4.3. Sympathetic Trunks

A prominent component of the sympathetic nervous system in all vertebrates consists of the paravertebral sympathetic trunks. The neurons of the sympathetic trunks derive from the cervicodorsal neural crest adjacent to somites 8–28; cells of the adrenal medulla arise from neural crest between somites 18–24; the cholinergic enteric ganglia and the ganglion of Remak arise from somitic levels 1–7, and also caudal to somite 28 (33).

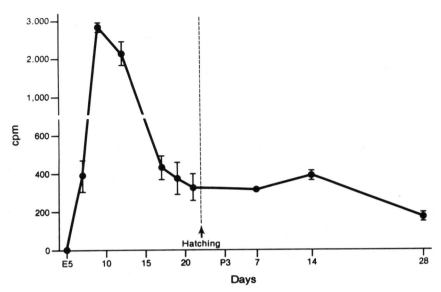

Fig. 1. [³H]Choline uptake (±SEM) in the developing chick atrium. The ordinate expresses neuronal uptake in cpm/mg tissue/10⁶ cpm available in the incubation well. The peak in choline uptake occurs just prior to the beginning of electrical transmission through the cardiac ganglia and is probably responsible for the accumulation of acetylcholine that is needed for chemical transmission at the terminals (reprinted from ref. *30*).

Recently, transplantation and extirpation of very precise regions of neural crest have been used to provide information regarding the origin and migratory pathways of precursors of the autonomic innervation to specific organs. Yip (*57*) has studied the formation of the paravertebral sympathetic chain ganglia using quail-chick chimeras. After transplanting single segments of quail neural tube and its overlying neural crest onto chick embryos of the same age, the embryos were allowed to develop to 7.5–12 d. The embryos were then sectioned and examined for the presence of quail cells in the sympathetic ganglia. Yip reported that in the cervical and thoracic regions, sympathetic ganglia arise primarily from the neural crest of the corresponding levels. Longitudinal migration of neural crest cells along the neural axis does occur, however, with cells migrating two segments rostrally and three segments caudally in the trunks. A single sympathetic trunk ganglion, therefore, may contain cells from as many as six segments of the neural crest. The majority of ganglion cells arise from the same level of neural crest; decreasing numbers of cells are contributed from more distant segmental levels.

Direct visceral branches of postganglionic neurons located in the cervicothoracic portion of the sympathetic trunks provide the sympathetic innervation of the heart. Postganglionic neurons contain catecholamines acting as neurotransmitters that can be detected using formaldehyde-induced fluorescence. Sympathetic trunk formation is first seen in the upper thoracic region as segmental aggregates of fluorescent cells that form bilaterally dorsolateral to the aorta at about d 3.5 of incubation (15,27). By d 4, these aggregates are present in cervical through lumbar levels, but remain segmental in arrangement with no continuity between segments. At d 4.5, the aggregates have become continuous cords of cells and are designated primary trunks. Segmental swellings are present near the intersegmental arteries. In 4.5–5-d embryos, cells extend dorsolaterally just caudal to each intersegmental artery toward the ventral root. These cells contribute to the formation of the secondary trunks. The primary trunks become longitudinally discontinuous and rapidly disappear as the secondary trunks form at d 5.5–6. Using fluorescence histochemistry for catecholamines, both the primary and secondary trunks can be visualized shortly after their development.

Ignarro and Shideman (20) have shown that tyrosine hydroxylase is present in whole embryos on the first day of incubation, dopa-decarboxylase is present on the second day, and dopamine-β-hydroxylase appears on the fourth day. Ignarro and Shideman (21) have also reported uptake of injected tritiated epinephrine (E) and norepinephrine (NE) from the yolk, which contains 25 ng NE/g protein and 195 ng E/g protein. To determine whether the visible catecholamine fluorescence in primary trunk cells is caused by synthesis or uptake, chick embryos were treated with α-methyl-p-tyrosine (AMT), a synthesis inhibitor, and desmethylimipramine (DMI), an uptake inhibitor (23). Treatment of chick embryos with AMT during the period when the primary and secondary trunks first appear caused a total loss of NE and E concentrations. Treatment with DMI causes no loss of NE or E concentrations. These results indicate that synthesis alone is responsible for NE and E that are normally present in the primary and secondary trunks in the chick embryo from d 3.5–5.

Although neuronal cell processes interconnect the primary and secondary trunks at 5.5 d, the neurons in the secondary trunks do not make peripheral connections until later, as described below. Secondary trunk formation is essentially complete at d 6, and by d 8 of incubation the secondary sympathetic trunks consist of homogeneous aggregations of brightly fluorescent cells. Between 8 and 16 d, the ganglia lose most of the fluorescence observed at d 8 (*28*); however, aggregations of small cells within the ganglia remain brightly fluorescent. Also, the adrenal medullary cells remain brightly fluorescent throughout this period. By d 16 the ganglion cells have regained some of their earlier fluorescence. Small intensely fluorescent (SIF) cells with short processes can be distinguished from principal ganglion cells. These SIF cells are most numerous in the lumbar ganglia, but are present in thoracic and cervical ganglia and appear to be the same cells that retained fluorescence, whereas the ganglion cells were nonfluorescent at d 9–15.

The ultrastructural development of sympathetic ganglia and catecholamine-containing cells has been studied by several investigators, frequently with emphasis placed on the characterization of catecholamine-storage vesicles. Luckenbill-Edds and van Horn (*36*) have described catecholaminergic cells of the sympathetic trunks in both adults and embryonic chicks. In contrast to adult principal sympathetic neurons that contain predominantly small dense-cored vesicles (SDCV), 40–60 nm in diameter, embryonic (d 7–8) ganglia contain neuroblasts that commonly have large dense-cored vesicles (LDCV), 100–200 nm in diameter. The LDCV appear to migrate into developing processes as differentiation continues. The neuroblasts then develop SDCV at d 13–15 of incubation, and gradually resemble small adult principal sympathetic neurons. Granule cells, characterized by sparse cytoplasm and clumped chromatin, have vesicles greater than 200 nm in diameter and are rarely found in 7–8-d embryos. After increasing in size and numbers, granule cells are no longer seen beyond embryonic d 13–15. Although Kirby et al. (*28*) have reported the presence of SIF cells, cells with ultrastructural characteristics similar to mammalian SIF cells have not been seen in the sympathetic trunks of late chick embryos and adults (*36*).

Rothman et al. (*46*) and Cohen (*11*) have shown that some catecholamine-containing cells in the sympathetic trunks of chick embryos retain the ability to undergo mitosis. Therefore, presumptive sympathetic neuroblasts do not have to withdraw from the cell cycle in order to synthesize catecholamines. In addition, the precursors of sympathetic ganglion cells specifically take up NE (*23,46*). This is evidence that adrenergic characteristics of both an ultrastructural (vesicle formation) and functional (uptake of NE) nature are attributable to neuroblasts of the sympathetic trunks.

Sprouting from the principal ganglion cells is first seen at d 6–8 of incubation and occurs along the entire length of the trunks. The branches of the superior cervical ganglia are all directed cranially along the carotid arteries. The one major branch seen leaving the trunks in the lower cervical and thoracic regions appears bilaterally in both formaldehyde-induced fluorescence and silver preparations at the level of the first thoracic ganglion. This branch, the thoracic cardiac nerve, is first seen at d 7–7.5. The thoracic cardiac nerves accompany the T1 ventral ramus and the first intercostal nerve just medial to the first rib. At d 8, the sympathetic cardiac nerves are found in the saccopleural membrane. On d 9, these nerves make contact with the vagus, and can be seen in the bulbar region of the heart on d 10. Small fluorescent cells are seen along the length of the sympathetic cardiac nerves from the first thoracic ganglion to the bulbar region of the heart.

Fluorescent nerve processes can be seen in atrial stretch preparations on d 10 of incubation, and in the ventricles on d 11 (*17, 28*). These fluorescent nerves are present in the atria in approximately 25% of d-10 hearts. The number of fluorescent nerves in the heart continues to increase throughout the remainder of the incubation period. [^{3}H]Norepinephrine uptake studies in the atrium confirm that ingrowth of adrenergic nerves begins on d 10–11, and specific neuronal uptake increases throughout the embryonic period to peak on d 19 (*48*). Between d 19 and 21, there is a decrease in neuronal uptake of [^{3}H]-NE, followed by a more gradual increase from incubation d 21 to posthatching d 14 (*see* Fig. 2).

Although sympathetic innervation arrives in the heart on d 10, field stimulation of intramural cardiac nerves does not evoke a positive inotropic response until embryonic d 16 (*18*). This lag in functional adrenergic transmission might occur for various reasons, including an insufficient density of adrenergic nerves and immature axon terminals not capable of manufacturing or releasing adequate amounts of neurotransmitter prior to d 16.

4.4. Catecholamines in the Developing Heart

Even though the sympathetic innervation is not present in the heart until the middle of the embryonic period, NE and E can be detected in the heart as early as d 3 of incubation (*21*). Dopa is first

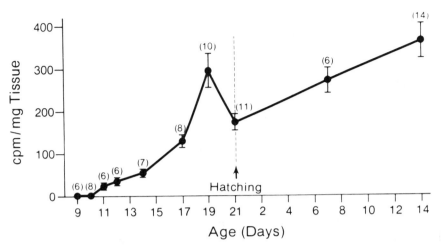

Fig. 2.   Specific neuronal uptake of [³H]-NE by adrenergic nerves in chick atria (in vitro). Specific neuronal uptake is defined as total uptake minus uptake in the presence of desmethylimipramine. The number of individual hearts used for each point is in parentheses. Each point represents the mean ±SEM (reprinted from ref. *48*).

detected in the heart on d 4, and dopamine appears 2 d later. The E content (Fig. 3A) in whole hearts of chick embryos is higher than the NE content (Fig 3B), except just prior to hatching, when NE surpasses E and remains higher throughout the first 2 wk of posthatching (*48*). In contrast to Ignarro and Shideman, we could detect only negligible amounts of NE during embryonic d 10–13 (*see* Fig. 3B).

Callingham and Cass (*8*) reported finding a greater proportion of E than NE in the hearts of young chicks and suggested that E may play a role in sympathetic transmission in the chick. More recently, DeSantis et al. (*14*) provided evidence that both NE and E act as sympathetic neurotransmitters in the hearts of adult chickens. The literature indicates that E is present in greater concentrations than NE in the hearts of embryonic and newborn chicks and that this relationship reverses in the adult chicken (*8, 14,21,48*). Some of the catecholamines present in the heart may be located in nonneuronal catecholamine-storing structures that have been described in both embryonic and mature chicken hearts (*5,15,28*). Circulating catecholamines from the yolk (*22*) probably contribute to the pool of cardiac catecholamines. Although the role of catecholamines in presympathetic embryogenesis remains an enigma, the evidence in support of their presence

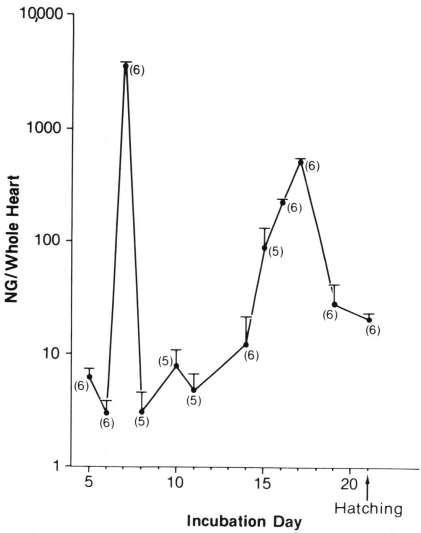

Fig. 3A. EPI (A) and NE (B) (*see* next page) content in whole hearts of chick embryos. Catecholamines were measured via HPLC and integration of chromatogram peaks. Each graph is plotted on a semi-logarithmic scale. the number of individual hearts used for each point is in parentheses. Each point represents the mean ± SEM (reprinted from ref. *48*).

early in development infers a nontransmitter function (*10*). Even the notochord has been shown to be a site of transient catecholamine synthesis and storage (*26,53*). Further studies are

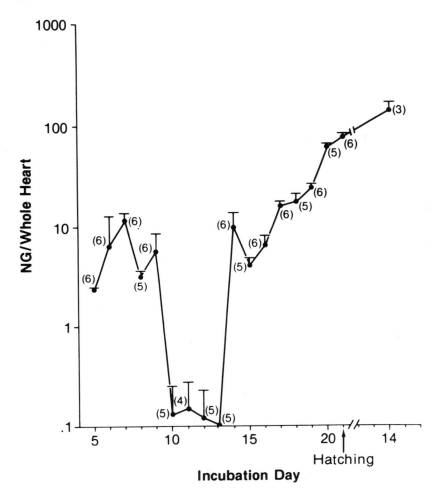

Fig. 3B.

necessary to elucidate what may be a critical link between catecholamines and cellular differentiation in embryonic development.

5. Receptor Development

5.1. Muscarinic Receptors

The chronotropic and inotropic effects of cholinergic innervation in the heart are mediated via muscarinic cholinergic receptors that are located on the myocardial membrane and on the terminals of postganglionic cholinergic and adrenergic neurons.

Atropine-sensitive cholinergic neuroeffector transmission begins in the chick heart on d 12 of incubation, although pretreatment with physostigmine allows transmission to begin on d 10. However, the chick heart responds to acetylcholine long before functional parasympathetic innervation occurs. Acetylcholine can arrest pacemaker activity in chick hearts from 44 h to 21 d of development. However, there is a progressive increase in sensitivity to acetylcholine as incubation proceeds.

Using [³H]quinuclidinyl benzilate (QNB), the development of muscarinic acetylcholine receptors in chick heart was studied from incubation d 5 through 20. At d 5, [³H]-QNB binds to a single population of receptors. The concentration of muscarinic acetylcholine receptors increases by 25–30% between d 5 and 9 (*see* Fig. 4). After 9 d of development, ventricular receptor content

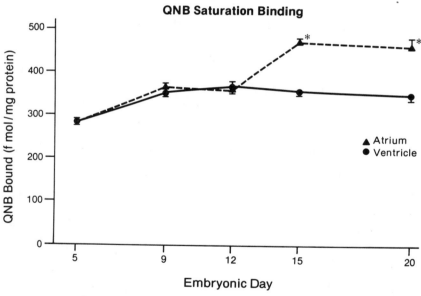

Fig. 4. [³H]-QNB binding to muscarinic receptors in chick atria and ventricles on d 5, 9, 12, 15, and 20 of incubation. Binding in whole heart is indicated for incubation d 5. The concentration of muscarinic receptors increased between d 5 and 9. Between d 12 and 15 of incubation, the atrial concentration of muscarinic receptors increased by about 30%, whereas the ventricular concentration did not change. Statistically significant differences ($p < 0.05$) in binding site densities between atrium and ventricle are indicated by asterisks (reprinted from ref. 24).

does not increase, plateauing at about 350 fmol/mg protein. Although receptor concentrations in atrium and ventricle were equivalent on d 12 of incubation, by d 15 the receptor concentration was 25–40% higher in atrium than in ventricle.

Carbamylcholine binding can be described by a two-receptor population model, although the possibility of additional sites is not ruled out. High-affinity ($K_d = 0.08$–0.13 μM) and low-affinity ($K_d = 4$ μM) components are observed. The fraction of receptors displaying high-affinity carbamylcholine binding rises 20% between d 12 and 15, from 50 and 52% to 61 and 63% in atrium and ventricle, respectively (24).

This time period corresponds to the natural elaboration of functional parasympathetic innervation of the heart.

5.2. β-Adrenergic Receptors

Catecholamines act via α and β receptors. The variation in responses elicited in different tissues is determined by the relative numbers of these receptor sites (43). The β-adrenergic receptors are coupled to adenylate cyclase and a guanine nucleotide regulatory protein. The activation of adenylate cyclase mediates a variety of intracellular physiological and metabolic functions by catalyzing synthesis of cAMP at the inner face of the plasma membrane. The mechanism of action of cAMP is complex and involves the initiation of a cascading series of protein phosphorylation reactions. Most, if not all, of the actions of catecholamines at β-adrenergic receptor sites appear to require activation of adenylate cyclase, and the consequent increase in the intracellular concentrations of cAMP. Hence, catecholamines influence a large number of cellular events via receptor activation, not all of which have been elucidated.

The β-adrenergic receptors have been more thoroughly characterized than other receptor types, but it is known that other plasma membrane receptors are also coupled to second messengers such as cAMP, Ca^{2+}, phosphoinositol breakdown, and others not yet defined (34). Most of the known receptors are studied by direct radioligand-binding assays.

The β-adrenergic receptors have been characterized using various radioligands, including [^{3}H]dihydroalprenolol (2, 47) and (^{125}I)iodocyanopindolol (38). McCarty et al. (37) have reported that β-adrenergic receptors are present and functional in the hearts of very early (d 4) chick embryos. Hence, receptors are present in cardiac tissue prior to innervation. To study the influence of sympathetic innervation on receptor development, we have measured β-adrenergic receptors in chick hearts prior to and following the arrival of sympathetic nerves in the heart (49).

Using (^{125}I)L-pindolol as a probe (synthesized according to Minneman, personal communication), receptors were measured in hearts from embryos of 7, 9, 12, 15, 18, and 20 d of incubation (*see* Fig. 5). Hearts were pooled as required to obtain sufficient tissue for the binding assay. We defined specific β-adrenergic receptor binding as total (^{125}I)L-pindol ol binding, minus binding measured in the presence of 50 μM isoproterenol. The β-adrenergic receptor concentration was highest on embryonic d 9 (664 fmol/mg protein), and declined to 119 fmol/mg protein on embryonic d 20. This indicates an 80% decrease in adrenergic receptor density over the same period of embryonic development in which the sympathetic nerves arrive in the heart (d 10–11) and functional adrenergic transmission begins (d 16).

6. Lesions of Autonomic Innervation to the Heart

6.1. Surgical Lesions

Previously, denervation of the heart has been achieved by transection of specific cardiac nerves, cardiac transplantation, or with chemicals such as 6-hydroxydopamine. Obtaining a model for the study of cardiac development in the absence of autonomic innervation is problematic for several reasons. Peripheral surgical denervation is not feasible in the very young chick embryo and chemical denervation lacks target specificity and might affect other cellular functions. In order to produce an animal model that is deficient in autonomic innervation to the heart, microsurgery is necessary to remove or destroy the neural crest precursors of the postganglionic autonomic nerves to the heart. This type of lesion may be produced with precision and a minimum of variables (such as nonspecific drug effects or surgical trauma) affecting the developing target organ.

Parasympathetically aneural hearts can be produced by removing the area of neural crest that seeds the cardiac ganglia. Extirpation or cauterization of the neural fold (containing neural

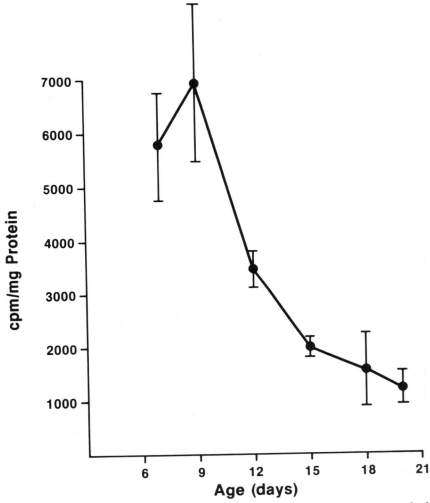

Fig. 5. (^{125}I)L-pindolol binding to β-adrenergic receptors in whole chick hearts on d 7, 9, 12, 15, 18, and 20 of incubation. The β-adrenergic receptor density decreases 80% from d 9 to 20 (reprinted from ref. 49).

crest) over somites 1–3 at stage 9 results in a 60–80% decrease in cardiac ganglion cells. Choline uptake in the atrium is decreased by 60%, whereas acetylcholine concentration in whole heart is decreased by 60–70%. Thus, although a complete parasympathetic denervation is not possible, the depletion of parasympathetic innervation may be sufficient to test several questions about interactions of these nerves in the developing heart.

Whereas the parasympathetic postganglionic innervation to the heart has been localized to a precise region of the neural crest, the sympathetic postganglionic innervation appears to arise from a more diffuse area. To define the origin of sympathetic cardiac innervation, we have performed [³H]-NE uptake studies using the atria of hearts from embryos that have had various somitic levels of the neural crest removed. Our data correlates well with the transplantation work of Yip (57), in that several somite levels of neural crest must be removed in order to obtain a maximum lesion of cardiac sympathetic innervation (29). Obviously a dispersed area of neural crest cells is contributing to the sympathetic trunk ganglia that innervate the heart.

6.2. Chemical Lesions

Chemical sympathectomy has been performed using drugs such as guanethidine sulfate, 6-hydroxydopamine (6-OHDA), and reserpine. Administration of guanethidine is reported to result in the permanent destruction of sympathetic ganglion cells, whereas 6-OHDA causes specific degeneration of sympathetic axon terminals and reserpine depletes the adrenergic neuron of vesicular stores of catecholamines (4,16).

Most of the studies of the effects of 6-OHDA on the adrenergic nerves of the domestic fowl have been performed on chicks several weeks after hatching (4,6). In an attempt to study the influence of adrenergic nerves on β-adrenergic receptor development, we administered 6-OHDA to chicks *in ovo* on embryonic d 13–19. [³H]-NE uptake studies on atria and HPLC determination of catecholamines in whole hearts were performed on 20-d embryos. β-Adrenergic receptors were measured using (^{125}I)L-pindolol in 6-OHDA-treated embryos on d 20. Although 6-OHDA treatment produced a 67% decrease in [³H]-NE uptake (in vitro) in atria, and a 53% decrease in NE concentration in whole hearts, no change in density of β-adrenergic receptors in whole hearts was detected (49). Although it is possible that adrenergic nerves have no influence on β-adrenergic receptor development, 6-OHDA might have been ineffective in altering receptor density because treatment was begun so late in development.

7. Conclusions

We have attempted to consolidate several facets of the development of the autonomic innervation to the chick heart. It is obvious from this discussion that autonomic terminals are probably present in the heart well in advance of their potential to control inotropic and chronotropic events in the heart. Thus, the potential exists for the autonomic terminals to influence developmental patterns. Gaps in our knowledge remain, particularly with regard to the interactions of parasympathetic and sympathetic nerves within the target organ. Combined biochemical and ultrastructural investigations of the developing synaptic terminals are necessary to further our understanding of autonomic interactions within the heart. Patterns of synaptic contact may change with differing functional states of the neuronal release and uptake mechanisms. This level of research demands close attention to the functional capabilities of maturing autonomic nerve terminals, in addition to morphology.

Acknowledgments

Supported by NIH Grant HD 17063. We would like to take this opportunity to thank Robert S. Aronstam and Kenneth P. Minneman for their help with several of these projects, as well as Martin R. Kirby for editorial comments and Pat O'Meara for typing the manuscript.

References

1. Abraham, A. (1969) Microscopic Innervation of the Heart and Blood Vessels, in *Vertebrates Including Man* Pergamon, Oxford, pp. 64–85.
2. Alexander, R. W., J. B. Galper, E. J. Neer, and T. W. Smith (1982) Noncoordinate development of β-adrenergic receptors and adenylate cyclase in chick heart. *Biochem. J.* **204,** 825–830.
3. Baumel, J. J. (1975) Nervous System, in *The Anatomy of the Domestic Animals* (R. Getty, ed.) Saunders, Philadelphia, pp. 2020, 2034, 2054.
4. Bennett, T., G. Burnstock, J. L. S. Cobb, and T. Malmfors (1970) An ultrastructural and histochemical study of the short-term effects of 6-hydroxydopamine on adrenergic nerves in the domestic fowl. *Br. J. Pharmac.* **38,** 802–809.
5. Bennett, T. and T. Malmfors (1970) The adrenergic nervous system of the domestic fowl [*Gallus domesticus* (L)]. *Z. Zellforsch. mikrosk. Anat.* **106,** 22–50.

6. Bennett, T., T. Malmfors, and J. L. S. Cobb (1973) Fluorescence histochemical observations on degeneration and regeneration of noradrenergic nerves in the chick following treatment with 6-hydroxydopamine. *Z. Zellforsch. mikrosk. Anat.* **142**, 103–130.
7. Black, I. B. (1978) Regulation of autonomic development. *Ann. Rev. Neurosci.* **1**, 183–214.
8. Callingham, B. A. and R. Cass (1965) Catecholamine levels in the chick. *J. Physiol.* **176**, 32p–33p.
9. Caserta, M. and L. L. Ross (1978) Biochemical and morphological studies of synaptogenesis in the avian sympathetic cell column. *Br. Res.* **144**, 241–255.
10. Claycomb, W. C. (1976) Biochemical aspects of cardiac muscle differentiation. *J. Biol. Chem.* **251**, 6082–6089.
11. Cohen, A. (1974) DNA synthesis and cell division in differentiating avian adrenergic neuroblasts, in *Wenner-Gren Center International Symposium Series*, vol. 22 (K. Fuxe, L. Olson, and Y. Zotterman, eds.) Pergamon, Oxford, pp. 359–370.
12. Cohen, D. H., A. M. Schnall, R. L. MacDonald, and L. H. Pitts (1970) Medullary cells of origin of vagal cardioinhibitory fibers in the pigeon. Anatomical studies of peripheral vagus nerve and the dorsal motor nucleus. *J. Comp. Neurol.* **140**, 299–320.
13. D'Amico-Martel, A. and D. M. Noden (1983) Contributions of placodal and neural crest cells to avian cranial peripheral ganglia. *Am. J. Anat.* **166**, 445–468.
14. De Santis, V. P., W. Längsfeld, R. Lindmar, and K. Löffelholz (1975) Evidence for noradrenaline and adrenaline as sympathetic transmitters in the chicken. *Br. J. Pharmac.* **55**, 343–350.
15. Enemar, A., B. Falck, and R. Hakanson (1965) Observations on the appearance of norepinephrine in the sympathetic nervous system of the chick embryo. *Dev. Biol.* **11**, 268–283.
16. Eranko, O. and L. Eranko (1971) Histochemical evidence of chemical sympathectomy by guanethidine in newborn rats. *Histochem. J.* **3**, 451–456.
17. Higgins, D. and A. Pappano (1979) A histochemical study of the ontogeny of catecholamine-containing axons in the chick embryo heart. *J. Molec. Cell. Cardiol.* **11**, 661–668.
18. Higgins, D. and A. J. Pappano (1981) Development of transmitter secretory mechanisms by adrenergic neurons in the embryonic chick heart ventricle. *Dev. Biol.* **87**, 148–162.
19. Horstadius, S. (1956) *The Neural Crest: Its Properties and Derivatives in the Light of Experimental Research.* Oxford, London.
20. Ignarro, L. J. and F. E. Shideman (1968) Catechol-O-methyl transferase and monoamine oxidase activities in the heart and liver of the embryonic and developing chick. *J. Pharmac. Exp. Ther.* **159**, 29–37.

21. Ignarro, L. J. and F. E. Shideman (1968) Appearance and concentration of catecholamines and their biosynthesis in embryonic and developing chick. *J. Pharmac. Exp. Ther.* **159**, 38–48.
22. Ignarro, L. J. and F. E. Shideman (1968) Norepinephrine and epinephrine in the embryo and embryonic heart of the chick: uptake and subcellular distribution. *J. Pharmac. Exp. Ther.* **159**, 49–58.
23. Kirby, M. L. (1978) Drug modification of catecholamine synthesis and uptake in early chick embryos. *Br. Res.* **149**, 443–451.
24. Kirby, M. L. and R. S. Aronstam (1983) Atropine-induced alterations of normal development of muscarinic receptors in the embryonic chick heart. *J. Molec. Cell. Cardiol.* **15**, 685–696.
25. Kirby, M. L., T. F. Gale, and D. E. Stewart (1983) Neural crest cells contribute to normal aorticopulmonary septation. *Science* **220**, 1059–1061.
26. Kirby, M. L. and S. A. Gilmore (1972) A fluorescence study on the ability of the notochord to synthesize and store catecholamines in early chick embryos. *Anat. Rec.* **173**, 469–478.
27. Kirby, M. L. and S. A. Gilmore (1976) A correlative histofluorescence and light microscopic study of the formation of the sympathetic trunks in chick embryos. *Anat. Rec.* **186**, 437–450.
28. Kirby, M. L., J. W. McKenzie and T. A. Weidman (1980) Developing innervation of the chick heart: a histofluorescence and light microscopic study of sympathetic innervation. *Anat. Rec.* **196**, 333–340.
29. Kirby, M. and D. E. Stewart (1984) Adrenergic innervation of the developing chick heart: neural crest ablations to produce sympathetically aneural hearts. *Am. J. Anat.* **171**, 295–305.
30. Kirby, M. L. and D. E. Stewart (1983) Neural crest origin of cardiac ganglion cells in the chick embryo: identification and extirpation. *Dev. Biol.* **97**, 433–443.
31. Kirby, M. L., T. A. Weidman and J. W. McKenzie (1980) An ultrastructural study of the cardiac ganglia in the bulbar plexus of the developing chick heart. *Dev. Neurosci.* **3**, 174–184.
32. LeDouarin, N. and M. A. Teillet (1973) The migration of neural crest cells to the wall of the digestive tract in avian embryo. *J. Embryol. Exp. Morph.* **30**, 31–48.
33. LeDouarin, N. M. and M. A. Teillet (1974) Experimental analysis of the migration and differentiation of neuroblasts of the autonomic nervous system and of neurectodermal mesenchymal derivatives, using a biological cell marking technique. *Dev. Biol.* **41**, 162–184.
34. Lefkowitz, R. J. (1982) Biochemical mechanisms of hormone receptor-effector coupling. *Fed. Proc.* **41**, 2662–2663.
35. Levi-Montalcini, R. (1950) The origin and development of the visceral system in the spinal cord of the chick embryo. *J. Morph.* **86**, 253–283.

36. Luckenbill-Edds, L. and C. van Horn (1980) Development of chick paravertebral sympathetic ganglia. I. Fine structure and correlative histofluorescence of catecholaminergic cells. *J. Comp. Neurol.* **191,** 65–76.

37. McCarty, L. P., W. C. Lee, and F. E. Shideman (1960) Measurement of the inotropic effects of drugs on the innervated and noninnervated embryonic chick heart. *J. Pharmac. Exp. Ther.* **129,** 315–321.

38. Minneman, K. P. (1983) Peripheral catecholamine administration does not alter cerebral β-adrenergic receptor density. *Brain Res.* **264,** 328–331.

39. Noden, D. M. (1980) The Migration and Cytodifferentiation of Cranial Neural Crest Cells, in *Current Trends in Prenatal Craniofacial Development* (Pratt and Christiansen, eds.) Elsevier, North Holland, pp. 3–25.

40. Noden, D. M. (1983) The role of the neural crest in patterning of avian cranial skeletal, connective, and muscle tissues. *Dev. Biol.* **96,** 144–165.

41. Noden, D. M. (1984) Craniofacial development: new views on old problems. *Anat. Rec.* **208,** 1–13.

42. Pappano, A. J. (1977) Ontogenetic development of autonomic neuroeffector transmission and transmitter reactivity in embryonic and fetal hearts. *Pharmac. Rev.* **29,** 3–33.

43. Prosser, C. L. (1974) Smooth muscle. *Ann. Rev. Physiol.* **36,** 503–535.

44. Rickenbacher, J. and E. Müller (1979) The development of cholinergic ganglia in the chick embryo heart. *Anat. Embryol.* **155,** 253–258.

45. Romanoff, A. L. (1960) *The Avian Embryo: Structural and Functional Development* MacMillan, New York, pp. 712–715.

46. Rothman, T. P., M. D. Gershon, and H. Holtzer (1978) The relationship of cell division to the acquisition of adrenergic characteristics by developing sympathetic ganglion cell precursors. *Dev. Biol.* **65,** 322–341.

47. Scarpace, P. J. and I. B. Abrass (1982) Desensitization of adenylate cyclase and down regulation of beta adrenergic receptors after in vivo administration of beta agonist. *J. Pharmac. Exp. Ther.* **223,** 327–331.

48. Stewart, D. E. and M. L. Kirby (1985) Endogenous tyrosine hydroxylase activity in the developing chick heart: A possible source of extraneuronal catecholamines. *J. Mol. Cell. Cardiol.* **17,** 389–398.

49. Stewart, D. E., M. L. Kirby, and R. S. Aronstam (1986) Regulation of β-adrenergic receptor density in the non-innervated and denervated embryonic chick heart. *J. Mol. Cell. Cardiol.* (in press).

50. Tennyson, V. M. (1970) The Fine Structure of the Developing Nervous System, in *Developmental Neurobiology* (W. Himwich, ed.) Thomas, Springfield, Illinois, pp. 47–116.
51. Terni, T. (1924) Richerche anatomiche sul sistema nervoso autonomo degli uccelli. *Arch. Ital. Anat. Embryol.* **20,** 433–510.
52. Tosney, K. W. (1982) The segregation and early migration of cranial neural crest cells in the avian embryo. *Dev. Biol.* **89,** 13–24.
53. Wallace, J. A. (1982) Monoamines in the early chick embryo: demonstration of serotonin synthesis and the regional distribution of serotonin-concentrating cells during morphogenesis. *Am. J. Anat.* **165,** 261–276.
54. Wetzel, G. T. and J. H. Brown (1983) Relationships between choline uptake, acetylcholine synthesis, and acetylcholine release in isolated rat atria. *J. Pharmac. Exp. Ther.* **226,** 343–348.
55. Windle, W. F. and M. F. Austin (1936) Neurofibrillar development in the central nervous system of chick embryos up to 5 days' incubation. *J. Comp. Neurol.* **63,** 431–463.
56. Wright, L. L. (1981) Time of cell origin and cell death in the avian dorsal motor nucleus of the vagus. *J. Comp. Neurol.* **199,** 125–132.
57. Yip, J. W. (1983) Formation of paravertebral sympathetic chain ganglia in chick embryos (abstract). *Neurosci.* **9,** 937.

Chapter 6

Development of Autonomic Innervation in Mammalian Myocardium

Howard L. Cohen

1. Introduction

Although the myocardium has an inherent rhythmicity, auto-nomic regulation of this pattern is necessary in order to meet the physiological demands placed upon the organism (*14,32,82*). The ontogeny of autonomic innervation of the myocardium involves a highly complex series of processes, still not well understood, involving axonal growth, terminal branching, neurotransmitter synthesis, elaboration of mechanisms for transmitter storage, re-lease, uptake, and inactivation, receptor growth, and develop-ment of myocardial responsivity (*36,65,69,90*). Each of these, in turn, may be subjected to modifications by circulating hormones (*64,83*), trophic influences (*10,12,67*), and presynaptic nerve im-pulses (*6,10,26*). Whereas the general developmental pattern is common to most mammals, the maturation of cardiac regulation manifests itself with marked species differences in its temporal course. The following review examines the development of those mechanisms underlying functional autonomic innervation of mammalian myocardium.

2. Sympathetic Innervation of the Myocardium

Sympathetic innervation of the myocardium originates in neurons whose cell bodies are found in the intermediolateral columns of spinal segments T1–T8 (47). Their axons, the preganglionic fibers, exit the spinal cord primarily through the white rami communicantes of spinal segments T1–T6 (85), but may also exit at levels C5–C8 (130). The fibers then enter the paravertebral chain of ganglia. The cells of origin of cardiac postganglionic neurons can be located in the superior, middle, or inferior stellate cervical ganglia, and there appear to be significant species differences (1, 2,56). In canines, physiologically identified cardiac nerves were found to originate primarily in the middle cervical ganglion, with fewer soma in the stellate ganglion and even less in the superior cervical ganglion (1,77); in felines, cardiac postganglionic neurons appear to be located primarily in the lateral half of the stellate ganglion (56); and in rats, they are apparently localized to the caudal pole of the superior cervical ganglion (13). However, although the postganglionic sympathetic cardiac nerves usually have some degree of functional specificity (3,96,97,102), little topographic organization of cell bodies within the ganglia has been reported, suggesting that preganglionic sympathetic neurons project to more than one cardiac nerve (1,77).

In most mammals, sympathetic innervation of the myocardium is poorly developed at birth (24,31,36,38,61,71,73,107,127), with some exceptions, e.g., guinea pig (51,69,94) and sheep (65, 104,134). Cardiac responses mediated by the sympathetic nervous system might be present, although reduced, undergoing gradual postnatal maturation (31,71) (Fig. 1). For example, in neonatal rabbit (36,127) and puppy (71,73), reduced chronotropic responses were detected following sympathetic nerve stimulation. Possible factors in the reduced neonatal response might include decreases in both quantal release and reuptake of neurotransmitter (40). Whereas species differences have often been observed in the temporal course of sympathetic fiber growth into the myocardium (65,89), in contrast, several species (e.g., fetal lamb, rat, and guinea pig) share a common innervation pattern in that the fibers appear to grow into the heart through larger nerve trunks, rather than to form *in situ* from structures within embryonic or fetal myocardium (65). Typically, cardiac sympathetic innervation is first distributed in the tunica adventitia surrounding

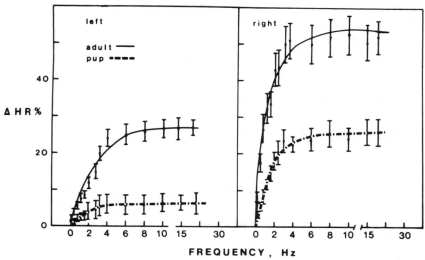

Fig. 1. Changes in heart rate as a function of the stellate stimulation frequency in adult dogs and puppies (reprinted, with permission, from ref. *71*).

the pulmonary and coronary arteries in the epicardium and then into the connective tissue between the myocardial muscle bundles (*34,65,89,90,136*). In several species, (e.g., rat, guinea pig, and rabbit) the sinoatrial (SA) node is the first region to be innervated (primarily by the right cardiac sympathetic nerve) with a progression to the atrioventricular (AV) junction, and then to the apex of the heart (*69*); this pattern of innervation is consistent with other studies showing that the atria are innervated at the same time as, or prior to, the ventricles (*24,51,69,90*).

In general, the density of sympathetic innervation in the neonate is low at birth (*4,24,36,127*), increasing to adult levels postnatally, and eventually forming a varicose plexus that ramifies along cells of both atrial and ventricular myocardium (*36,89,90*). The regional specificity of these distributions is such that there are varying degrees of refractory period-shortening depending on which sympathetic cardiac nerve is stimulated (*60,61,102,136*). Thus, stimulation of the left ventrolateral cardiac nerve decreased refractory periods by a maximum of 28 ms on the posterior and inferior left ventricular wall and posterolateral right ventricular wall; stimulation of the left ventromedial cardiac nerve reduced refractory periods by as much as 15 ms on the anterior surface of the myocardium and the posterolateral right and left ventricular walls (*60,61*). Although the magnitude of refractory period-shortening following cardiac sympathetic nerve stimulation generally increases with age, the results may be quite variable in the neonate (e.g., neonatal dog) (*60,61*).

3. General Criteria for Functional Synapses

The presence of functional synaptic connections cannot be implied, even when nerve terminals appear morphologically mature, contain synaptic vesicles, are totally or partially free of Schwann cell coverings, and are adjacent to myocardial cells (*89*). For example, in rat, there was an initial delay between the appearance of cholinergic neurons in the myocardium and the detection of choline acetyltransferase [ChAT, an enzyme necessary for the synthesis of acetylcholine (ACh)] and a second delay, between the onset of ChAT activity and the onset of neurotransmission (*75*). Similarly, at adrenergic neuroeffector junctions, several days may elapse between the presence of catecholamine-containing vesicles and the ability of the nerve terminals to release their neurotransmitter (*7,110*). The techniques used to assess functional innervation at cardiac adrenergic synapses include: (a) measurement of muscle twitch tension as a function of transmitter release following electrical nerve stimulation, (b) blocking the twitch response with propranolol (a β-blocker), guanethidine (a blocker of catecholamine release), or reserpine (a depleter of endogenous catecholamines), and (c) measuring the accumulation, retention, and release of norepinephrine (NE) (*49*). Similarly, the absence of fluorescent staining, indicating that nerves are either not yet synthesizing or storing NE or that synthesis and storage is at such low levels as to yield immeasurable reaction product, further supports nonfunctional sympathetic innervation (*49,69*).

4. Development of Sympathetic Neurons

The development of the immature postganglionic sympathetic neuron is characterized by variations in the fluorescent intensity of its cell body; fluorescence is low in early developmental stages, increases rapidly during fetal life, and decreases near birth (*16,17*). Mechanisms for uptake and storage of NE are present in adrenergic neurons even before the first appearance of fluorescence (*98*). This latter phenomenon manifests species differences, with detection occurring at 11, 13, and 14 d, and 7 wk gestation in mouse, rat, rabbit, and human, respectively (*16,17*). Neuronal growth is evidenced at this time by increased size and by development of long, smooth, fluorescent axons (*16,17*). Indications that the fluorescence observed is a result of catecholamines is supplied by the fact that tissue pretreatment with either reserpine or 6-hydroxydopamine (6-OHDA, which selectively destroys postsynaptic adrenergic terminals) prevents the fluorescent reaction product (*10,26,107,129*).

In mature adrenergic nerves containing adequate amounts of catecholamines, highly intense fluorescence can be observed over their entire length, whereas in developing nerves with lower catecholamine concentrations, evidence for catecholamines may not be seen (*16,17,24*). The absence of fluorescent nerves in the fetus or neonate must be interpreted carefully because: (a) there are limitations in the resolution of the histofluorescence technique, (b) there is greater difficulty in processing both fetal and neonatal material for fluorescence (*24,89,90*), (c) the fibers may not be synthesizing, storing, or releasing NE (*69*), or (d) the number of storage vesicles may be low (*4*). At this time it is unlikely that there would be functional sympathetic innervation (*69*). Increases in fluorescence might reflect an increase in cardiac sympathetic innervation; e.g., (a) the number of nerves per gram of tissue (*24*), (b) an increase in NE concentration stored in the neuronal vesicles (*24,69*), or (c) increased efficiency of the uptake mechanism (*4*). In the mature organism, adrenergic axons have varicose swellings specialized for transmitter release (*24*). These varicosities have also been observed in the terminal portions of cholinergic nerve fibers (*32,36,89,90*). Early in development, a significant proportion of the measured fluorescent NE is found in the preterminal nerve trunks rather than in the varicosities, suggesting that neurotransmitter is not in close proximity to the adrenergic receptors of the myocardial cells (*36*). However, during the maturation of both peripheral cholinergic and adrenergic neurons, neurotransmitter release may be detected, although ultrastructural studies reveal few or no regions in the nerve terminals specialized for this function, i.e., varicosities (*36,49*). With maturation, there is an increase in the intensity of the fluorescence, reflecting increases in: (a) the number of nerve terminals forming the ground plexus, (b) the NE content of the nerve terminals, and (c) interneuronal reuptake of NE (*4,24,36,65*). Neuronal development does not terminate with the onset of functional innervation because additional increases in NE content, reuptake, and fiber diameter have been observed (*91,104*).

5. Nerve Growth Factor

One of the primary contributors to sympathetic nerve ontogeny is nerve growth factor (NGF). Nerve growth factor is a neuropeptide that appears to be essential for both the normal development and maintenance of adrenergic nerves, although there is some evidence for an age-related decrease in dependence (*10,12,44,59,67,123*). It is present in both embryonic and fetal sympathetic neurons and in adrenergically innervated tissue (*10,12,67*). When given exogenously to either the mature or developing organism, NGF causes greater than normal innervation, as well as increased fiber diameter and terminal density (*10,44,67*). The effector organs appear to be the source of NGF and, through it, regulate the growth of the nerves providing their innervation (*10,12,67*). There is evidence that NGF can be taken up by adrenergic nerve terminals and transported to the cell body via retrograde axonal transport (*10,59,67*). In the cell body, it is involved in the synthesis of tyrosine hydroxylase (TH, the rate-limiting enzyme in catecholamine synthesis) and dopamine β-hydroxylase (DBH, which converts dopamine to NE) (*123*). For example, treatment with NGF antisera caused abnormalities in adrenergic nerve development and function, leading to degeneration of both cell bodies and nerve terminals (*67,123*). Removal of the effector organ (*27*) acted to prevent maturation of the neurons destined to innervate that structure and also led to disturbances in the normal ontogeny of both TH and dopa decarboxylase (DDC, which converts L-Dopa to dopamine). This supports the hypothesis that the effector organ is the source of NGF, and its removal leads to abnormalities in adrenergic growth and function.

6. Norepinephrine

The NE found in postganglionic sympathetic fibers is synthesized primarily in the nerve endings, but also in the cell body, and is stored in vesicles either at the nerve endings or in the varicosities

(70). Following its action on the postsynaptic membrane, the inactivation of NE is primarily through reuptake into the nerve ending, a process involving an active transport mechanism that accounts for about 75% of the inactivation (55,105,106). It has been shown in several species, including swine (34,120), rabbit (36), and rat (24,55,107) that endogenous myocardial NE is very low at birth and increases rapidly during the first postnatal weeks (24,36,55,127,135) (Fig. 2). For example, endogenous NE in rat myocardium at birth is 5% of the adult value and increases to 80% by 3 wk of age (55). In contrast, in sheep, fetal concentrations of myocardial NE increase between several weeks preterm to about 3 d postnatally, at which time they are approximately at adult levels (34). In human fetal myocardium, NE synthesis has been detected as early as 13 wk (42).

7. Norepinephrine Uptake Mechanisms

Uptake is a function of the availability of sites for both axonal transmembrane transport and vesicular catecholamine storage (43). Immaturity of uptake may be reflected in the ease with which [^{3}H]-NE can be washed out of myocardial tissue (43). In the

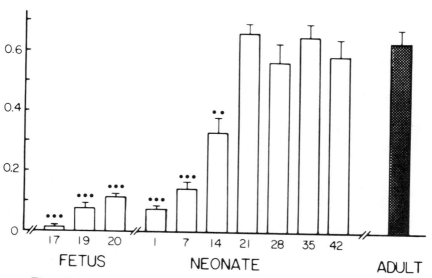

Fig. 2. Norepinephrine concentrations during development of the mouse heart (reprinted, with permission, from ref 135).

fetus, the ability to accumulate endogenous NE is low (*105,106*), but may be present in developing adrenergic neurons prior to the first appearance of endogenous NE in a specific tissue of the fetus or neonate (*38,55*). In some myocardial regions, e.g., right atrium, the ability to accumulate [^{3}H]-NE developed more rapidly than endogenous NE and appeared fully functional at birth; in other regions, e.g., left ventricular apex, the ability to accumulate [^{3}H]-NE paralleled the appearance of endogenous NE (*38*). Since it appears that uptake depends upon: (a) the integrity and transport capabilities of the axonal membrane of the postganglionic sympathetic neuron and (b) the development of the apparatus and enzymes necessary for synthesis and storage of NE, uptake is perhaps a more accurate index of the density of sympathetic innervation than would be the determination of levels of endogenous NE (*38*). In neonates of most species, uptake shows age-related increases (Fig. 3) (*7,16,43,55,95,105–107*) and ultimately will depend upon: (a) increased cellular efficiency in accu-

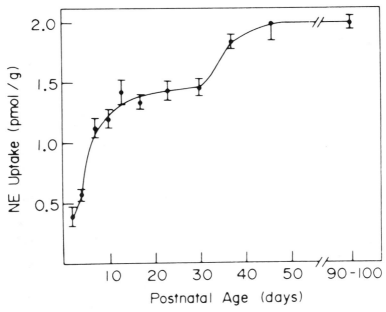

Fig. 3.   Ontogeny of 1–[^{3}H]norepinephrine uptake into rat heart synaptic vesicles expressed per gram of heart weight (reprinted, with permission, from ref. *7*).

mulating NE, (b) increased numbers of both fibers and intraneuronal vesicles, and (c) increased vesicular uptake *(4,7, 38,55)*. There is some evidence to suggest that early in development uptake is primarily into the neuronal space, whereas later in development NE is stored in vesicles *(4)*. Furthermore, the ontogeny of the uptake mechanism is marked by fluctuations in both transmembrane transport and storage, i.e., throughout the length of the axon vs concentrated in the varicosities, as evidenced by variations in fluorescent intensity *(4,43,107)*. In the human fetus, uptake and storage mechanisms appear during the second trimester; fluorescent terminals have been identified in coronary arteries by 12–13 wk, and in the myocardium itself after 14–15 wk *(106)*. In rat, [^{3}H]-NE uptake is present in both atria and ventricles by 1 d of age; in addition, uptake can be blocked by reserpine *(4)*. Although adult uptake levels have been reported to be present by 7 d in atria and 14 d in ventricle *(4)*, other studies in the rat report maturation of uptake continuing until 6 wk postnatally *(43)*. Many investigators have observed that uptake mechanisms develop first in the atria and then in the ventricles and follow the cephalocaudal progression of sympathetic myocardial innervation *(4,90)*.

8. Synaptic Vesicles

Synaptic vesicles, the sites of synthesis, reuptake, and storage of neurotransmitters, originate in the cell body and migrate down the axon to the nerve terminals *(4,70)*. The integral role of vesicles in the synthesis, storage, and release of NE indicates that their development is an important determinant of synaptic maturation *(7)*. There is an excellent correlation between the establishment of functional sympathetic innervation of the myocardium and the development of synaptic vesicles within postganglionic noradrenergic nerve terminals *(7,114,116)*. In the neonate, synaptic efficacy is associated with both vesicular maturation, i.e., increased uptake and storage ability, and increased number of vesicles *(4,7)*. In rat, vesicles appear in isolated atria by about 4–7 d postnatally; they mature in two stages that occur approximately at 1 and 4–7 wk *(4,7)*. The majority of vesicles appear after 30 d of age, well after the establishment of functional sympathetic neurotransmission *(7)*. It has been demonstrated that the ontogeny of [^{3}H]-NE uptake by isolated synaptic vesicle preparations closely parallels the development and storage capability of intact synaptic terminals (7,64).

9. Neonatal Sympathetic Dominance

Although, as has been stated, functional sympathetic innervation of the myocardium is generally immature in the neonate and myocardial catecholamine concentrations at birth might be only 5% of that of the adult (e.g., in rat) (43,55), early sympathetic dominance in the control of myocardial function is still possible. Two factors contribute to this possibility: (a) a six-fold increase in the sensitivity of the neonatal myocardium to NE, as compared to the adult and (b) levels of circulating catecholamines that may be up to 30 times higher in the neonate than in the adult (34,40,54). The increased sensitivity persists for several days postnatally and appears to be identical to that observed in the chronically sympathectomized adult myocardium (29,31,36). The increased levels of circulating catecholamines, which may be a result of the contribution of the adrenal medulla to early sympathetic control, have been verified by: (a) greater decreases in both heart rate and blood pressure were observed after bilateral adrenalectomy in neonatal dogs than were seen in adults (40) and (b) ventricular responses to prolonged sympathetic nerve stimulation were greater in the intact than in the adrenalectomized preparation (31) (Fig. 4). However, the latter decrease in response magnitude is species specific, e.g., present in the neonatal dog, but not in the neonatal sheep (28). Regardless of whether the neonate has been adrenalectomized or not, response magnitudes in neonates were generally less than those in adults (31). Age-related decreases in sympathetic dominance probably reflect such factors as: (a) decreases in circulating catecholamines, (b) down-regulation of myocardial receptors, and (c) further maturation of functional parasympathetic innervation (29–31,54).

10. Presynaptic Influences on Myocardial Development

Alterations in sympathetic input to the myocardium, e.g., changes in the temporal course of functional neurotransmission, have been observed to retard or accelerate myocardial growth and development (26). For example, during periods of myocardial

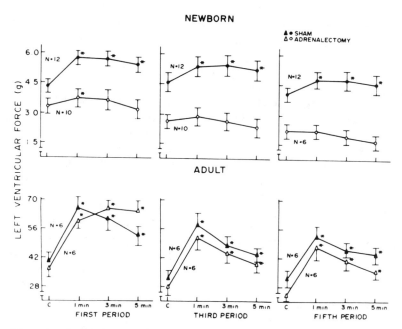

Fig. 4. Left ventricular force in newborn (top) and adult (bottom) dogs prior to stimulation (C, control) and after 1, 3, and 5 minutes of left stellate ganglion stimulation in 1st, 3rd, and 5th stimulation periods (reprinted, with permission, from ref. *31*).

growth characterized by rapid proliferation of cells and increased synthesis of RNA, DNA, and protein, administration of sympathomimetics or exposure to high concentrations of circulating catecholamines may produce decreased DNA synthesis or accelerated protein synthesis (*6*). Another measure of the relationship between sympathetic input and myocardial growth is the activity of myocardial ornithine decarboxylase (ODC, a rate-limiting enzyme in the synthesis of polyamines) (*79*). Its activity is stimulated under conditions in which myocardial growth is induced and depressed as growth ceases or as the number of growing or actively dividing cells declines (*79,114,115*). Thus, hypothyroidism or hyperthyroidism will cause either decreased or increased ODC activity, respectively (*115*). Similarly, prenatal exposure to ethanol, which accelerates the onset of functional sympathetic neurotransmission, will produce premature ODC responses when compared with nonexposed controls (*8*). These results suggest the involvement of basal sympathetic tone in the regulation of myocardial growth and development (*6,26*). However, in contrast are the published results concerning the effects of guanethidine administration (*7*). Guanethidine, a neurotoxic, adrenergic-blocking agent, produces a long-term peripheral sympathectomy with deficits in NE uptake that persist even after treatment is terminated (*6,129*). However, even with a reduced sympathetic nerve input there was normal myocardial growth, as evidenced by the ratio between heart and body weight and by the lack of a significant effect on RNA and protein synthesis, suggesting that basal sympathetic tone is not essential for normal myocardial tissue growth, differentiation, and maturation (*6*).

As indicated, myocardial growth and development can be influenced directly by sympathomimetics and indirectly by agents altering the time course of sympathetic nerve ontogeny. However, when these substances are administered systemically, it is difficult to separate the direct from the indirect influences on the myocardium (*26*). In one study (*26*), 6-OHDA was administered intracisternally prior to onset of tonic sympathetic myocardial control. This treatment destroys central catecholaminergic terminals, leaving peripheral sympathetic terminals intact, thus, permitting testing of the hypothesis that alterations in basal sympathetic tone alone can influence cardiac growth and cellular development. At 2 or more wk postnatally, when tonic sympathetic control is established, the 6-OHDA-treated animals had significant decreases in the ratio of heart to body weight, in the level of ODC activity, and in the resting heart rate (Fig. 5). The latter deficit was restored to normal following vagotomy. In contrast to the aforementioned deficits, sympathetic nerve function was normal, as indicated by [^{3}H]-NE uptake. Thus, it appears that central catecholaminergic lesions that alter the level of basal sympathetic tone without destroying peripheral sympathetic terminals can also alter myocardial growth.

11. Parasympathetic Innervation of the Myocardium

Although it had once been believed that the dorsal motor nucleus (DMN) of the 10th nerve was the site of origin of cardiac vagal

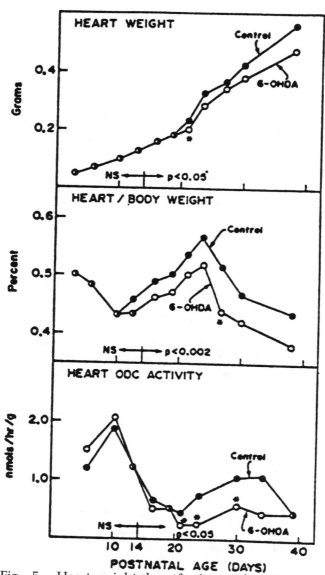

Fig. 5.   Heart weight, heart/body weight ratio, and cardiac ODC
basal activity in developing control rats and rats treated with 6-OHDA ic
(reprinted, with permission, from ref. 26).

preganglionic motoneurons (CVPM) (*18,117*), recent anatomical (*52,53,86,122*) and electrophysiological (*39,119*) evidence supports the view that in many mammals nucleus ambiguus (NA) and/or ventrolateral nucleus ambiguus (VLNA) (*52,53*) are the primary sites of origin of CVPM. This latter region, consisting of neurons located in a tightly packed cluster ventrolateral to the main cell column, has been identified in rat (*86*), cat (*122*), dog (*52*), and swine (*53*) and may represent a special subdivision of NA concerned with cardiac control in these species. The axons of the vagal preganglionic neurons course through the carotid sheath toward the atria and ventricles (*48,68,90*). In several species, including dog, rabbit, and baboon, they are accompanied by sympathetic cardiac fibers and together form the vagosympathetic trunk (*48,68,90*); a substantial degree of intermixing of both fiber types has been observed (*2*). Vagal efferent nerves have been observed approaching the heart over the posterior walls of the atria, reaching the ventricular myocardium via the ventricular septum, as well as the epicardium (*50,136*). The terminals of both sympathetic and parasympathetic neurons are often observed in close proximity in the myocardium, providing a structural basis for sympathetic–parasympathetic interactions (*72,90*).

Although it has long been known that vagal fibers richly innervate the atria and specialized conduction system (i.e., His bundle, bundle branches, and Purkinje fibers, and the AV and SA nodes) (*28,31,48*), evidence for their innervation of the ventricles is much more recent and comes from (a) staining for acetylcholinesterase (AChE, which hydrolyzes ACh), (b) the presence of ChAT (which catalyzes ACh production in parasympathetic nerves), (c) the presence of muscarinic cholinergic receptors in ventricular myocardium, and (d) physiological studies demonstrating alterations in both automaticity and contractility to vagal stimulation (*48,68,136*). At present, there is little evidence for vagal regulation of ventricular contractility in the neonate (*28*). This may reflect the fact that parasympathetic activity may modulate cardiac performance by altering the quantity of NE released by a given level of sympathetic activity. However, in the neonate, sympathetic innervation is incomplete and NE concentrations are low (*36*).

In many species, e.g., kittens, puppies, and rabbits, parasympathetic innervation of neonatal myocardium is more mature than sympathetic innervation, having undergone most of its development during the fetal period (28,71). For example, in neonatal puppies (28) and guinea pigs (127) there is no significant difference in the density of cholinergic fibers in the SA and AV nodes, or in the atria and ventricles, between the neonate and adult. However, although there is little apparent difference between neonate and adult in the anatomy of autonomic innervation of the heart, notable differences in function have been observed; e.g., in the neonate, (a) the (−)-chronotropic effect of vagal nerve stimulation is smaller (Fig. 6) (28,71,73,78) and (b) both vagotomy (15,40) and atropine infusion (99,108) produce smaller elevations in heart rate.

12. Acetylcholine

The parasympathetic transmitter (ACh) decreases myocardial automaticity and inotropy by selectively increasing membrane

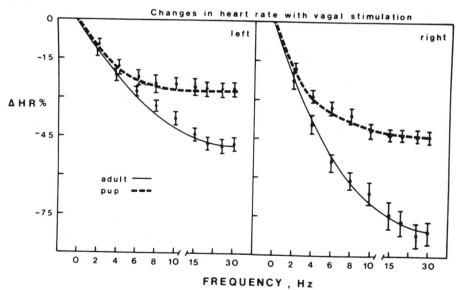

Fig. 6. Changes in heart rate as a function of the vagal stimulation frequency in adult dogs and puppies (reprinted, with permission, from ref. 71).

conductance of potassium in myocardial pacemaker cells of the SA and AV nodes and in atrial and ventricular muscle cells producing both hyperpolarization and a marked decrease in the rate of diastolic depolarization (76,82,124). It has been observed that the ability of elevated external potassium to suppress automaticity increases with developmental age (25). Myocardial muscarinic receptors are present early in development, indeed, prior to vagal innervation, as indicated by the ability of ACh to inhibit spontaneous myocardial contractions and the effect of atropine on these responses (46,113,132). Myocardial pacemaker tissue generally shows age-related increases in sensitivity to the inhibitory effects of ACh, although it is difficult to determine whether the increased sensitivity is a result of increased innervation density and/or increased muscarinic receptor number or to lower levels of AChE (46,91,132). Additionally, there are species differences in the sensitivity of pacemaker tissue to ACh, so that, although it is less sensitive in fetal guinea pig than in the adult, it is more sensitive in neonatal rat and rabbit pups (127) and equally sensitive in fetal lamb and adult sheep (34).

Cholinergic neurons in the fetus and neonate have a relatively low ability to synthesize ACh from choline and acetyl coenzyme A, and the actual onset of synthesis remains to be determined (91,113). However, age-related increases in ChAT result in corresponding increases in ACh concentration, leading to the establishment of tonic cholinergic regulation of the heart (75,113) (Fig. 7). The onset of tonic vagal cardiac activity may be documented by both significant increases in heart rate following injection of atropine into the mother or fetus (99,108) and the bradycardia observed following vagal stimulation (28,126,127) (Fig. 8). In each case the fetal or neonatal responses were significantly smaller than the adult responses. However, it should be noted that the bradycardia elicited by vagal nerve stimulation only indicates that a functional neuroeffector junction exists, not that it is active in vivo (127).

13. Acetylcholinesterase

Acetylcholinesterase, which hydrolyzes ACh to choline and acetic acid (76), shows significant species and age-related differences in its activity, distribution, and concentration in the myocardium (111,112,125,126). True AChE appears in the cytoplasm of

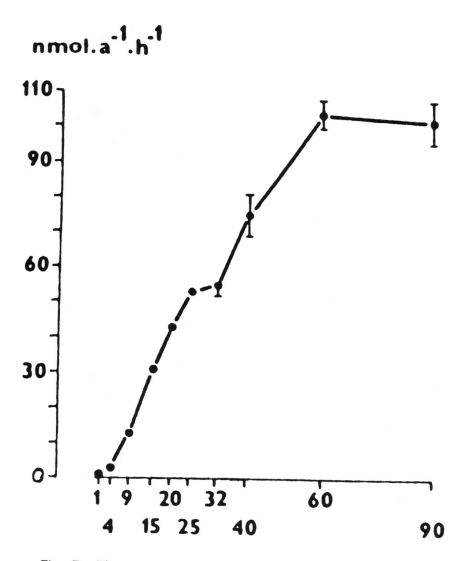

Fig. 7. The synthesis of ACh in homogenates of heart atria from rats of different ages (reprinted, with permission, from ref. *113*).

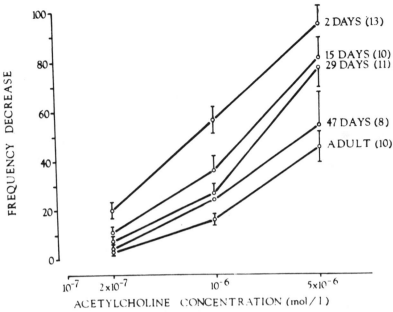

Fig. 8. Negative chronotropic effects of acetylcholine on isolated
atria of rats of different ages (reprinted, with permission, from ref. 126).

myocardial cells early in fetal development in human, rabbit, and
guinea pig, preceding functional innervation of the myocardium
(45,57,95,111,112). In rat, the first appearance of AChE in cardiac
ganglia is at 4 d postnatally; at 15 d, fibers containing the enzyme
can be identified in the SA node (81). Because there is a good correlation among ACh, AChE, and choline acetylase concentrations
throughout both the peripheral and central nervous system, it is
quite probable that one can assume a correlation between AChE
distribution and cholinergic innervation density in the neonate.
This relationship has been observed in the SA and AV nodes,
both of which have dense cholinergic innervation. Although
AChE is present in all cholinergic neurons, some investigators be-
lieve that the only criteria for adequately identifying the phenotype of a neuron as cholinergic would be high concentrations of
either ACh and/or choline acetylase (70). In comparisons between
two species during the neonatal period (111,112), the AChE
measured in human atrial appendage was over four times greater
than that measured in puppy. However, although its distribution
pattern was nonuniform and showed wide differences between

the species, postnatal maturation tended to minimize the quantitative differences, i.e., the concentration of right atrial AChE manifested age-related increases in human and age-related decreases in canine. The studies also demonstrated that, by using the criterion of AChE concentration, parasympathetic innervation of puppy myocardium is immature at birth, developing rapidly with increasing postnatal age. These studies also verified earlier findings that: (a) the human neonate has a high level of myocardial enzyme activity at birth, (b) the AChE concentration in atrial appendage is greater in infants than in adults, (c) there are greater concentrations of AChE in all areas of infant atria than are found in corresponding areas in the adult, and (d) the degree of change in AChE concentration is less dramatic in the left atrium when compared to the right.

14. Sympathetic–Parasympathetic Interactions

As described in "Parasympathetic Innervation of the Myocardium," (see Section 11), the anatomic basis for sympathetic–parasympathetic interactions is determined by the close proximity among adrenergic nerve terminals, cholinergic nerve terminals, and myocardial cells (72,90). Interactions may involve either presynaptic or postsynaptic mechanisms (72,90). For presynaptic interaction, the ACh released from vagal nerve terminals must inhibit the release of NE from nearby sympathetic postganglionic nerve terminals. The magnitude of the interaction depends, in part, upon the proximity between adrenergic and cholinergic terminals; the closer the terminals, the smaller the quantity of ACh hydrolyzed before binding postsynaptically to a muscarinic receptor. For postsynaptic interaction (72,90), both adrenergic and cholinergic neurotransmitters must act upon the same myocardial cell. This interaction may result in changes in the cyclic nucleotides cAMP and cGMP. There is evidence to indicate that cGMP, which increases (intracellularly) in response to muscarinic receptor stimulation, may antagonize the inotropic effects mediated by cAMP in response to β-adrenoceptor stimulation (68,72,121).

There is substantial evidence of significant interactions between cardiac vagal and cardiac sympathetic nerves in the adult of most species. The degree of vagal inhibition of cardiac rhythmicity elicited by stimulation of the peripheral end of the cervical vagus nerve is dependent upon prevailing levels of sympathetic activity (*68,72*). For example, vagal stimulation at 8 Hz will elicit a greater decrease in heart rate (greater than 40 bpm) carried out in the presence of cardiac sympathetic nerve stimulation (at 4 Hz). The converse has also been observed.

Sympathetic–parasympathetic interactions have also been demonstrated in neonatal myocardium, suggesting that the underlying functional mechanisms have been established during the fetal period. In some species, the magnitude of the interactive effects may be less in the neonate than in the adult because of incomplete development of sympathetic nerves, parasympathetic nerves, and/or the interactive mechanisms (*72*). In contrast, in a comparison of interactions in both neonatal and adult dogs, no significant differences were found, indicating that in canine, interactive mechanisms are functional in the early neonatal period (*72*).

15. Myocardial Beta Adrenoceptors

In the developing organism there is evidence for the ontogeny of β-adrenoceptor function and/or the numerical increase in receptors, as may be observed in the increases in both sensitivity and magnitude of chronotropic and inotropic responses to catecholamines, sympathomimetics, and their antagonists. However, it is also possible that the smaller responses simply reflect an immature myocardial structure (*23,35,93*). For example, in a comparison of ultrastructural elements in papillary muscle of neonatal vs adult cat (*23*), the neonate had more noncontractile elements [e.g., sarcoplasm 36.5–20.9 (mean volume fractions)], fewer and more disorganized myofibrils (34.1–47.8), and a lower mitochondrial content (15.9–22.2).

The ontogeny of β-adrenoceptors occurs at different rates in different mammalian species (*37*); differences are also observed in the final receptor number in the adult (*80*). However, in some species [e.g., mouse (*132*) and sheep (*22*)] there is evidence that fetal β-adrenoceptors may be fully functional before sympathetic innervation of the myocardium is completed (*19,66*). Isoproterenol- (ISO, a specific β-adrenoceptor agonist) induced elevations of cAMP, which can be blocked by propranolol, have been observed. Apparently, these receptors have the same sensitivity to exogenous NE and ISO as those reported for the adult (*34,92*). On the other hand, responses to stimulation of sympathetic cardiac nerves might not occur until functional innervation can be documented. For example, in rat, the (+)-inotropic response to sympathetic nerve stimulation was not present until 2–3 wk postnatally (*74*).

In human fetal myocardium at 9 and 10 wk, respectively, (+) inotropic and chronotropic responses to epinephrine (E) have been observed in ventricular muscle (*41*). In embryonic rat heart, (+)-chronotropic responses to E and ISO have been detected by 10–10.5 d gestation (*74*); the same time that spontaneous myocardial contractions are first observed (*46,100*). These responses, present prior to functional innervation, can also be blocked by propranolol (*37*).

In most species, the time at which functional β-adrenergic components are incorporated into the myocardial cell membrane, and the rates of both β-adrenoceptor ontogeny and proliferation, have not yet been accurately determined (*92*). As a consequence, experimental results may be inconsistent. For example, in two studies examining both [^{3}H]-dihydroalprenolol ([^{3}H]-DHA, a β-adrenoceptor antagonist) binding and (+)-chronotropic responses in fetal mouse myocardium, one (*62*) found [^{3}H]-DHA binding and (+)-chronotropic responses at 9 d, and the other (*19*) found [^{3}H]-DHA binding, but no (+)-chronotropic responses at 13 d, even though binding was at 14% of adult values. Among many possible factors contributing to the lack of agreement between different investigators are: species differences, differences in the membrane fraction used, sympathetic nerve growth, hormonal influences, and cellular proliferation and differentiation (*80,84,109*). Using [^{3}H]-DHA binding as a criterion, some studies have found that during ontogeny there may be increases (*19,109*), decreases (*11,84,101*), or no change at all (*22,131*) in the number and/or density of binding sites. One study (*63*) used triiodothyronine, which apparently first increases and then decreases β-adrenoceptor number, in order to demonstrate the relationship between receptor density and the (+)-chronotropic effects of ISO. Others (*19,20,37*) found that concentration- (receptor density) dependent inotropic responses to ISO could be elicited before functional sympathetic innervation was established. However, although the elevated heart rate response to ISO may have reflected increased receptor density, it is also possible that maturation of other membrane properties and/or coupling mechanisms and/or increased affinity may have been the precipitating factor(s). Whereas increased receptor density may be associated with increased receptor affinity (*109*), the rate of increase in the latter may lag behind that of the former (*19,20*). Furthermore, there may also be increases in receptor density with no concommitant changes in affinity (*19,101*).

At birth, or early in the postnatal period, responses that initially manifested age-related increases in magnitude, reflecting changes in receptor density, receptor affinity, and coupling mechanisms, may soon decrease in magnitude from their prenatal levels. This decrease has been described for several species (*19*) and may represent down-regulation of β-adrenoceptors as the possible result of the establishment of functional sympathetic innervation, i.e., catecholamine release and the resultant increased stimulation of β-adrenoceptors by catecholamines (*37, 103*). Finally, it has also been reported that even with rapid increases in tissue catecholamine concentrations, there may be no detectable changes in either β-adrenoceptor number or function, indicating that variations in endogenous NE may not be the most important factor in their perinatal maturation (*109*).

16. AC–cAMP

Adenylate cyclase–cyclic 3′,5′-adenosine monophosphate (Ac–cAMP) is involved in the mediation of β-adrenoceptor responses to stimulation by catecholamines (*20,121*). Most, if not all, actions of catecholamines at β-adrenoceptors appear to require Ac activation with subsequent increases in intracellular cAMP production, protein kinase activation, and calcium conductance (*9,92*). In fetal tissue not yet possessing functional sympathetic innervation, the relationship between β-adrenoceptor stimulation and Ac activation can be demonstrated by blocking the ISO-induced increase in cAMP with propranolol (*9,92*). In human fetal myocardium, catecholamine-induced activation of Ac has been observed as early as 5 wk postconception (*88*). The activation of Ac is apparently mediated by a β-adrenoceptor because transfused propranolol, but not phentolamine (an α-adrenoceptor antagonist), can block the activation (*92*). The cAMP response to catecholamines apparently also develops prenatally in rabbit, since it is present as early as 1 d postnatally and is also blocked by propranolol (*9*). However, whereas β-adrenoceptor ontogeny during the fetal period may bring receptor density near to adult levels, Ac activity matures more slowly and may be at only 50% of adult levels (*20*). Although ISO may stimulate Ac in fetal myocardium, it may not initiate a heart rate response should the excitation–contraction coupling system in myocardium be immature (*9,20,35*). In rats, both Ac and cAMP-dependent protein kinase activity increased above adult levels through prenatal and early postnatal periods and then fell to adult levels (*87,118*). In rabbit, however, ISO elicited age-related increases in inotropic response and, in contrast, elicited age-related decreases in both cAMP and protein kinase activation (*9,93*). Other investigators have tried to determine whether the (+)-inotropic effects of catecholamines are causally related to the concurrent accumulation of cAMP; there is some evidence that indicates maximal contractile tension is not attained until exposure to ISO or NE produces a maximal increase in cAMP concentration. On the other hand, it has been reported that although the cAMP response to E increased from 2–20 d in neonatal rat heart, the (+)-inotropic responses did not occur until the appearance of cAMP-dependent protein kinase at 20 d (*5*). In the neonatal dog, which has a greater density of β-adrenoceptors and stronger Ac coupling than the adult, there is a smaller (+)-inotropic response to ISO. Possible factors contributing to this reduced responsivity include: (a) activation of cAMP dependent protein kinase, (b) protein phosphorylation, and (c) contractile protein activity (*101*).

17. Myocardial α-Adrenoceptors

Studies of myocardial α-adrenoceptors, like those of myocardial β-adrenoceptors, must consider both age and species differences. For example, whereas myocardial α-adrenoceptors apparently exist in mice and rats (*84,128,133,135*), their presence in guinea pig (*58,128*) is controversial. They have been detected in fetal sheep (*21,66*) and neonatal dogs (*33*), although they might not be present in ewes (*21*) and adult dogs (*33,128*). In general, myocardial α-adrenoceptors have been detected prior to sympathetic innervation (*135*), increase in number (*135*), but not in affinity (*84*), through fetal and early neonatal periods, and then, like β-adrenoceptors, down regulate to adult levels in response to functional sympathetic innervation and the increased release of NE (*135*).

18. Myocardial Cholinergic Receptors

Muscarinic receptors have been identified in myocardial pace-maker cells prior to vagal innervation of the myocardium (*132*) and ACh inhibits myocyte contractions (*62*). In human fetal myocardium, ACh inhibits the SA node as early as 10 wk gestation (*41*). Studies using [^{3}H]-quinuclidinyl benzoate ([^{3}H]-QNB, a cholinergic muscarinic antagonist) binding to examine developmental changes in receptor number have observed both increases and decreases. When binding was expressed as mol of [^{3}H]-QNB/mg tissue, there was an age-related increase through both fetal and early neonatal periods (*103*). In contrast, when binding was expressed per μm^2 of cell surface, there was an initial increase in the early fetal period followed by a decrease continuing into the early postnatal period (*62*).

Acknowledgments

The author's research is supported by NIH Grant #HL-20864, awarded to Dr. Phyllis M. Gootman. The author wishes to thank Dr. Nancy M. Buckley for her review of this manuscript and for her most helpful suggestions.

References

1. Armour, J. A. and Hopkins, D. A. (1981) Localization of sympathetic postganglionic neurons of physiologically identified cardiac nerves in the dog. *J. Comp. Neurol.* **202**, 169–184.
2. Armour, J. A. and Hopkins, D. A. (1984) Anatomy of the Extrinsic Efferent Autonomic Nerves and Ganglia Innervating the Mammalian Heart, in *Nervous Control of Cardiovascular Function* (Randall, W. C., ed.), Oxford University, New York.
3. Armour, J. A. and Randall, W. C. (1975) Functional anatomy of canine cardiac nerves. *Acta Anat.* **91**, 510–528.
4. Atwood, G. F. and Kirshner, N. (1976) Postnatal development of catecholamine uptake and storage of the newborn rat heart. *Dev. Biol.* **49**, 532–538.
5. Au, T. L. S., Collins, G. A., and Walker, M. J. A. (1980) Rate, force, and cyclic adenosine 3',5'-monophosphate responses to (−) adrenaline in neonatal rat heart tissue. *Br. J. Pharmacol.* **69**, 601–608.

6. Bareis, D. L., Morgan, R. E., Lau, C., and Slotkin, T. A. (1981) Maturation of sympathetic neurotransmission in the rat heart. IV. Effects of guanethidine-induced sympathectomy on neonatal development of synaptic vesicles, synaptic terminal function, and heart growth. *Dev. Neurosci.* **4,** 15–24.

7. Bareis, D. L. and Slotkin, T. A. (1980) Maturation of sympathetic neurotransmission in the rat heart. I. Ontogeny of the synaptic vesicle uptake mechanism and correlation with development of synaptic function. Effects of neonatal methadone administration on development of synaptic vesicles. *J. Pharmacol. Exp. Ther.* **212,** 120–125.

8. Bartolome, J. V., Schanberg, S. M., and Slotkin, T. A. (1981) Premature development of cardiac sympathetic neurotransmission in the fetal alcohol syndrome. *Life Sci.* **28,** 571–576.

9. Beck, N. and Park, M. K. (1981) Maturation-related changes in catecholamine-dependent cyclic AMP and protein kinase in the rabbit myocardium. *Dev. Pharmacol. Ther.* **3,** 129–138.

10. Bevan, J. A., Bevan, R. D., and Duckles, S. P. (1980) Adrenergic Regulation of Vascular Smooth Muscle, in *Handbook of Physiology,* Section 2, *The Cardiovascular System,* vol. II, *Vascular Smooth Muscle,* (Bohr, D. F., Somlyo, H. P., Sparks, H. V., Jr., eds.), American Physiological Society, Bethesda, Maryland.

11. Bhalla, R. C., Sharma, R. V., and Ramanathan, S. (1980) Ontogenetic development of isoproterenol subsensitivity of myocardial adenylate cyclase and beta adrenergic receptors in spontaneously hypertensive rats. *Biochem. Biophys. Acta* **632,** 497–506.

12. Black, I. B. (1978) Regulation of Autonomic Development, in *Annual Review of Neuroscience 1* (Cowan, M. W., Hall, Z. W., Kandel, E. R., eds.), Annual Reviews, Inc., Palo Alto, California.

13. Bowers, C. W. and Zigmond, R. E. (1979) Localization of neurons in the rat superior cervical ganglion that project into different postganglionic trunks. *J. Comp. Neurol.* **185,** 381–392.

14. Broadley, K. J. (1982) Review: Cardiac adrenoceptors. *J. Auton. Pharmacol.* **2,** 119–145.

15. Buckley, N. M., Gootman, P. M., Yellin, E. L., and Brazeau, P. (1979) Age-related cardiovascular effects of catecholamines in anesthetized piglets. *Circ. Res.* **45,** 282–292.

16. Burnstock, G. (1978) Do some sympathetic neurons synthesize and release both noradrenaline and acetylcholine? *Progr. Neurobiol.* **11,** 205–222.

17. Burnstock, G. (1980) Cholinergic and Purinergic Regulation of Blood Vessels, in *Handbook of Physiology,* vol. II, *Vascular Smooth Muscle,* (Bohr, D. F., Somlyo, A. P., and Sparks, H. V., Jr., eds.), American Physiological Society, Bethesda, Maryland.

18. Calaresu, F. R. and Cottle, M. K. (1965) Origins of cardiomotor fibres in the dorsal nucleus of the vagus in the cat: A histological study. *J. Physiol.* **176,** 252–260.

19. Chen, F. C. M., Yamamura, H. I., and Roeske, W. R. (1979) Ontogeny of mammalian myocardial beta-adrenergic receptors. *Eur. J. Pharmacol.* **58,** 255–264.
20. Chen, F. C. M., Yamamura, H. I., and Roeske, W. R. (1982) Adenylate cyclase and beta adrenergic receptor development in the mouse heart. *J. Pharmacol. Exp. Ther.* **222,** 7–13.
21. Cheng, J. B., Cornett, L. E., Goldfein, A., and Roberts, J. M. (1980) Alpha adrenergic receptor is present in fetal, but not in adult sheep myocardium. *Fed. Proc.* **39,** 693.
22. Cheng, J. B., Goldfein, A., Cornett, L. E., and Roberts, J. A. (1981) Identification of beta-adrenergic receptors using [³H]-dihydroalprenolol in fetal sheep heart: Direct evidence of qualitative similarity to the receptors in adult sheep heart. *Pediatr. Res.* **15,** 1083–1087.
23. Cullen, M., Sheridan, D., and Tynan, M. (1977) Postnatal ultrastructural development of the cat myocardium. *J. Physiol.* **265,** 16P–17P.
24. de Champlain, J., Malmfors, T., Olson, L., and Sachs, Ch. (1970) Ontogenesis of peripheral adrenergic neurons in the rat: Pre- and postnatal observations. *Acta Physiol. Scand.* **80,** 276–288.
25. De Haan, R. L. (1970) The potassium-sensitivity of isolated embryonic heart cells increases with development. *Dev. Biol.* **23,** 226–240.
26. Deskin, R., Mills, E., Whitmore, W. L., Seidler, F. J., and Slotkin, T. A. (1980) Maturation of sympathetic neurotransmission in the rat heart. VI. The effect of neonatal central catecholaminergic lesions. *J. Pharmacol. Exp. Ther.* **215,** 342–347.
27. Dibner, M. D. and Black, I. B. (1976) The effect of target organ removal on the development of sympathetic neurons. *Brain Res.* **103,** 93–102.
28. Downing, S. E. (1979) Baroreceptor Regulation of the Heart, in *Handbook of Physiology,* Section 2, *The Cardiovascular System,* vol. I, *The Heart.* (Berne, R. M., Sperelakis, N., and Geiger, S. R., eds.), American Physiological Society, Bethesda, Maryland.
29. Downing, S. E., Hellenbrand, W. E., Lee, J. C., and Nudel, D. B. (1978) Adrenal contribution to coronary regulation in the newborn. *Am. J. Physiol.* **234,** 173–179.
30. Egbert, J. R. and Katona, P. G. (1980) Development of autonomic heart rate control in the kitten during sleep. *Am. J. Physiol.* **238,** 829–835.
31. Erath, H. G., Jr., Boerth, R. C., and Graham, T. P., Jr. (1982) Functional significance of reduced cardiac sympathetic innervation in the newborn dog. *Am. J. Physiol.* **243,** 20–26.
32. Fambrough, D. M. (1976) Development of Cholinergic Innervation of Skeletal, Cardiac, and Smooth Muscle, in *Biology of Cholinergic Function.* (Goldberg, A. M. and Hanin, I., eds.), Raven, New York.

33. Felder, R. A., Calcagno, P. L., Eisner, G. M., and Jose, P. A. (1982) Ontogeny of myocardial adrenoceptors. II. Alpha adrenoceptors. *Pediatr. Res.* **16,** 340–342.

34. Friedman, W. F. (1972) The intrinsic physiologic properties of the developing heart. *Prog. Cardiovasc. Dis.* **15,** 87–111.

35. Friedman, W. F. Physiological Properties of the Developing Heart, in *Pediatric Cardiology,* vol. 6, (Anderson, R. H., Macartney, F. J., Shinbourne, E. A., and Tynan, M., eds.), Churchill Livingstone, Robert Stevenson House, Edinburgh, Scotland (in press).

36. Friedman, W. F., Pool, P. E., Jacobowitz, D., Seagren, S. C., and Braunwald, E. (1968) Sympathetic innervation of the developing rabbit heart: Biochemical and histochemical comparisons of fetal, neonatal, and adult myocardium. *Circ. Res.* **23,** 25–32.

37. Frieswick, G. M., Danielson, T., and Shideman, F. E. (1979) Adrenergic inotropic responsiveness of embryonic chick and rat hearts. *Dev. Neurosci.* **2,** 276–285.

38. Gauthier, P., Nadeau, R. A., and de Champlain, J. (1975) The development of sympathetic innervation and the functional state of the cardiovascular system in newborn dogs. *Can. J. Physiol. Pharmacol.* **53,** 763–776.

39. Geis, G. S., Kozelka, J. W., and Wurster, R. D. (1981) Organization and reflex control of vagal cardiomotor neurons. *J. Auton. Nerv. Syst.* **3,** 437–450.

40. Geis, W. P., Tatooles, C. J., Priola, D. V., and Friedman, W. F. (1975) Factors influencing neurohumoural control of the heart in the newborn dog. *Am. J. Physiol.* **238,** 1685–1689.

41. Gennser, G. and Nilsson, E. (1970) Response to adrenaline, acetylcholine, and change of contraction frequency in early human foetal hearts. *Experientia* **26,** 1105–1107.

42. Gennser, G. and Studnitz, W. (1975) Noradrenaline synthesis in human fetal heart. *Experientia* **31,** 1422–1424.

43. Glowinski, J., Axelrod, J., Kopin, I. J., and Wurtman, R. J. (1964) Physiological disposition of H^3-norepinephrine in the developing rat. *J. Pharmacol. Exp. Ther.* **146,** 48–53.

44. Greene, L. A. (1977) Quantitative in vitro studies on the nerve growth factor (NGF) requirement of neurons. I. Sympathetic neurons. *Dev. Biol.* **58,** 96–105.

45. Gyevai, A. (1969) Comparative histochemical investigations concerning prenatal and postnatal cholinesterase activity in the heart of chickens and rats. *Acta Biol. Acad. Sci. Hung.* **20,** 253–262.

46. Hall, E. K. (1957) Acetylcholine and epinephrine effects on the embryonic rat heart. *J. Cell. Comp. Physiol.* **19,** 187–200.

47. Henry, J. L. and Calaresu, F. R. (1972) Distribution of cardioacceleratory sites in intermediolateral nucleus of the cat. *Am. J. Physiol.* **222,** 700–704.

48. Higgins, C. B., Vatner, S. F., and Braunwald, E. (1973) Parasympathetic control of the heart. *Pharmacol. Rev.* **25,** 119–155.
49. Higgins, D. and Pappano, A. J. (1981) Development of transmitter secretory mechanisms by adrenergic neurons in the embryonic chick heart ventricle. *Dev. Biol.* **87,** 148–162.
50. Hirsch, E. F. (1971) The Innervation of the Vertebrate Heart. Charles C. Thomas, Springfield, Illinois.
51. Hoar, R. M. and Hall, J. L. (1970) The early pattern of cardiac innervation in the fetal guinea pig. *Am. J. Anat.* **128,** 499–508.
52. Hopkins, D. A. and Armour, J. A. (1982) Medullary cells of origin of physiologically identified cardiac nerves in the dog. *Brain Res. Bull.* **8,** 359–365.
53. Hopkins, D. A., Gootman, P. M., DiRusso, S. M., and Zeballos, M. E. (1984) Brainstem cells of origin of the cervical vagus and cardiopulmonary nerves in the neonatal pig (*Sus scrofa*). *Brain Res.* **306,** 63–72.
54. Ishii, K., Shigenobu, K., and Kasuya, Y. (1982) Postjunctional supersensitivity in young rat heart produced by immunological and chemical sympathectomy. *J. Pharmacol. Exp. Ther.* **220,** 209–215.
55. Iversen, L. L., de Champlain, J., Glowinski, J., and Axelrod, J. (1967) Uptake, storage, and metabolism of norepinephrine in tissues of the developing rat. *J. Pharmacol. Exp. Ther.* **157,** 509–516.
56. Kalia, M. and Mesulam, M. M. (1980) Brainstem projections of sensory and motor components of the vagus complex in the cat. II. Laryngeal, tracheobronchial, pulmonary, cardiac, and gastrointestinal branches. *J. Comp. Neurol.* **193,** 467–508.
57. Karczmar, A. G., Srinivasan, R., and Bernsohn, J. (1973) Cholinergic Function in the Developing Fetus, in *Fetal Pharmacology.* (Boreus, L., ed.), Raven, New York.
58. Karliner, J. S., Barnes, P., Hamilton, C. A., and Dollery, C. T. (1979) Alpha one adrenergic receptors in guinea pig myocardium: Identification by binding with a new radioligand (^{3}H) Prazosin. *Biochem. Biophys. Res. Commun.* **90,** 142–150.
59. Korsching, S. and Thoenen, H. (1983) Nerve growth factor in sympathetiic ganglia and corresponding target organs of the rat: Correlation with density of sympathetic innervation. *Proc. Natl. Acad. Sci. USA* **80,** 3513–3516.
60. Kralios, F. A., Martin, L., Burgess, M. J., and Millar, C. K. (1975) Local ventricular repolarization changes due to sympathetic nerve branch stimulation. *Am. J. Physiol.* **228,** 1621–1626.
61. Kralios, F. A. and Millar, C. K. (1978) Functional development of cardiac sympathetic nerves in newborn dogs: Evidence for asymmetrical development. *Cardiovasc. Res.* **12,** 547–554.
62. Lane, M. A., Sastre, A., Law, M., and Salpeter, M. (1977) Cholinergic and adrenergic receptors on mouse cardiocytes in vitro. *Dev. Biol.* **57,** 254–269.

63. Lau, C. and Slotkin, T. A. (1979) Accelerated development of rat sympathetic neurotransmission caused by neonatal triiodothyronine administration. *J. Pharmacol. Exp. Ther.* **208,** 485–490.

64. Lau, C. and Slotkin, T. A. (1980) Maturation of sympathetic neurotransmission in the rat heart. II. Enhanced development of presynaptic and postsynaptic components of noradrenergic synapses as a result of neonatal hyperthyroidism. *J. Pharmacol. Exp. Ther.* **212,** 126–130.

65. Lebowitz, E. A., Novick, J. S., and Rudolph, A. M. (1972) Development of myocardial sympathetic innervation in the fetal lamb. *Pediatr. Res.* **6,** 887–893.

66. Lee, J. C., Fripp, R. R., and Downing, S. E. (1982) Myocardial responses to alpha-adrenoceptor stimulation with methoxamine hydrochloride in lambs. *Am. J. Physiol.* **242,** 405–410.

67. Levi-Montalcini, R. and Angeletti, P. U. (1968) The nerve growth factor. *Physiol Rev.* **48,** 534–569.

68. Levy, M. N. and Martin, P. J. (1979) Neural Control of the Heart, in *Handbook of Physiology,* Section 2, *The Cardiovascular System,* vol. I, *The Heart.* (Berne, R. M., Sperelakis, N., and Geiger, S. R., eds.), American Physiological Society, Bethesda, Maryland.

69. Lipp, J. M. and Rudolph, A. M. (1972) Sympathetic nerve development in the rat and guinea pig heart. *Biol. Neonate* **21,** 76–82.

70. Loggie, J. M. H. (1982) Growth and Development of the Autonomic Nervous System, in *Scientific Foundations of Paediatrics.* (Davis, J. A. and Dobbing, J., eds.), University Park, Baltimore, Maryland.

71. Mace, S. E. and Levy, M. N. (1983) Neural control of heart rate: A comparison between puppies and adult animals. *Pediatr. Res.* **17,** 491–495.

72. Mace, S. E. and Levy, M. N. (1983) Autonomic nervous control of heart rate: Sympathetic–parasympathetic interactions and age-related differences. *Cardiovasc. Res.* **17,** 547–552.

73. Mace, S. E., Levy, M. N., and Liebman, J. (1980) The immaturity and asymmetry of heart rate responses to autonomic stimulation in anesthetized puppies. *Pediatr. Res.* **14,** 448.

74. Mackenzie, E. and Standen, N. B. (1980) The postnatal development of adrenoceptor responses in isolated papillary muscles from rat. *Pfluegers Arch.* **383,** 185–187.

75. Marvin, W. J., Hermesmeyer, K., McDonald, R. I., Roskoski, L. M., and Roskoski, R., Jr. (1980) Ontogenesis of cholinergic innervation in the rat heart. *Circ. Res.* **46,** 690–695.

76. Mayer, S. E. (1980) Neurohumoral Transmission and the Autonomic Nervous System, in *The Pharmacological Basis of Therapeutics.* (Gilman, A. G., Goodman, L. S., and Gilman, A., eds.) Macmillan, New York.

188   *Cohen*

77. McGill, M., Hopkins, D. A., and Armour, J. A. (1982) Physiological studies of canine sympathetic ganglia and cardiac nerves. *J. Auton. Nerv. Syst.* **6**, 157–171.
78. Mills, E. (1978) Time course for development of vagal inhibition of the heart in neonatal rats. *Life Sci.* **23**, 2717–2720.
79. Miska, S. P., Kimmel, G. L., Harmon, J. R., and Webb, P. (1983) Ontogeny of cardiac ornithine decarboxylase and its beta adrenergic responsiveness in the rat. *J. Pharmacol. Exp. Ther.* **220**, 419–423.
80. Mukherjee, A., Haghani, Z., Brady, J., Bush, L., McBride, W., Buja, L. M., and Willerson, J. T. (1983) Differences in myocardial alpha and beta adrenergic receptor number in different species. *Am. J. Physiol.* **245**, 957–961.
81. Navaratnam, F. (1965) The ontogenesis of cholinesterase activity within the heart and cardiac ganglia in man, rat, rabbit, and guinea pig. *J. Anat.* **99**, 459–467.
82. Noble, D. (1975) Chronotropic Actions of Autonomic Nervous Transmitters, in *The Initiation of the Heartbeat*. Clarendon, Oxford.
83. Noguchi, A. and Whitsett, J. A. (1983) Ontogeny of alpha-1 adrenergic receptors in the rat myocardium: Effects of hypothyroidism. *Eur. J. Pharmacol.* **86**, 43–50.
84. Noguchi, A., Whitsett, J. A., and Dickman, L. (1981) Ontogeny of myocardial alpha-1 adrenergic receptor in the rat. *Dev. Pharmacol. Ther.* **3**, 179–188.
85. Norris, J. E., Foreman, R. D., and Wurster, R. D. (1974) Responses of the canine heart to stimulation of the first five ventral thoracic roots. *Am. J. Physiol.* **227**, 9–12.
86. Nosaka, S., Yamamoto, T., and Yasunaga, K. (1979) Localization of cardioinhibitory preganglionic neurons within the rat brainstem. *J. Comp. Neurol.* **186**, 79–82.
87. Novak, E., Drummond, G. I., Skala, J., and Hahn, P. (1972) Developmental changes in cyclic AMP, protein kinase, phosphorylase kinase, and phosphorylase in liver, heart, and skeletal muscle of the rat. *Arch. Biochem. Biophys.* **150**, 511–518.
88. Palmer, G. C. and Dail, W. D., Jr. (1975) Appearance of hormone sensitive adenylate cyclase in the developing human heart. *Pediatr. Res.* **9**, 98–103.
89. Papka, R. E. (1978) Development of innervation to the atrial myocardium of the rabbit. *Cell. Tissue Res.* **194**, 219–236.
90. Papka, R. E. (1981) Development of innervation to the ventricular myocardium of the rabbit. *J. Mol. Cell. Cardiol.* **13**, 217–228.
91. Pappano, A. J. (1977) Ontogenetic development of autonomic neuroeffector transmission and transmitter reactivity in embryonic and fetal hearts. *Pharmacol. Rev.* **29**, 3–33.
92. Pappano, A. J. (1981) Adrenoceptors and Adrenergic Mechanisms in the Embryonic and Fetal Heart, in *Adrenoceptors and*

Catecholamine Action. vol. 1, (Kunos, G., ed.), John Wiley and Sons, New York.

93. Park, M. K., Sheridan, P. H., Morgan, W. W., and Beck, N. (1980) Comparative inotropic response of newborn and adult rabbit papillary muscles to isoproterenol and calcium. *Dev. Pharmacol. Ther.* **1**, 70–82.

94. Pavlovic, M. (1982) Tonic sympathetic influences on the guinea pig heart during postnatal development. *Proc. 12th Cong. Un. Yug. Physiol. Soc.* 159–161.

95. Rakusun, K. (1984) Cardiac Growth, Maturation, and Aging, in *Growth of the Heart in Health and Disease.* (Zak, R., ed.), Raven, New York.

96. Randall, W. C. and Armour, A. J. (1974) Regional vagosympathetic control of the heart. *Am. J. Physiol.* **227**, 444–452.

97. Randall, W. C., Armour, J. A., Geis, W. P., and Lippincott, D. B. (1972) Regional cardiac distribution of the sympathetic nerves. *Fed. Proc.* **31**, 1199–1208.

98. Read, J. B. and Burnstock, G. (1969) A method for the localization of adrenergic nerves during early development. *Histochemie* **20**, 197–200.

99. Renou, P., Newman, W., and Wood, C. (1969) Autonomic control of fetal heart rate. *Am. J. Obstet. Gynecol.* **105**, 949–953.

100. Robkin, M. A., Shepard, T. H., and Dyer, D. C. (1976) Autonomic receptors of the early rat embryo heart: Growth and development. *Proc. Soc. Exp. Biol. Med.* **151**, 799–803.

101. Rockson, S. G., Homcy, C. J., Quinn, P., Manders, W. T., Haber, E., and Vatner, S. F. (1981) Cellular mechanisms of impaired adrenergic responsiveness in neonatal dogs. *J. Clin. Invest.* **67**, 319–327.

102. Rogers, M. C., Abildskov, J. A., and Preston, J. B. (1973) Cardiac effects of stimulation and block of the stellate ganglion. *Anesthesiology* **39**, 525–533.

103. Roeske, W. R. and Yamamura, H. I. (1978) Maturation of mammalian myocardial muscarinic receptors. *Life Sci.* **23**, 127–132.

104. Rudolph, A. M. and Heymann, M. A. (1974) Fetal and neonatal circulation and respiration. *Ann. Rev. Physiol.* **36**, 187–207.

105. Saarikoski, S. (1977) Development of noradrenaline uptake in the human foetal heart. *Experientia* **33**, 251–252.

106. Saarikoski, S. (1983) Functional development of adrenergic uptake mechanisms in the human fetal heart. *Biol. Neonate* **43**, 158–163.

107. Sachs, Ch., de Champlain, J., Malmfors, T., and Olson, L. (1970) Postnatal development of noradrenaline uptake in adrenergic nerves: In vitro isotope studies of different rat tissues with or without pretreatment with drugs. *Eur. J. Pharmacol.* **9**, 67–79.

108. Schifferli, P. Y. and Caldeyro-Barcia, R. (1973) Effects of Atropine and Beta Adrenergic Drugs on the Heart of the Human Fetus, in *Fetal Pharmacology* (Boreus, L., ed.), Raven, New York.

190   *Cohen*

109. Schumacher, W., Mirkin, B. L., and Shepard, J. R. (1984) Biological maturation and beta adrenergic effectors: Development of beta adrenergic receptors in rabbit heart. *Mol. Cell. Biochem.* **58,** 173–181.
110. Seidler, F. J. and Slotkin, T. A. (1979) Presynaptic and postsynaptic contributions to ontogeny of sympathetic control of heart rate in the pre-weanling rat. *Br. J. Pharmacol.* **65,** 431–434.
111. Sinha, S., Keresztes-Nagy, S., and Frankfater, A. (1976) Studies on the distribution of cholinesterases: Activity in the human and dog heart. *Pediatr. Res.* **10,** 754–758.
112. Sinha, S., Yelich, M. R., Keresztes-Nagy, S., and Frankfater, A. (1979) Regional distribution of acetylcholinesterase in the right atria of humans and dogs. *Pediatr. Res.* **13,** 1217–1221.
113. Slavikova, I. and Tucek, S. (1982) Postnatal changes of the tonic influence of the vagus nerves on the heart rate, and of the activity of choline acetyltransferase in the heart atria of rats. *Physiol. Bohemoslov.* **31,** 113–120.
114. Slotkin, T. A. (1979) Ornithine decarboxylase as a tool in developmental neurobiology. *Life Sci.* **24,** 1623–1630.
115. Slotkin, T. A., Johnson, A., Whitmore, W. L., and Slepetis, R. J. (1984) Ornithine decarboxylase and polyamines in developing rat brain and heart: Effects of perinatal hypothyroidism. *Intl. J. Dev. Neurosci.* **2,** 155–161.
116. Slotkin, T. A., Smith, P. G., Lau, C., and Bareis, D. L. (1980) Functional Aspects of Development of Catecholamine Biosynthesis and Release in the Sympathetic Nervous System, in *Biogenic Amines in Development.* (Parvez, H. and Parvez, S., eds.), Elsevier, Amsterdam.
117. Smolen, A. J. and Truex, R. C. (1977) The dorsal motor nucleus of the vagus nerve of the cat: Localization of preganglionic neurons by quantitative histological methods. *Anat. Rec.* **189,** 555–566.
118. Sperelakis, N. and Pappano, A. J. (1983) Physiology and pharmacology of developing heart cells. *Pharmacol. Ther.* **22,** 1–39.
119. Spyer, K. M. (1981) Neural organization and control of the baroceptor reflex. *Rev. Physiol. Biochem. Pharmacol.* **88,** 23–124.
120. Stanton, H. C. and Mueller, R. L. (1975) Onotgenesis of catecholamines and some related enzymes in heart and spleen of swine (*Sus domesticus*). *Comp. Biochem. Physiol.* **50,** 171–176.
121. Stull, J. and Mayer, S. E. (1979) Biochemical Mechanisms of Adrenergic and Cholinergic Regulation of Myocardial Contractility, in *Handbook of Physiology,* Section 2, *The Cardiovascular System,* vol. I, *The Heart.* (Berne, R. M., Sperelakis, N., and Geiger, S. R., eds.), American Physiological Society, Bethesda, Maryland.
122. Sugimoto, T., Itoh, K., Mizuno, N., Nomura, S., and Konishi, A. (1979) The site of origin of cardiac preganglionic fibers of the vagus nerve: An HRP study in the cat. *Neurosci. Lett.* **12,** 53–58.

123. Thoenen, H. and Barde, Y. -A., (1980) Physiology of nerve growth factor. *Physiol. Rev.* **60,** 1284–1335.
124. Trautwein, W. (1963) Generation and conduction of impulses in the heart as affected by drugs. *Pharmacol. Rev.* **15,** 277–332.
125. Vlk, J. (1979) Postnatal development of postganglionic parasympathetic neurons in the heart of the albino rat. *Physiol. Bohemoslov.* **28,** 561–568.
126. Vlk, J. (1981) Cholinesterases and postnatal development of the negative chronotropic effects of acetylcholine in albino rats. *Physiol. Bohemoslov.* **30,** 497–503.
127. Vlk, J. and Vincenzi, F. F. (1977) Functional autonomic innervation of mammalian cardiac pacemaker during the perinatal period. *Biol. Neonate* **31,** 19–26.
128. Wei, J. W. and Sulakhe, P. V. (1979) Regional and subcellular distribution of beta and alpha adrenergic receptors in the myocardium of different species. *Gen. Pharmacol.* **10,** 263–267.
129. Weiner, N. (1980) Drugs That Inhibit Adrenergic Nerves and Block Adrenergic Receptors, in *The Pharmacological Basis of Therapeutics.* (Gilman, A., Goodman, L. S., and Gilman. A. G., eds.) Macmillan, New York.
130. Weisman, G. G., Jones, D. S., and Randall, W. C. (1966) Sympathetic outflows from cervical spinal cord in the dog. *Science* **152,** 381–382.
131. Whitsett, J. A. and Darovec-Beckman, C. (1981) Developmental aspects of beta adrenergic receptors and catecholamine-sensitive adenylate cyclase in rat myocardium. *Pediatr. Res.* **15,** 1363–1369.
132. Wildenthal, K. (1973) Maturation of responsiveness to cardioactive drugs: Differential effects of acetylcholine, norepinephrine, theophylline, tyramine, glucagon, and dibutyryl cyclic AMP on atrial rate in hearts of fetal mice. *J. Clin. Invest.* **52,** 2250–2258.
133. Williams, R. S. and Lefkowitz, R. J. (1978) Alpha adrenergic receptors in rat myocardium. *Circ. Res.* **43,** 721–727.
134. Woods, J. R., Jr., Dandavino, A., Nuwayhid, B., Brinkman, C. R., III, and Assali, N. S. (1978) Cardiovascular reactivity of neonatal and adult sheep to autonomic stimuli during adrenergic depletion. *Biol. Neonate* **34,** 112–120.
135. Yamada, S., Yamamura, H. I., and Roeske, W. R. (1980) Ontogeny of mammalian cardiac alpha-1 adrenergic receptors. *Eur. J. Pharmacol.* **68,** 217–221.
136. Zipes, D. P., Martins, J. B., Ruffy, R., Prystowsky, E. N., Elharrar, V., and Gilmour, R. F., Jr. (1981) Roles of Autonomic Innervation in the Genesis of Ventricular Arrhythmias, in *Disturbances in Neurogenic Control of the Circulation.* American Physiological Society, (Abboud, F. M., Fozzard, H. A., Gilmore, J. P., and Reis, D. J., eds.), American Physiological Society, Bethesda, Maryland.

Chapter 7

Autonomic Effects in the Developing Heart

Robert F. Reder, Ofer Binah, and Peter Danilo, Jr.

1. Introduction

Embryonic and fetal hearts have been studied for over 300 yr. Initially, the effects on the rate of impulse generation of various physical factors (e.g., temperature) and chemicals were observed. Later, cardiac electrophysiologic properties, including the characteristics of transmembrane action potentials, were investigated. Changes related to age and development were observed in these studies, the vast majority of which utilized the chick embryo and, to a lesser extent, the rat embryo. Although the heart of the human embryo or fetus has been studied, it has not been possible to perform detailed electrophysiological studies over the entire gestational period.

In this chapter we will describe (1) the changes in the transmembrane action potential that occur during ontogeny and the neonatal period; and (2) certain aspects on the relationship between development and the effects of autonomic agents on cardiac electrophysiologic characteristics.

193

2. Transmembrane Action Potential Characteristics of Embryonic Hearts

Because of the ease with which the chick embryo can be studied, the ontogeny of its transmembrane action potential has been well characterized. By the second day (postfertilization) the resting membrane potential of embryonic chick myocardial cells is -35 to -40 mV $(9,31,34)$, increasing to between -60 and -70 mV by days 4–7 $(13,18,22,38,41)$. Between the 16th and the 19th day, it reaches a maximum value of -70 to -80 mV $(13,18,22,38,41)$. Because experimental conditions have differed, some caution is required in evaluating the data on ontogenetic changes in resting membrane potential. Differing extracellular potassium concentrations can cause differences in resting potential, and secondarily, in action potential amplitude, overshoot, and maximum upstroke velocity ($\dot{V}_{max}$). Action potential amplitude, overshoot, and $\dot{V}_{max}$ increase concurrently with, and perhaps secondarily to, the increasing resting membrane potential. For embryonic chick ventricle, action potential amplitude increases from approximately 50 mV at the second embryonic day to between 90 and 115 mV at or near the time of hatching (19 d). Overshoot increases over the range of 11 to 19 mV at day 2, to approximately 33 mV at hatching. For $\dot{V}_{max}$, values of from 7 to 20 V/s have been reported at day 2 $(21,34)$ and values greater than 200 V/s near the time of hatching $(18,22)$. Some uncertainty exists regarding the upstroke of the action potential, which may change from one dependent on "slow" (sodium as well as calcium) channels, to one that is dependent on "fast" (sodium) channels.

In contrast to the depolarization phase of the action potential of the developing heart, relatively little information is available concerning the voltage-time course of repolarization. Vleugels et al. (38) found little change in duration (measured to 90% repolarization) between days 7 and 19; values were 154 and 165 ms, respectively. Similarly, Yeh and Hoffman (41) found that (at a stimulus cycle length of 600 ms) action potential duration was unchanged (130 ms) from days 6 through 19.

Because the mammalian embryo is technically more difficult to obtain and study, as well as more expensive, its cellular electrophysiology is not as completely characterized as that of the chick. In the rat, between days 10 and 21 postconception (total gestation $= 22$ d) the resting membrane potential of ventricular myocardium increases from -48 mV to -82 mV, and $\dot{V}_{max}$ increases from 5–8 V/s to approximately 50 V/s (2).

Recently, we have reported the developmental changes in electrophysiologic properties of the canine ventricular specialized conducting system (12). The dog has a total gestation time of 60–63 d. For fetal Purkinje fibers obtained from just after implantation (approximately day 18) to just prior to natural birth, a linear relationship exists between fetal crown–rump length (CRL) and maximum diastolic potential. For least-developed (CRL = 6–8 mm) hearts, maximum diastolic potential is approximately -68 mV; whereas for fibers from near-term hearts (CRL = 150–160 mm), it is approximately -80 mV. During this time, action potential amplitude increases from approximately 98 to 120 mV, and $\dot{V}_{max}$ from 200 to 450 V/s. The action potential duration of fetal canine Purkinje fibers, apparently unlike that of either the rat or the chick myocardium, increases as development progresses. At a stimulus cycle length of 500 ms, action potential duration (measured to 50% repolarization) increases from 65 to 240 ms during gestation. Figure 1 summarizes the changes that occur in transmembrane action potential characteristics during the latter two-thirds of gestation.

A limited number of studies of cellular electrophysiology of the human fetal heart have been reported (17,20,36). Available data suggest that the resting membrane potential of midtrimester human fetal ventricular myocardium is relatively high (-81 mV) (17). Action potential amplitude is approximately 95–120 mV at this time, and the duration of the action potential is similar to that of the adult tissue (36).

3. Developmental Changes in Ionic Currents—Embryonic and Fetal Hearts

Studies to determine developmental changes in ionic currents responsible for the resting and action potentials have utilized alterations in the ionic composition of the extracellular fluid, the addition of blockers of various ionic currents to the superfusion solutions, and the estimation of intracellular ionic activities using ion-sensitive microelectrodes. Most studies have been performed using the chick embryo.

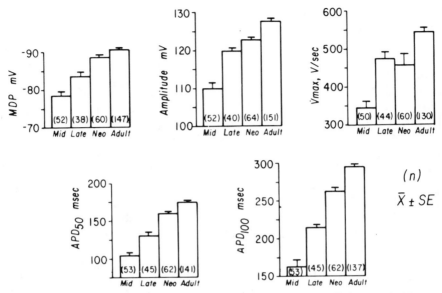

Fig. 1. The effect of development on transmembrane action potential characteristics of canine cardiac Purkinje fibers. In the top half of this figure are shown values for maximum diastolic potential, MDP (left), action potential amplitude (center), and maximum action potential upstroke velocity, $\dot{V}_{max}$ (right). The bottom half illustrates the effect of development on action potential duration measured to 50% repolarization, APD_{50} (left), and full repolarization, APD_{100} (right). In all panels are shown four bars representing the means $\pm$ SEM for midgestation, late gestation, neonatal (0–10 d of age), and adult Purkinje fibers. The number of observations comprising the mean is shown in parentheses.

The resting membrane potential of embryonic heart is dependent on the gradient of potassium ions across the cell membrane. For older chick embryos, at or near the time of hatching, the relationship between $[K^+]_o$ and resting membrane potential is similar to that for the adult; that is, an approximately 50–60-mV change occurs per decade change in $[K^+]_o$ (34). For less-developed hearts, the slope is lower: 30–40 mV/decade for 2–3-d hearts and 40–50 mV/decade for 4–7 d hearts. The derivations at lower $[K^+]_o$ from values predicted from the Nernst equation are greatest in the least developed hearts, in which resting potentials are relatively low. Younger hearts apparently also tolerate a higher $[K^+]_o$ before loss of excitability ensues (34). Membrane permeability to K^+ (P_{K^+})

derived from estimated values of $[K^+]_i$ (intracellular potassium ion concentration) suggest that the ratio P_{Na}/P_K is higher in less developed hearts: 0.2, 0.1, and 0.01 for 3-, 5-, and 15-d ventricles, respectively. Similar changes in P_{Na}/P_K have been reported by others (4,16,23, 34). Although decreases in $[K^+]_i$ related to development have been reported (23–24), other studies suggest that $[K^+]$ increases (16,21, 28). Because of the many technical limitations involved in the estimation of intracellular ions, however, it seems preferable to directly measure ionic activity.

Fozzard and Sheu (16) studied the chick embryo using both K^+-selective glass microelectrodes and K^+-selective liquid ion exchange microelectrodes. They found that the intracellular activity of K^+ ($a^i K^+$) increased from 71 mM at day 4 to approximately 90 mM at day 18. In contrast, the activity of intracellular Na$^+$ ($a^i Na^+$) decreased from 12.5 to 7.0 mM. Using these values to calculate equilibrium potentials for K^+ and Na$^+$, the following were reported: V_K increased from -73.3 mV at day 4 to -79.3 mV at day 18, whereas V_{Na} increased from 57.0 to 71.8 mV. Fozzard and Sheu suggest that a changing activity of a sodium-potassium exchange pump can explain the concomitant increase in $a^i K^+$ and fall in $a^i Na^+$. Developmental changes in Na$^+$-K$^+$-ATPase have been reported for embryonic chick ventricle (21,35).

Alterations in external $[Na^+]$, and $[Ca^{2+}]$, and agents that either enhance (catecholamines) or depress (Mn^{2+}, tetrodotoxin, verapamil) fast and slow inward currents have been used to study the upstroke of embryonic and fetal action potentials. In general, it is believed that early in cardiac development the action potential upstroke is more dependent on an inward current moving through "slow channels," whereas later in development, inward current moves primarily through "fast channels" in the chick (13,18,22,31,33,34,41), rat (2,9), dog (12), and human (20). Yeh and Hoffman (41) found that reduction of $[Na^+]_o$ to 50% of normal reduced $\dot{V}_{max}$ in chick ventricles to the same extent at 6 and 19 d. Substitution of choline, tetraethylammonium chloride (TEA), or tetramethylammonium chloride (TMA) for Na$^+$ results in the generation of "all or none" action potentials during the first 4 d of incubation (19). Substitution of Na$^+$ with choline or TMA inhibits action potentials after day 6. In hearts younger than 6 d, tetrodotoxin (TTX), 1 mg/L, reduces amplitude and $\dot{V}_{max}$; complete abolition of the action potential, however, does not occur. Hearts greater than 10 d of age (postfertilization) lose the ability to generate action potentials in the presence of TTX. Those younger hearts in which TTX did not abolish the action potential responded to Mn^{2+} (a blocker of inward Ca^{2+}) with a complete loss of excitability. Subsequent studies (31,34) confirmed these results, showing that TTX (as high as 20 mg/L) had no effect on the rate of spontaneously generated action potentials or on $\dot{V}_{max}$ of action potentials from hearts younger than 3 d. Lanthanum (La^{2+}), 1mM, a putative blocker of inward Ca^{2+} current, was without effect in these younger hearts, suggesting that despite TTX-insensitivity, action potential upstroke was still dependent on Na^+, probably moving through a "slow channel." Removal of extracellular Na^+ abolished excitability (31). The relationships between $[Na^+]_o$ and action potential overshoot, and $[Na^+]_o$ and $\dot{V}_{max}$ have been reported to change markedly during embryonic life (34). With development, the curves relating $[Na^+]_o$ to each of these parameters shift upward and to the left. For overshoot, the curves are linear between $[Na^+]_o = 40$–50 mM for 4- and 17-d embryonic ventricles (slope ~60 mV/decade). Above 50 mM, the curve for day-4 hearts attains a plateau and the slope approaches zero, whereas for the older hearts, the slope remains greater. Similar qualitative changes occur for $\dot{V}_{max}$.

Pappano (25) found that action potentials from atria of ≥ 12-d chick embryos showed rapid reductions in amplitude and $\dot{V}_{max}$ following exposure to TTX, $5 \times 10^{-8}M$. In contrast, the excitability of atria from 6-d hearts was not blocked by TTX, although amplitude and $\dot{V}_{max}$ were reduced. Reduction of $[Na^+]_o$ had different effects on the overshoot of atrial action potentials in 6- and 18-d embryos; the slope of the line relating overshoot to Na^+ was ~60 mV/decade in 18-d embryos and only 10–20 mV/decade in 6-d atria, suggesting that the Na^+ electrode characteristics of the membrane were augmented during increasing development. For atria exposed to TTX, it was found that divalent cations (Ca^{2+}, Ba^{2+}, Sr^{2+}) were less effective in restoring excitability in more developed embryonic atria—an effect attributed to a decline in the ratio of inward current carried by divalent cations to that carried by outward K^+ (26). An increase in gK (potassium conductance) related to development has been reported for embryonic hearts (4,34).

For fetal canine Purkinje fibers, TTX has effects on the action potential that are equivalent throughout development. TTX, 7×10^{-7} to $1.6 \times 10^{-5}M$, decreases action potential amplitude and $\dot{V}_{max}$ in a concentration-dependent manner. The lack of developmentally related effects of TTX on fetal canine fibers in these studies could be a result of species differences, tissue differences, or the inability to study canine hearts earlier than 18 d postconception. Iijima and Pappano (*18*) found that the sensitivity to TTX of $\dot{V}_{max}$ of embryonic chick ventricle was equivalent between the 4th and 18th d of development. Similar results were reported by Marcus and Fozzard (*22*), who also found that the hearts of the embryonic and adult chicken are equally sensitive to TTX.

Inward Ca^{2+} currents have been studied in fetal canine Purkinje fibers (*12*) using verapamil, a blocker of the slow inward current. Equivalent effects of verapamil on fetal Purkinje fibers from early, middle, and late gestation were observed. The depression of $\dot{V}_{max}$ was greatest at verapamil $1 \times 10^{-5}M$. Action potential amplitude was also reduced by verapamil. Fibers from the least- and most-developed hearts were equivalent in their sensitivity to verapamil, although there was no effect on fibers from midgestation. Verapamil also decreased the duration of the plateau to an equivalent degree in all age groups. The effects of verapamil on $\dot{V}_{max}$ and action potential amplitude suggests that either these parameters depend to some extent on a slow inward current (Ca^+ or Na^+), or that verapamil reduces the rapid inward Na^+ current. Although there is evidence that verapamil blocks the movement of Na^+ through a slow channel in embryonic chick ventricle (*32*), the existence of such a current in fetal canine Purkinje fibers has not been determined.

4. Neonatal Hearts

Various animal models have been used to study the effects of age on the cardiac transmembrane action potential. Cavoto et al. (*6*) found that the resting potential of electrically stimulated rat atria increased from -74 mV at 1 mo of age to -82 mV at 3 mo, whereas action potential amplitude increased from 83 to 90 mV. Increases of $\dot{V}_{max}$ in young rat apparently do not parallel the increase in membrane potential. $\dot{V}_{max}$ in 1-mo-old rats was 445 V/s, and in 3-mo-old rats, 308 V/s.

Neonatal (0–10 d of age) canine cardiac Purkinje fibers continue the developmental changes in the transmembrane action potential begun during fetal life (*28*). Mean values of action potential characteristics recorded for neonatal and adult Purkinje fibers can vary slightly from study to study, but overall there are statistically significant differences between these age groups (Fig. 1). Action potential amplitude increases from approximately 120 mV in the term fetus to 123 mV within the first 10 d of life. $\dot{V}_{max}$ is statistically equivalent over this time, being approximately 474 V/s in the late-gestation fetus and 456 V/s in the neonate. Action potential duration (to full repolarization) increases from approximately 220 to 263 ms. When considering action potential duration it should be kept in mind that for neonatal Purkinje fibers, duration is equivalent in proximal subendocardial and free-running portions, and decreases in distal locations (37). In contrast, action potential duration in the adult Purkinje fiber is greatest in distal segments, at the ventricular muscle–Purkinje fiber junction (25,37).

We have studied the effects of modifiers of ionic currents underlying the canine Purkinje fiber action potential as a function of development (28). TTX decreases the amplitude and duration of the action potential, as well as $\dot{V}_{max}$ of neonatal and adult fibers. However, the neonate requires somewhat less TTX than the adult to depress the action potential. The duration of neonatal action potentials is significantly decreased by TTX $7 \times 10^{-7}M$, whereas that of the adult requires a higher threshold concentration, $1.5 \times 10^{-6}M$. Action potential amplitude of neonatal fibers required about one-fifth as much TTX as adults for a threshold reduction. $\dot{V}_{max}$ of neonatal fibers significantly decreased by TTX $>1.5 \times 10^{-6}M$, whereas $\dot{V}_{max}$ of adult fibers required a slightly greater threshold concentration of $3.1 \times 1\text{-}^{-6}M$. For all three variables (amplitude, duration, and $\dot{V}_{max}$), the maximum effect of TTX was greater in the neonate than in the adult. The reasons for these age-dependent effects of TTX are not known. A developmental increase in the number and/or density of Na^+ channels is suggested by the increase in $\dot{V}_{max}$ occurring between neonatal and adult life, but there are no data to support this hypothesis. Because the relationship between membrane potential and $\dot{V}_{max}$ (i.e., membrane responsiveness) is equivalent in neonatal and adult fibers, it is likely that fast channel kinetics do not change significantly. The effect of TTX on action potential duration suggests that there is a background inward Na^+ current operating during repolarization (8). This current is more sensitive to TTX than that of the action potential upstroke, both in adults (8,28,37) and neonates (28). Voltage clamp studies of adult sheep Purkinje fibers have shown that a significant sodium current is operative in plateau potentials (i.e., a "window current") and that this current is probably more sensitive to the effects of tetrodotoxin than is the current underlying the upstroke of the action potential (1). The fact that the duration of fetal Purkinje fiber action potential is more sensitive to tetrodotoxin than $\dot{V}_{max}$ suggests that this "window current" exists relatively early in fetal life.

Repolarizing currents have been studied in neonatal and adult Purkinje fibers (28) using a variety of blocking agents, including verapamil and tetraethylammonium (TEA) chloride. The effect of verapamil on the amplitude of slow response action potentials (produced in Na^+-free, Ca^{2+}-enriched solutions) is greater in the adult than the neonatal Purkinje fiber. These results suggest that the adult Purkinje fiber has a larger inward Ca^{2+} current than does that of the neonate, which is attenuated by blocking agents. In contrast, sensitivity to tetrodotoxin is greater in the neonate than in the adult.

Thus, despite the lack of precise quantification of age-related changes in ionic conductances, it is possible to detect changes in the sensitivities in blocking agents of cardiac transmembrane action potentials and to hypothesize that age and development may be responsible for changes in cellular electrophysiology.

5. Effects of Autonomic Nervous System Agonists and Antagonists

Ontogenetic changes in cardiac autonomic innervation and responsiveness have been studied extensively, primarily in the embryonic chick heart [*see* review by Pappano, (27)].

We have studied Purkinje fibers obtained from canine fetal hearts during early (18–30 d, postconception), middle (31–45 d), and late (46–63 d) gestation. Fibers were superfused with a physiologic salt solution containing 4 m*M* K^+. Over this gestational interval the spontaneous automatic discharge rate of isolated Purkinje fibers increases from 0 during early gestation to approximately 10 beats/min during late gestation. The spontaneous rate of discharge of fibers from near-term fetuses is equivalent to that of the neonate. For adult fibers, spontaneous rate is approximately 12 beats/min and is statistically equivalent to that of the neonate as well as the near-term fetus.

6. Effects of Sympathetic Agonists and Antagonists

Responsiveness of the ventricular specialized conducting system to adrenergic agonists is also a function of ontogenesis (*11*). Purkinje fibers from canine fetal hearts initially show little response to epinephrine. During early gestation, relatively high (>1 × $10^{-6}M$) epinephrine concentrations are required to increase spontaneous rate. These increases are of a much lower magnitude than those that occur in near-term fetal hearts. With development, responsiveness increases and becomes similar to that seen for neonatal fibers. Of further interest is the development in fibers from near-term fetal hearts of a negative chronotropic response to epinephrine. That is, low concentrations of epinephrine actually decrease spontaneous rate. Thus, not only are there quantitative changes in the dose–response relationship for epinephrine and spontaneous rate, but a qualitative change also occurs—a negative chronotropic response to epinephrine in addition to the positive response at higher concentrations.

The response to adrenergic agonists and antagonists of neonatal and adult Purkinje fibers has been studied extensively (*30*). Significant differences exist between the two age groups with respect to both the negative and positive chronotropic response to adrenergic stimulation. The agonists studied were isoproterenol, a relatively pure beta-agonist, and epinephrine, an agonist with both alpha- and beta-agonist properties. Initial studies with adult Purkinje fibers yielded two distinct response patterns. Approximately two-thirds of adult fibers responded to low concentrations of epinephrine (1 × 10^{-11} to 1 × $10^{-8}M$) with a decrease in rate followed by an increase in rate at higher epinephrine concentrations. This type of response was termed a biphasic response. A second group of adult Purkinje fibers responded to all concentrations of epinephrine with increases in rate only; this response was called monophasic. These two groups of adult fibers differed further from each other in that the maximum positive chronotropic response of the biphasic group was approximately 75% above control, whereas that of the monophasic group was 250% above control. Blockade of beta-adrenergic receptors by propranolol and alpha-adrenergic blockade by phentolamine suggested that the negative chronotropic effect of epinephrine was alpha-adrenergic mediated, and the positive chronotropic effect was beta-adrenergic mediated. In contrast to the group of biphasic adult fibers only one-half of the fibers from neonatal hearts showed a biphasic response. The maximum positive chronotropic effect of biphasic neonatal fibers was significantly greater than for adults—250 vs 75%, respectively, above control. However, for those fibers responding with only an increase in spontaneous rate (i.e., monophasic) the responses were statistically equivalent. Figure 2 illustrates these differences between neonatal and adult Purkinje fibers. These age-dependent effects of adrenergic amines on canine Purkinje fiber spontaneous rate require further study to determine the underlying mechanism.

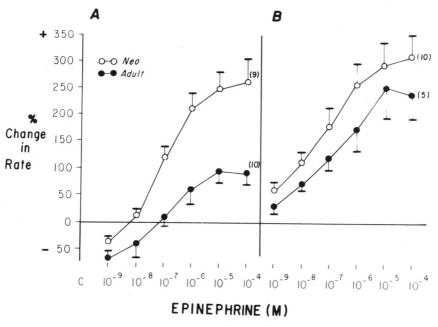

EPINEPHRINE (M)

Fig. 2. Effects of epinephrine on the spontaneous rate of neonatal and adult Purkinje fibers. Panel A shows the biphasic effect of epinephrine on fibers from both age groups. Note the greater increase in spontaneous rate of neonatal fibers. The number of fibers responding in this biphasic manner is shown in parentheses next to each concentration-response curve. Panel B illustrates the monophasic effect of epinephrine on neonatal and adult fibers. In contrast to the biphasic curve, these curves are statistically equivalent (from ref. *30*, by permission of the American Heart Association, Inc.).

Recent studies of canine Purkinje fibers (29) have shown that in the first 2 d of life, alpha-adrenergic stimulation (via phenylephrine) results in an increase in spontaneous rate in some fibers. Blockade of alpha-adrenergic receptors by phentolamine prevents this effect (Fig. 3). In contrast propranolol, a beta-adrenergic receptor blocker, was without effect (Fig. 4), suggesting that the alpha-receptor component of catecholamine responsiveness undergoes marked qualitative changes during the first 2 d of neonatal life. Although the underlying reasons for such a change are presently unknown, an increase in cardiac innervation does occur, as is indicated by the increase in norepinephrine content of neonatal hearts that occurs during this time interval. Norepinephrine content increases from approximately 100 mg/g tissue (wet wt) at day 2 to 1600 mg/g at day 10 (29).

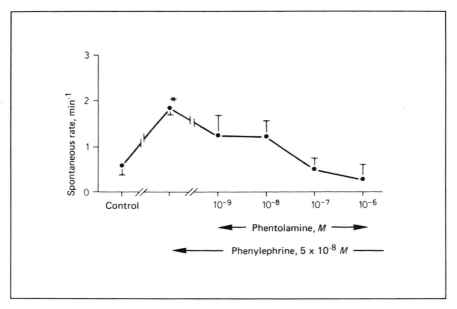

Fig. 3. Response of six neonatal fibers showing increased automaticity in the presence of $5 \times 10^{-8}M$ phenylephrine to superfusion with graded concentrations of phentolamine. The axes are similar to those in the previous figures. A significant increase in rate (is induced in these fibers and is blocked by phentolamine. Data are presented as mean $\pm$SEM (from ref. 29, by permission).

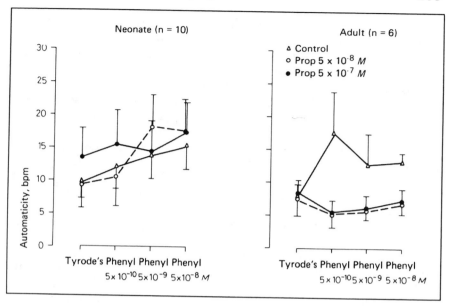

Fig. 4. Effect of phenylephrine and propranolol on automaticity of adult and neonatal Purkinje fibers. Abscissa: concentrations of phenylephrine. Ordinate: automaticity (impulses/min). The curves represent the response to phhenylephrine, in the absence and presence of propranolol, 5×10^{-8} and $5 \times 10^{-7}M$. Mean $\pm$SEM are presented. The increase in rate in the presence of phenylephrine alone is significant ($p < 0.0005$). Propranolol, $5 \times 10^{-7}M$, significantly attenuates the response to phenylephrine in the adults ($p < 0.005$), but not the neonates (from ref. *29*, by permission).

Additional support for the hypothesis that sympathetic innervation for the manifestation of alpha-adrenergic effects is derived from recent preliminary experiments on cultured myocytes (*14*). In spontaneously beating rat ventricles, alpha-stimulation induced by phenylephrine causes a decrease in rate. Monolayer cultures of rat ventricular myocardial cells, on the other hand, respond to alpha-adrenergic stimulation with only an increase in spontaneous rate. Of greater interest is the observation that in cells from 2-d-old rat hearts, cocultured with rat sympathetic ganglion cells plus nerve growth factor, alpha-adrenergic stimulation decreases spontaneous rate. From these preliminary data, it is plausible to consider that the negative chronotropic response of cardiac cells to adrenergic amines requires the presence of sympathetic nerves. Differences in adrenergic receptor characteristics, such as receptor number, density, affinity, or specificity, are likely to occur during development. There may also be differences in the relative rates of development of alpha- and beta-receptors during fetal and neonatal life. The presence of alpha-receptors in cardiac tissues has been amply demonstrated (7,15,39,40). These receptors have been demonstrated during fetal life and have been reported to decrease in number during ontogenesis in the fetal lamb (7). For the dog, specific binding of various ligands by the alpha-receptor has been demonstrated in the neonate (1–5 wk of age) but not in the adult (15). For canine Purkinje fibers in which the magnitude of the alpha-adrenergic receptor-mediated negative chronotropic effect is greater in the adult than in the neonate, one might expect to demonstrate alpha-receptors more easily in the adult.

Preliminary data from studies of alpha-receptors in canine myocardium indicate that there are $alpha_1$ receptors in both neonatal and adult hearts (3). There may be two separate binding sites distinguishable on the basis of affinity. One site has a high affinity and low density, the other has a lower affinity and higher density. For both age groups, the high affinity/low density site is equivalent. The lower affinity, higher density site differs, however. For the adult, K_d = 790 ± 120 pM; β_{max} = 460 ± 90 8 fmol/mg, whereas for neonate these values are 1870 ± 140 pM and 1.13 ± 24 pmol/mg. (K_d = affinity constant; β_{max} = receptor density.) Although the data are preliminary and pertain to myocardium rather than Purkinje fibers, they indicate that characteristics of alpha-receptors change with development and suggest that differences in responsiveness to adrenergic agonists that occur in developing hearts may result from changes in receptor properties.

7. Effects of Parasympathetic Agonists and Antagonists

The effects of parasympathetic agonists on fetal and embryonic hearts have been extensively reviewed (27). The effects of this class of drugs on the developing ventricular conducting system have been less well characterized. For adult cardiac Purkinje fibers, the major effects of acetylcholine on action potential duration are variable and depend, at least in part, on the species of animal studied, on $[K^+]_o$, and on the rate of stimulation.

Neonatal canine cardiac Purkinje fibers respond differently than adult fibers to acetylcholine (10). At low concentrations (1×10^{-11} to $1 \times 10^{-9}M$), spontaneous rate increases, whereas higher concentrations ($>1 \times 10^{-9}M$) have a negative chronotropic effect. Both the negative and positive effects are blocked by atropine, suggesting that there is a common muscarinic receptor. The possibility of an acetylcholine-induced release of endogenous catecholamine was estimated by the use of propranolol. Neither aspect of the chronotropic response of neonatal Purkinje fibers was altered by beta-blockade with propranolol. Similarly, the alpha-blocking agent, phentolamine, was without effect. The possibility that histamine may have been released secondary to exposure of neonatal fibers to acetylcholine was eliminated when it was found that histamine, 1×10^{-9} to $1 \times 10^{-4}M$, was without effect on spontaneous rate. Thus, the likelihood is very low that the positive chronotropic effect of acetylcholine in the neonate is mediated by other than a muscarinic receptor. Although the ionic mechanisms for the positive chronotropic effect of acetylcholine on neonatal Purkinje fibers are not known, such an effect has been studied and demonstrated in adult sheep Purkinje fibers, which also develop a positive chronotropic response to acetylcholine (5). Here, the positive chronotropic effect results from a decrease in potassium conductance. Whether this occurs in neonatal canine fibers remains to be determined.

8. Conclusions

It is clear from the preceding discussion that the interaction between the developing autonomic nervous system and the developing heart are complex and incompletely understood. Species variation, as well as the difficulty in obtaining uniformly and accurately "timed" tissue, increase the complexity. With this in mind, it can, however, be stated that embryonic, fetal, and neonatal cardiac electrophysiologic parameters are modified by changing characteristics of cellular structure, membrane and organelle function, receptor characteristics, and autonomic nervous system development and ingrowth. With the tools currently available, it is difficult to dissect out the specific alterations resulting exclusively from the developmental changes in the autonomic nervous system; however, with continued efforts, these changes will no doubt become better defined.

Acknowledgments

Parts of the studies described herein were supported by USPHS HD-13063 and HL-28958. Dr. Danilo is the recipient of a Research Career Development Award 1-KO-4-00853.

References

1. Atwell, D., L. Cohen, D. Eisner, M. Ohba, and C. Ojeda (1979) The steady state TTX-sensitive ("window") sodium current in cardiac Purkinje fibers. *Pfluegers Arch.* **379,** 137–142.
2. Bernard, C. (1975) Establishment of Ionic Permeabilities of the Myocardial Membrane During Embryonic Development of the Rat, in *Developmental and Physiological Correlates of Cardiac Muscle* (M. Lieberman and T. Sano, eds.) Raven, New York, pp. 169–184.
3. Buchthal, S. D., L. E. Kupfer, J. P. Bilezikian, and P. Danilo (1984) Identiification of α-adrenergiic receptors in adult and neonatal canine ventricles. *Fed. Proc.* (Abstract) **43,** 690.
4. Carmeliet, E. E., C. R. Horres, M. Lieberman, and J. F. Vereecke (1973) Developmental aspects of potassium flux and permeability of the embryonic chick heart. *J. Physiol.* **254,** 673–692.
5. Carmeliet, E. and J. Ramon (1980) Effect of acetylcholine on time-independent currents in sheep cardiac Purkinje fibers. *Pflugers Arch.* **387,** 207–216.
6. Cavoto, F. V., G. J. Kelliher, and J. Roberts (1974) Electrophysiological changes in the rat atrium with age. *Am. J. Physiol.* **226,** 1293–1297.
7. Cheng, J. B., L. E. Cornett, A. Goldfein, and J. M. Roberts (1980) Decreased concentration of myocardial α-adrenergic with increasing age in fetal lambs. *Br. J. Pharmacol.* **70,** 515–517.
8. Coraboeuf, E., E. Deroubaix, and A. Coulombe (1979) Effect of tetrodotoxin on action potentials of the conducting system in the dog. *Am. J. Physioll.* **236,** H561–H567.
9. Couch, J. R., T. C. West, and H. E. Hoff (1969) Development of the action potential of the prenatal rat heart. *Circ. Res.* **24,** 19–31.
10. Danilo, P., M. R. Rosen, and A. J. Hordorf (1978) Effects of acetylcholine on the ventricular conducting system on neonatal and adult dogs. *Circ. Res.* **43,** 777–784.
11. Danilo, P., R. Reder, J. Mill, and R. Petrie (1979) Developmental

changes in cellular electrophysiologic characteristics and catecholamine content of fetal hearts. *Circulation* **59,60,** II–50.

12. Danilo, P., R. F. Reder, O. Binah, and M. J. Legato (1984) Fetal canine cardiac Purkinje fibers: electrophysiology and ultra-structure. *Am. J. Physiol.* **246,** H250–H260.

13. DeHaan, R. L., T. F. McDonald, and H. G. Sachs (1975) Development of Embryonic Chick Heart Cells In Vitro, in *Developmental and Physiological Correlates of Cardiac Muscle* (M. Lieberman and T. Sano, eds.) Raven, New York, p. 161.

14. Drugge, E. D., M. R. Rosen, and R. R. Robinson (1985) Neuronal regulation of the cardiac alpha adrenergic chronotropic response. *Circ. Res.* **57,** 415–423.

15. Felder, R. A., P. L. Calcagno, G. M. Eisner, and P. A. Jose (1982) Ontogeny of myocardial adrenoceptors. II. Alpha adrenoceptors. *Ped. Res.* **16,** 340–342.

16. Fozzard, H. A. and S. S. Sheu (1980) Intracellular potassium and sodium activities of chick ventricular muscle during embryonic development. *J. Physiol.* **306,** 579–586.

17. Gennser, G. and E. Nilson (1970) Excitation and impulse conduction in human fetal heart. *Acta Physiol. Scand.* **79,** 305–320.

18. Iijima, T. and A. J. Pappano (1979) Ontogenetic increase in the maximal rate of rise of the chick embryonic heart action potential. Relationship to voltage, time and tetrodotoxin. *Circ. Res.* **44,** 359–367.

19. Ishima, Y. (1968) The effect of tetrodotoxin and sodium substitution on the action potential in the course of development of the embryonic chicken heart. *Proc. Japan. Acad.* **44,** 170–175.

20. Janse, M. J., R. H. Anderson, F. J. L. van Capell, and D. Durrer (1976) A combined electrophysiological and anatomical study of the human heart. *Am. Heart J.* **91,** 556–562.

21. Klein, R. L. (1963) The induction of a transfer adenosine triphosphate phosphohydrolase in embryonic chick heart. *Biochem. Biophy. Acta* **73,** 488–498.

22. Marcus, N. C. and H. Fozzard (1981) Tetrodotoxin sensitivity in the developing and adult chick heart. *J. Mol. Cell. Cardiol.* **13,** 335–340.

23. McDonald, T. F. and R. L. DeHaan (1973) Ion levels and membrane potential in chick heart tissue and cultured cells. *J. Gen. Physiol.* **61,** 89–109.

24. Myerburg, R. J., J. W. Stewart, and B. F. Hoffman (1970) Electrophysiological properties of the canine peripheral AV conducting system. *Circ. Res.* **26,** 361–378.

25. Pappano, A. J. (1972) Action potentials in chick atria. Increased susceptibility to blockade by tetrodotoxin during embryonic development. *Circ. Res.* **31,** 379–338.

26. Pappano, A. J. (1976) Action potentials in chick atria. Ontogenetic

changes in the dependence of tetrodotoxin-resistant action potentials on calcium, strontium and barium. *Circ. Res.* **39,** 99–105.

27. Pappano, A. J. (1977) Ontogenetic development of autonomic neuroeffect or transmission and transmitter reactivity in embryonic and fetal hearts. *Pharmacol. Rev.* **29,** 3–33.

28. Reder, R. F., D. S. Miura, P. Danilo, and M. R. Rosen (1981) The electrophysiological properties of neonatal and adult canine cardiac Purkinje fibers. *Circ. Res.* **48,** 658–668.

29. Reder, R., P. Danilo, and M. R. Rosen (1984) Developmental changes in adrenergic effects on canine Purkinje fiber automaticity. *Dev. Pharmacol. Ther.* **7,** 94–108.

30. Rosen, M. R., J. P. Hordof, J. P. Illvento, and P. Danilo (1977) Effects of adrenergic amines on electrophysiological properties and automaticity of neonatal and adult canine Purkinje fibers. *Circ. Res.* **40,** 390–400.

31. Shigenobu, K. and N. Sperelakis (1971) Development of sensitivity to TTX of chick embryonic hearts with age. *J. Mol. Cell. Cardiol.* **3,** 271–286.

32. Shigenobu, K., J. A. Schneider, and N. Sperelakis (1974) Verapamil blockade of slow Na^+ and Ca^{++} response in myocardial cells. *J. Pharmacol. Exp. Ther.* **190,** 280–288.

33. Shimizu, Y. and K. Tasaki (1966) Electrical excitability of developing cardiac muscle in chick embryos. *Tohoku J. Exp. Med.* **88,** 49–56.

34. Sperelakis, N. and K. Shigenobu (1972) Changes in membrane properties of chick embryonic hearts during development. *J. Gen. Physiol.* **60,** 430–453.

35. Sperelakis, N. (1972) (Na^+, K^+)-ATPase activity of embryonic chick heart and skeletal muscle as a function of age. *Biochem. Biophys. Acta* **266,** 230–237.

36. Tuganowski, W. and A. Cekanski (1971) Electrical activity of a single fiber of a human embryonic heart. *Pflugers Arch.* **323,** 21–26.

37. Untereker, W. J., J. Danilo, and M. R. Rosen (1984) Developmental changes in action potential duration, refractoriness and conduction in the canine ventricular conducting system. *Ped. Res.* **18,** 53–58.

38. Vleugels, A., E. Carmeliet, S. Bosteels, and M. Zaman (1976) Differential effects of hypoxia with age on the chick embryonic heart. *Pflugers Arch.* **365,** 159–166.

39. Wei, J. W. and P. V. Sulakhe (1979) Regional and subcellular distribution of beta and alpha adrenergic receptors in the myocardium of different types. *Gen. Pharmacol.* **10,** 263–267.

40. Yamada, S., H. I. Yamamura, and W. R. Roeske (1980) Ontogeny of mammalian cardiac α_1-adrenergic receptors. *Eur. J. Pharmacol.* **68,** 217–221.

41. Yeh, B. K. and B. F. Hoffman (1968) The ionic basis of electrical activity in embryonic cardiac muscle. *J. Gen. Physiol.* **52,** 666–681.

Chapter 8

Development, Aging, and Plasticity of Perivascular Autonomic Nerves

Tim Cowen and Geoffrey Burnstock

1. Introduction

Neuromuscular relationships in different blood vessels show remarkable heterogeneity. Variability exists between similar blood vessels in different vascular beds, and between the same blood vessels in different species. Many features are involved, including nerve type and density, neuromuscular separation, pre- and postjunctional receptor type and distribution, and transmitter synthesis, degradation, and reuptake systems (for reviews, *see* refs. *7,17,18*).

Such heterogeneity in the adult, which seems to represent adaptations of perivascular nerve–muscle relationships to local physiological requirements, poses several questions regarding the development of these relationships. For instance, when does heterogeneity appear during development? How does aging affect the innervation of blood vessels? Are development and aging similar or separate processes? To what extent do vascular

211

nerve–muscle relationships in early development possess plasticity? And what are the regulatory mechanisms involved? Blood vessels, because of their relatively simple intrinsic nerve plexuses, with few, if any, intramural neurons and discrete localization of the nerves at the adventitial–medial border, provide a particularly suitable model for studying the development and aging of an autonomic neuroeffector system.

There are few studies of the development of perivascular nerves, and of those published the majority consider the development of adrenergic innervation of blood vessels (*31,33, 43,45,49,58,75,78*). Others deal with changes in nerves staining for acetylcholinesterase during development (*14*). Studies of changes in other putative neurotransmitters during development are in their infancy (*17,60,68*). In our own laboratory, quantitative histochemical methods (*22,23,26*) have been used recently to follow the age changes in sympathetic and other perivascular nerves (*25,42,67*). In vitro pharmacology has also been used in some of these studies to monitor changes in neuromuscular activity.

2. Perinatal Development

Aspects of the earliest stages of development of peripheral autonomic nerves that have been examined in recent years include the physiological maturation of neuromuscular reflexes during the perinatal period (*34,43,45,56,73,75,84*) and the structural changes that accompany these developments in perivascular nerves (*33,40,58,65,70,77,78*).

The first stage of the development of the noradrenergic nerve plexus viewed by fluorescence histochemistry consists of the outgrowth of smooth, faintly fluorescent nerves over fetal blood vessels (*25,27*) and into fetal gut (*65*). In rabbit blood vessels, this first phase of development of the sympathetic plexus takes place between 25 d *in utero* and birth (*see* Fig. 1). The development of varicosities takes place later; they are seen first in large blood vessels such as the carotid artery, in which varicosities are already present at 30 d *in utero* (Fig. 1) and later, often during the postnatal period, in smaller blood vessels such as the mesenteric artery (*25*). In rat vessels, fluorescent nerve bundles are visible by the 18th day of gestation, but varicosities do not appear until the postnatal period (*31*).

Two stages have been proposed in the formation of autonomic neuromuscular junctions (*10,13,15*). The first stage is the migra-

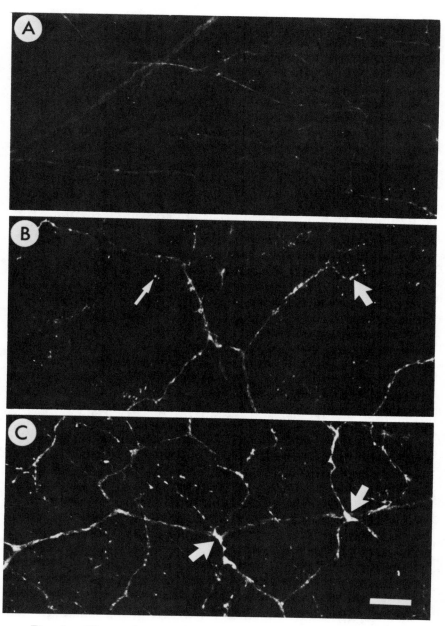

Fig. 1.   Photomicrographs of stretch preparations of the carotid artery at three age stages. Nerves demonstrated by glyoxylic acid fluorescence histochemistry. A, 25 d *in utero*, showing sparse, faintly fluorescent nerves. B, 30 d *in utero*, showing varicose terminal swellings (large arrow) and isolated, brightly fluorescent "varicosities" (small arrow). C, 1 d after birth, note the increase in nerve density, irregular bright areas of nonvaricose axonal fluorescence (large arrows), and the increasing numbers of varicosities. Bar = 25 μm (from ref. *25*).

tion of the neuron to its adult position in the body and the outgrowth of axons toward their target sites. This would match the appearance of smooth, faintly fluorescent nerve bundles between 15 and 25 d *in utero* in the large arteries of the rabbit (*25*). The second stage involves the "recognition" of the effector cells by the nerves and subsequent structural and chemical differentiation of the neuromuscular relationships. In studies of the formation of autonomic neuromuscular junctions in vitro, varicosities matured both in size and vesicular composition during the 24 h following "recognition" (*19,20,59*). A similar recognition process may take place in vivo, marking the transition from the initial period of axon outgrowth to the later process of varicosity formation. This stage occurs during the perinatal period in rabbit arteries (*25*).

In anterior eye-chamber transplants as well, the formation of varicosities depends upon contact with the target organ (*11,62*), although this may represent only the final stage in a gradual process of commitment of each neuron to its particular phenotype, since the earliest stages of neurochemical differentiation in neurons have been shown to depend on interactions with the local environment (*53*).

In developing perivascular plexuses, varicosities were seen only in fine nerve fibers at the perinatal stage (*25*). However, at about the same stage of development, bright patches of fluorescence appeared that could be distinguished from varicosities by their greater size and irregularity (Fig. 1). They were situated mainly in the larger nerve bundles and may therefore indicate the transport of noradrenaline in vesicles from the nerve cell body to the varicosities. Similar observations have been made in the early development of sympathetic nerves in several tissues of the chick (*36*) and rat (*31*). This provides supporting evidence for increased noradrenaline synthesis immediately following "recognition." The onset of noradrenaline synthesis in vivo may also be correlated with the loss of transmitter pluripotentiality shown in tissue culture during the development of neurons, and their dedication to a single neurotransmitter phenotype (*41,63*).

The state of maturation of the perivascular innervation at birth varies among different species. In the rat, much of this development takes place postnatally (*56*), whereas in the rabbit (*30*), cat (*50*), and dog (*43*), development takes place somewhat earlier, during the perinatal period. In the sheep and guinea pig, which are relatively mature at birth, development of the perivascular innervation occurs mainly *in utero* (*26,75*).

Nerve density and varicosity numbers increase dramatically over the perinatal period in rabbit arteries (25) (*see* Figs. 2,3, and compare Figs. 1B,C). From the day before to the day after birth, the nerve plexuses of the carotid and renal arteries approximately double in density. These changes match the physiological demonstration of rapid perinatal maturation of cardiovascular reflexes in the rabbit (34), and the in vitro demonstration in this laboratory of noradrenergic neuromuscular activity in the renal and carotid arteries of the 1-d-old rabbit (S. Griffith, personal observation). The importance of thermoregulation in the neonatal rabbit may provide adaptive pressure for this spurt of growth and maturation in the perivascular nerves of larger arteries in the rabbit that takes place earlier than in some other blood vessels and species (43,58).

Preliminary observations from a study of the early development of sympathetic and substance P-containing perivascular

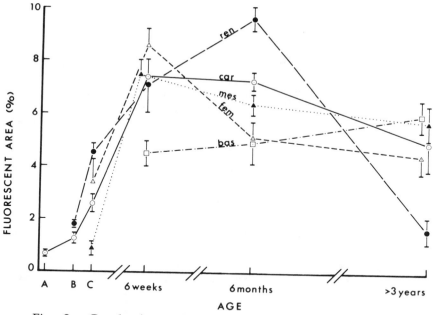

Fig. 2. Graph of age-related changes in fluorescent area (% of total surface area) of perivascular adrenergic nerves in five arteries of the rabbit. Abbreviations: ren, renal; car, carotid; mes, mesenteric; fem, femoral; bas, basilar. Nerve demonstrated by glyoxylic acid fluorescence histochemistry and measured by image analysis (from ref. 25).

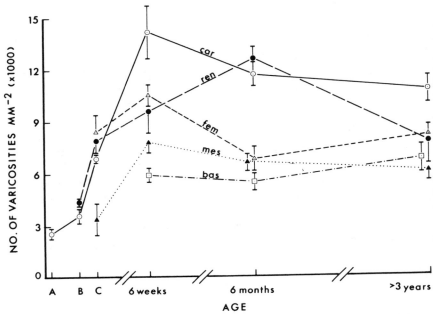

Fig. 3. Graph of age-related changes in numbers of varicosities ($\times 10^{-3}$ mm^{-2} surface area) of perivascular adrenergic nerves in five arteries of the rabbit. Abbreviations: ren, renal; car, carotid; mes, mesenteric; fem, femoral; bas, basilar. Nerve demonstrated by glyoxylic acid fluorescence histochemistry and measured by image analysis (from ref. 25).

nerves in the guinea pig suggest that peptide-containing axons and varicosities are formed in advance of the noradrenergic nerve plexus (Dhall et al., in preparation).

3. Postnatal Development

The first stage of postnatal growth of autonomic nerve plexuses follows a similar pattern of development in many blood vessels (78) and other tissues (31), although some nerve plexuses develop before others. In rabbit arteries we have found large increases during the first 6 wk following birth in the number of nerve fibers and varicosities and in the size of the component nerve bundles of the sympathetic nerve plexuses (Figs. 2,3). These increases occur both in terms of nerve density and the total numbers of nerves supplying a particular blood vessel (77).

In a recent study, the electrophysiological activity and fluorescence histochemistry of the sympathetic innervation of the rat mesentery were compared during early postnatal development (49). Fluorescence histochemistry showed a continuous increase in nerve density during the first 3 wk of postnatal life, whereas development of the electrophysiological responses was discontinuous. Responses were present from d 1 to d 4, but did not resemble those seen in the adult; from d 4, no responses to stimulation could be elicited until d 9, when responses reappeared as excitatory junction potentials similar to those seen in the adult.

During the later postnatal development of perivascular nerves, there is a gradual appearance by a process of divergent development of the remarkably heterogenous patterns of neuromuscular relationships seen in adults. Marked variations of nerve density have been demonstrated in different regions of the same vessel—for example, in the adult guinea pig mesenteric artery (22) and in the rabbit ear artery (46); also, contrasting innervation patterns and pharmacological activity have been shown in the same arteries in different species—for instance, the rabbit and guinea pig renal arteries (42).

The nerve density in many tissues, including many blood vessels, continues to increase in postnatal life (31,77). In some cases, nerve density continues to increase into adult life, for instance in the rabbit renal artery (Fig. 2) (25) and in the rat salivary gland (79). Measurements of the increase in length and circumference of a growing blood vessel (61) have been used in rabbit blood vessels to estimate the total increases in nerve fiber density over the whole blood vessel wall during postnatal development. These calculations show a 20-fold increase in nerve fiber density in the renal and femoral arteries and a 60-fold increase in the mesenteric artery (25).

However, there are a number of examples of blood vessels that exhibit falling nerve density during postnatal development. Doležel (32) showed a reduction of sympathetic nerve density in early postnatal life in the femoral artery of the dog. A further study showed that at the same stage of development, while nerve density was falling in the main trunk of the femoral artery, it increased in the peripheral branches supplying the musculature of the leg (33), suggesting a peripheral shift of vasomotor control. The guinea pig renal artery has a noradrenergic nerve plexus of medium density at birth that declines in density continuously between birth and adulthood (42). Figure 4 shows a comparison of

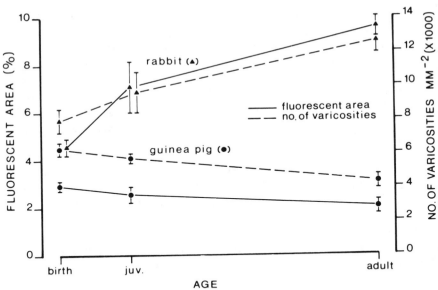

Fig. 4.   Postnatal changes in the density of perivascular adrenergic nerves of the renal arteries of the guinea pig (●) and rabbit (▲). Measurements of fluorescent area (–) and number of varicosities (---) were made using image analysis of stretch preparations treated with glyoxylic acid. Three ages were used; at birth; juvenile (guinea pig, 4 wk; rabbit, 6 wk); and adult (guinea pig, 4 mo; rabbit, 6 mo). Note that during postnatal development, both parameters decrease in the guinea pig and increase in the rabbit (from ref. 42).

the developmental changes in the perivascular sympathetic nerves of the guinea pig and rabbit renal arteries. The nerves of the guinea pig renal artery are doubly unusual in that they fail to respond in vitro to transmural nerve stimulation either at birth or in the adult, whereas the sympathetic nerves of the rabbit renal artery respond strongly to in vitro nerve stimulation (42), as do those of the rat renal artery (57).

Reductions of nerve density are also seen in the tail and saphenous arteries of the rat between about 15 and 30 d after birth (77). This study showed that the total nerve numbers in these blood vessels continued to increase over the period in question, demonstrating that the reductions in nerve density probably involved a spreading out of existing nerves with the growth in surface area of the vessel wall, rather than nerve loss. A substantial reduction in density of nerves and varicosities is also seen in the

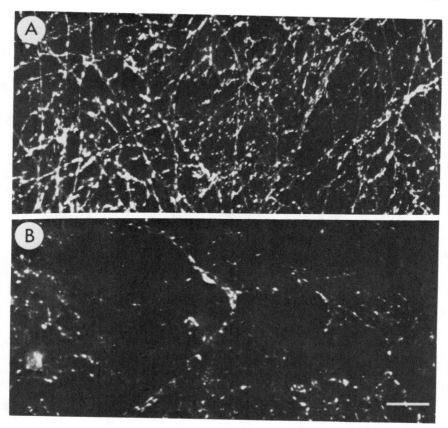

Fig. 5. Photomicrographs of stretch preparations of the femoral artery of the rabbit at two age stages. Nerves demonstrated by glyoxylic acid fluorescence histochemistry. A, 6 wk after birth. B, 6 mo after birth, nerve density and varicosity numbers have fallen significantly, and resemble the neonate. Bar = 25 μm (from ref. *25*).

rabbit femoral artery between 6 wk and 6 mo (Fig. 5), the nerve density at 6 mo resembling that seen at 1 d after birth. The artery continues to grow over this period; therefore the same explanation may apply, i.e., that the nerves are not lost, but become more spread out over the growing vessel wall. Again, a peripheral shift of vasomotor control may be involved.

Heterogeneity is characteristic of perivascular neuromuscular relationships in the adult and involves several features, including the receptors, neuromuscular cleft width, and transmitter synthesis, storage, and release, as well as nerve pattern and density. For example, receptor distribution and type vary between species, and even in particular blood vessels from the same species: α-adrenergic receptors in the cerebral vasculature have been shown to be relatively insensitive to noradrenaline compared with those in other blood vessels, such as the aorta (5,35); the thoracic aorta of the rabbit is responsive to β-adrenoceptor activation, whereas the abdominal aorta is not (39). There are as yet few clues about the developmental changes that give rise to these variations in the adult. However, it has been shown that β-adrenergic receptor activity in the rat and rabbit thoracic aorta declines with age (38,39). Receptors to several putative neurotransmitters in the vasculature have also been characterized: for example, purinoceptors (14,74) and 5-hydroxytryptamine receptors (2,80), but at present there is little knowledge of the changes in their activity or distribution during development.

Neuromuscular cleft width varies considerably in adult blood vessels, with important physiological consequences (7,21). Briefly, the neuromuscular cleft in all adult blood vessels is relatively wide, ranging from 80 to 1000 nm in muscular arteries, to 2 μm or more in elastic arteries (18,21). Response to nerve stimulation in small muscular arteries, in which there is a narrow neuromuscular cleft, is relatively rapid and is terminated partly by neuronal reuptake (uptake 1) of the transmitter and partly by uptake into muscle (uptake 2). In blood vessels with a wider neuromuscular gap, the response to nerve stimulation builds up slowly and is slowly terminated, largely by uptake 2 (5,12). Again, there is little information about developmental changes in these features. However, in cocultures of sympathetic nerves and vascular smooth muscle, neuromuscular gaps as narrow as 10–50 nm have been observed (19), and it may be speculated that the earliest associations in vivo of perivascular nerves and smooth muscle cells are considerably closer than those of the adult.

Changes in transmitter synthesis, storage, release, and breakdown have been shown to occur in the adult and during development, and both aspects have been reviewed in depth (7,9).

Nerve density is generally highest in muscular arteries and arterioles, and lower in elastic arteries, precapillaries, and veins (12). However, as some of the studies cited here have shown, there are numerous exceptions to these generalizations. The pattern of development in blood vessels also seems to be variable.

However, the results of several studies suggest that the perivascular nerves develop first in the larger blood vessels, that these larger vessels also seem to be more prone to postnatal reductions of nerve density, and that postnatal nerve loss in larger arteries may indicate a peripheral shift of vasomotor control to smaller blood vessels with increasing age. This process can also be seen in an evolutionary sense as a gradual adaptation of the nerve plexus in each locality and at each stage of development to local physiological requirements.

4. Old Age

Aging may be seen as part of a continuous developmental sequence of events in which each stage can act as a trigger for the onset of the next stage (37). Tissues vary in their responses to age, and the same tissue in different species may also vary in its responses; thus the cerebral vasculature of Sprague-Dawley rats seems to be different from that of humans in being relatively protected against the types of vascular disease and stroke observed in humans (82,83). Several studies have demonstrated reductions in monoamines with age in sympathetic ganglia (48,69) and in the noradrenergic terminal plexus (44,71), as well as in some areas of the brain (37,66). Reductions of perivascular catecholamine fluorescence with age have been reported in human gingival tissue (81) and temporal artery (12).

Comparisons between young and old adult rabbits show important changes in several areas of the circulation. Although there is a stabilization of the nerve population in the smaller arteries (mesenteric, femoral, and basilar), the larger elastic arteries (renal and carotid) show significant reductions of nerve density in old age, and in the renal artery nerve loss is dramatic, falling to 20% of the adult value (Fig. 6; and *see* Figs. 2,3) (25). It should be noted that the reduction of varicosity density in old age is not so marked (60% of the adult value). These results may be explained in part by the evidence already cited for reduced noradrenaline synthesis in old age. Noradrenaline levels are low in the nonvaricose regions of the nerve fiber and very high in varicosities (18,29). A general reduction of noradrenaline content would render the nonvaricose nerve fibers, in which noradrenaline content is already low, invisible in the fluorescence

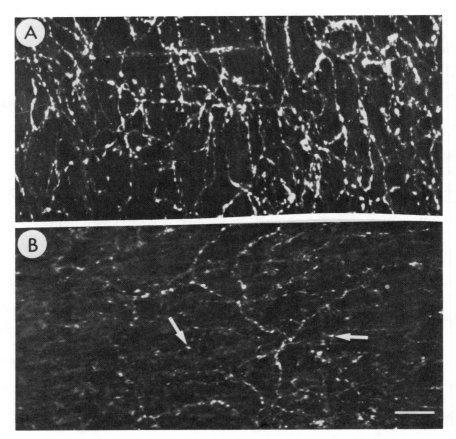

Fig. 6.   Photomicrographs of stretch preparations of the renal artery of the rabbit at two age stages. Nerves demonstrated by glyoxylic
acid fluorescence histochemistry. A, 6 mo, the nerves are dense, with
many varicosities. B, old age, nerves are very sparse and some varicosities appear isolated (small arrows). Background autofluorescence is relatively high. Bar = 25 μm (from ref. 25).

microscope, whereas the varicosities would be relatively unaf-
fected. However, reduced noradrenaline synthesis alone cannot
explain the dramatic reduction of nerve density seen in the renal
artery, which seems likely to involve, in addition, the loss of
nerve fibers and perhaps some neurons.

Substantial reductions of nerve density have also been shown
in old age in the noradrenergic and cholinesterase-positive nerve
populations in rabbit cerebral arteries (67), indicating that not

Table 1
Comparison of Noradrenergic Nerve Densities in the
Left Middle Cerebral (LMC) and Right Middle
Cerebral (RMC) Arteries of Young Adult and
Old Age Rabbits[a]

		Density of noradrenergic nerve bundles, expressed as fluorescent nerves mm^{-1} vessel circumference			
	n	Young adult	p	n	Old age
RMC	9	20 ± 1.9	<0.05	7	14 ± 1.6
LMC	8	18 ± 0.9	<0.001	8	10 ± 1.0
		Density of noradrenergic varicosities, expressed as fluorescent varicosities mm^{-1} vessel circumference			
	n	Young adult	p	n	Old age
RMC	9	27 ± 4.0	<0.05	7	15 ± 2.4
LMC	8	21 ± 2.7	<0.02	8	11 ± 2.0

[a]Nerves were demonstrated using fluorescence histo-chemistry on 6 μm cryostat transverse sections. Nerve densities are compared using 2-tailed Student's t-test (67).

only noradrenergic nerves are affected by old age. Comparison of the left and right middle cerebral arteries showed that the reduction of noradrenergic nerve density tended to be greater in the left than in the right artery (Table 1), whereas the reduction of cholinesterase-positive nerves was similar on both sides of the cerebral circulation (Table 2) (67). However, the total numbers of noradrenergic nerves in the left and right cerebral arteries did not change significantly in old age, suggesting that the reduction of nerve density could be caused by a spreading out of the existing nerve plexuses over the surface of the vessel wall, since the vessel wall continues to grow in surface area over the same period. This redistribution of nerves is similar to that discussed in relation to the rabbit femoral artery in early development, and to the reductions of nerve density coupled with increases in total numbers of nerves observed in postnatal life in rat muscular arteries (77). It seems possible that throughout life the nerve population in each locality is in a mobile equilibrium between nerve loss and new growth, mediated perhaps by a constant competitive demand between adjacent nerves for a target-released growth factor (47,64). Increasing separation of muscle cells by connective tissue in the

Table 2

Comparison of Acetylcholinesterase-Positive Nerve
Densities in the Left Middle Cerebral (LMC) and
Right Middle Cerebral (RMC) Arteries of
Young Adult and Old Age Rabbits[a]

	n	Young adult	p	n	Old age
RMC	11	21 ± 1.5	<0.01	7	14 ± 1.4
LMC	11	21 ± 0.8	<0.001	8	15 ± 1.2

[a]Nerves were demonstrated using acetylcholinesterase histochemistry on 6 μm cryostat transverse sections. Nerve densities are expressed as nerves mm^{-1} of vessel circumference and compared using Student's t-test (67).

aging artery wall (76) could, by this mechanism, lead to a spreading out of the terminal plexus such as we have observed. A small reduction with age in production of a growth factor by the target organ might lead to impairment of noradrenaline synthesis, although a larger reduction might produce degeneration of nerve fibers or even of nerve cell bodies.

5. Plasticity

Plasticity is a well-established feature in the embryonic development of the autonomic nervous system. Migrating neural crest cells, from the level that would normally differentiate into the adrenal medulla, can form cholinergic enteric neurons when transplanted into the appropriate region along the neuraxis (55); conversely, transplanted presumptive enteric neurons can form catecholamine-producing adrenal medullary cells (53,54). Even in adult life, plasticity has been demonstrated. For example, in the sympathetic innervation of the uterine blood vessels during late pregnancy, there is a switch from adrenergic vasoconstrictor control to cholinergic vasodilatation. This is accompanied by a reduction of noradrenaline levels and redistribution of staining for acetylcholinesterase associated with a 100-fold increased sensitivity to acetylcholine (3,4). Impairment of the noradrenaline-uptake mechanism (1) and ultrastructural evidence of axon degeneration (72) have also been demonstrated in the sympathetic innervation of the uterus wall of the pregnant guinea pig. Another example of plasticity in the adult is the increase in density of the adrenergic innervation to the dorsal penile vein accompanying reduction of vasoactive intestinal polypeptide (VIP) innervation of penile vessels during early streptozotocin-induced diabetes (*28*).

Studies in this laboratory have also demonstrated plasticity in peripheral sympathetic nerves. Chronic stimulation of the guinea pig hypogastric nerves in vivo has demonstrated an increase in neuronal noradrenaline reuptake associated with a reduced neuromuscular gap in the vas deferens (*51*). In addition, chronic stimulation of the hypogastric nerve has been shown to produce much faster reinnervation of a minced smooth muscle implant in the guinea pig vas deferens, compared with nonstimulated controls (*52*). Further evidence for the trophic effect of sympathetic nerves on the smooth muscle effector organ has been provided by investigators who have shown structural and functional changes in the muscle walls of blood vessels after chronic sympathectomy (*6,8*).

Crush-lesions of the nerves supplying the mesenteric and carotid arteries of the guinea pig were followed by a substantially faster regrowth rate of the mesenteric nerves (*24*), suggesting locally specific levels of neurotrophic activity. These experiments also indicated that the neurons supplying the carotid artery could expand their field of innervation when some nerves were prevented by scar tissue from regrowing into their target area.

In conclusion, the long-term regulation of autonomic neuromuscular relationships seems to involve more than one trophic system: continuity of nerve supply and nervous activity has been shown to be important in adult life, as have some hormonal systems. In early development, the muscle target organ may also play an important role, partly by the production of growth factors such as NGF, to "attract" nerves from a distance, and partly by selective "recognition" systems in effector cells to help determine the density and pattern of innervation of a particular organ.

6. Conclusions

There are still remarkably few studies of the development of perivascular nerves and neuromuscular relationships, and those reports that are available deal mainly with the ontogeny of the noradrenergic system. Future studies will need to investigate, in addition, the development of the newly discovered perivascular nerves, such as those containing 5-hydroxytryptamine, and the peptides substance P and VIP, as well as the possibility of trophic interactions during development between these different groups of nerves and their effector organs.

References

1. Alm, P., C. Owman, N-O. Sjöberg, and G. Thorbert (1979) Uptake and metabolism of [^{3}H]norepinephrine in uterine nerves of pregnant guinea pig. *Am. J. Physiol.* **236,** C277–C285.
2. Apperley, E., W. Feniuk, P. P. A. Humphrey, and G. P. Levy (1980) Evidence for two types of excitatory receptor for 5-hydroxytryptamine in dog isolated vasculature. *Br. J. Pharmacol.* **68,** 215–224.
3. Bell, C. (1968) Dual vasoconstrictor and vasodilator innervation of the uterine arterial supply in the guinea pig. *Circ. Res.* **23,** 279–289.
4. Bell, C. (1969) Fine structural localization of acetylcholinesterase at a cholinergic vasodilator nerve–arterial smooth muscle synapse. *Circ. Res.* **24,** 61–70.
5. Bevan, J. A. (1977) Some functional consequences of variation in adrenergic synaptic cleft width and in nerve density and distribution. *Fed. Proc.* **36,** 2439–2443.
6. Bevan, R. D. and H. Tsuru (1979) Long term denervation of vascular smooth muscle causes not only functional but structural change. *Blood Vessels* **16,** 109–112.
7. Bevan, J. A., R. D. Bevan, and S. P. Duckles (1980) Adrenergic Regulation of Vascular Smooth Muscle, in *Handbook of Physiology,* Section 2, The Cardiovascular System; Vol. II, Vascular Smooth Muscle (D. F. Bohr, A. P. Somlyo, and H. V. Sparks, eds.) American Physiological Society, Maryland, pp. 515–566.
8. Bevan, R. D., H. Tsuru, and J. A. Bevan (1983) Cerebral artery mass in the rabbit is reduced by chronic sympathetic denervation. *Stroke* **14,** 393–396.
9. Black, I. B. (1978) Regulation of autonomic development. *Ann. Rev. Neurosci.* **1,** 183–214.
10. Bunge, R., M. Johnson, and D. C. Ross (1978) Nature and nurture in development of the autonomic neuron. *Science* **199,** 1409–1416.
11. Burnstock, G. (1974) Degeneration and Oriented Growth of Autonomic Nerves in Relation to Smooth Muscle in Joint Tissue Cultures and Anterior Eye Chamber Transplants, in *Dynamics of Degeneration and Growth in Neurons* (K. Fuxe, L. Olson, and Y. Zotterman, eds.) Pergamon Press, Oxford and New York, pp. 509–519.
12. Burnstock, G. (1975) Innervation of vascular smooth muscle: histochemistry and electron microscopy. *Clin. Exp. Pharacol. Physiol.,* Suppl. **2,** 7–20.
13. Burnstock, G. (1978) Do Some Sympathetic Neurons Synthesize and Release Both Noradrenaline and Acetylcholine?, in *Progress in Neurobiology,* Vol. II, Pergamon Press, United Kingdom, pp. 205–222.

14. Burnstock, G. (1980) Cholinergic and Purinergic Regulation of Blood Vessels, in *Handbook of Physiology;* Section 2, The Cardiovascular System (D. F. Bohr, A. P. Somlyo, H. W. Sparks, and S. R. Geiger, eds.) American Physiological Society, Waverley Press, Baltimore, pp. 567–612.

15. Burnstock, G. (1981) Development of Smooth Muscle and Its Innervation, in *Smooth Muscle: An Assessment of Current Knowledge* (E. Bülbring, A. F. Brading, A. W. Jones, and T. Tomita, eds.) Edward Arnold, London.

16. Burnstock, G. and M. Costa (1975) *Adrenergic Neurons.* Chapman and Hall, London.

17. Burnstock, G. and S. G. Griffith (1983) Neurohumoral Control of the Vasculature, in *The Biology and Pathology of the Vessel Wall: A Modern Appraisal* (N. Woolf, ed.) Praeger, Eastbourne, pp. 15–40.

18. Burnstock, G., J. H. Chamley, and G. R. Campbell (1980) The Innervation of Arteries, in *Structure and Function of the Circulation,* Vol. I (C. J. Schwartz, N. T. Werthessen, and S. Wolf, eds.) Plenum Press, New York and London, pp. 729–767.

19. Chamley, J. H. and G. R. Campbell (1975) Trophic influences of sympathetic nerves and cyclic AMP on differentiation and proliferation of isolated smooth muscle cells in culture. *Cell Tiss. Res.* **161,** 497–510.

20. Chamley, J. H., G. R. Campbell, and G. Burnstock (1973) An analysis of the interactions between sympathetic nerve fibers and smooth muscle cells in tissue culture. *Develop. Biol.* **33,** 344–361.

21. Cowen, T. (1984) Functional and nonfunctional vascular neuromuscular transmission, a comparison of neuromuscular relationships and noradrenaline content in the renal, carotid, and mesenteric arteries of the guinea pig. *J. Neurocytol.* **13,** 369–392.

22. Cowen, T. and G. Burnstock (1980) Quantitative analysis of the density and pattern of adrenergic innervation of blood vessels. *Histochemistry* **66,** 19–34.

23. Cowen, T. and G. Burnstock (1982) Image analysis of catecholamine fluorescence. *Brain Res. Bull.* **9,** 81–87.

24. Cowen, T., D. E. M. MacCormick, W. D. Toff, G. Burnstock, and J. S. P. Lumley (1982b) The effect of surgical procedures on blood vessel innervation. A fluorescence histochemical study of degeneration and regrowth of perivascular adrenergic nerves. *Blood Vessels* **19,** 65–78.

25. Cowen, T., A. J. Haven, C. Wen-Qin, D. D. Gallen, F. Franc, and G. Burnstock (1982a) Development and aging of perivascular adrenergic nerves in the rabbit. A quantitative fluorescence histochemical study using image analysis. *J. Auton. Nerv. Syst.* **5,** 317–336.

26. Cowen, T., A. J. Haven, and G. Burnstock (1985) Image analysis of catecholamine fluorescence and immunofluorescence in studies

on blood vessel innervation, in Wenner Gren International Symposium *Quantitative Neuroanatomy in Transmitter Research,* in press.

27. Crowe, R. and G. Burnstock (1982) Small intensely fluorescent (SIF) cells and sympathetic nerves in the adult rabbit portal vein and during perinatal development. *Cell Tissue Res.* **227,** 601–607.

28. Crowe, R., J. Lincoln, P. F. Blacklay, J. P. Pryor, J. S. P. Lumley, and G. Burnstock, (1983) Vasoactive intestinal polypeptide-like immunoreactive nerves in diabetic penis: A comparison between streptozotocin-treated rats and man. *Diabetes* **32,** 1075–1077.

29. Dahlström, A., J. Häggendal, and T. Hökfelt (1966) The noradrenaline content of the varicosities of sympathetic adrenergic nerve terminals in the rat. *Acta Physiol. Scand.* **67,** 289–294.

30. Dawes, G. S., J. J. Handler, and J. C. Mott (1957) Some cardiovascular responses in fetal, newborn, and adult rabbits. *J. Physiol.* (Lond.) **139,** 123–136.

31. De Champlain, J., T. Malmfors, L. Olson, and C. Sachs (1970) Ontogenesis of peripheral adrenergic neurons in the rat: pre- and postnatal observations. *Acta Physiol. Scand.* **80,** 276–288.

32. Doležel, S. (1973) Über die Variabilität der adrenergen Innervation der grossen Gefässe. *Acta Anat.* **85,** 123–132.

33. Doležel, S., M. Gerova, and J. Gero (1974) Postnatal development of the sympathetic innervation in skeletal muscles of the dog. *Physiol. Bohemoslov.* **23,** 138–139.

34. Downing, S. E. (1960) Baroreceptor reflexes in newborn rabbit. *J. Physiol. (Lond.)* **150,** 201–213.

35. Edvinsson, L. and C. Owman (1974) Pharmacological characterization of adrenergic alpha- and beta-receptors mediating the vasomotor responses of cerebral arteries in vitro. *Circ. Res.* **35,** 835–849.

36. Enemar, A., B. Falck, and R. Håkanson (1965) Observations on the appearance of norepinephrine in the sympathetic nervous system of the chick embryo. *Acta Physiol. Scand.* **72,** 15–24.

37. Finch, C. E. (1973) Monoamine Metabolism in the Aging Male Mouse, in *Development and Aging in the Nervous System* (M. Rockstein, ed.) Academic Press, London, pp. 199–218.

38. Fleisch, J. H., H. M. Maling, and B. B. Brodie (1971) Further studies on the effect of aging on β-adrenoceptor activity of rat aorta. *Brit. J. Pharmacol.* **42,** 311–313.

39. Fleisch, J. H., H. M. Maling, and B. B. Brodie (1970) Beta-receptor activity in aorta: variation with age and species. *Circ. Res.* **26,** 151–162.

40. Friedman, W. J., P. E. Pool, D. Jacobowitz, S. C. Seagren, and E. Braunwald (1968) Sympathetic innervation of the developing rabbit heart. *Circ. Res.* **23,** 25–32.

41. Furshpan, E. J., P. R. MacLeish, P. H. O'Lague, and D. D. Potter (1976) Chemical transmission between rat sympathetic neurons and cardiac myocytes developing in microcultures: evidence for

cholinergic, adrenergic, and dual-function neurons. *Proc. Natl. Acad. Sci. USA* **73,** 4225–4229.

42. Gallen, D. D., T. Cowen, S. G. Griffith, A. J. Haven, and G. Burnstock (1982) Functional and nonfunctional perivascular nerve–smooth muscle transmission in the renal arteries of the rabbit and guinea pig: a developmental study. *Blood Vessels* **19,** 237–246.

43. Gauthier, P., R. A. Nadeau, and J. De Champlain (1975) The development of sympathetic innervation and the functional state of the cardiovascular system in newborn dogs. *Can. J. Physiol. Pharmacol.* **53,** 763–776.

44. Gey, K. F., W. P. Burkard, and A. Pletscher (1965) Variations of the norepinephrine metabolism of the rat heart with age. *Gerontologia* **11,** 1–11.

45. Gootman, P. M., N. Gootman, P. D. M. V. Turlapaty, A. D. Yao, B. J. Buckley, and B. M. Altura (1981) Autonomic Regulation of Cardiovascular Function in Neonates, in *Development of the Autonomic Nervous System* (Ciba Foundation Symposium 83), Pitman Medical, London, pp. 70–93.

46. Griffith, S. G., R. Crowe, J. Lincoln, A. J. Haven, and G. Burnstock (1982) Regional differences in the density of perivascular nerves and varicosities, noradrenaline content, and responses to nerve stimulation in the rabbit ear artery. *Blood Vessels* **19,** 41–52.

47. Hendry, I. A. and C. E. Hill (1980) Retrograde axonal transport of target tissue-derived macromolecules. *Nature* **287,** 647–649.

48. Hervonen, A., A. Vaalasti, M. Partanen, L. Kanerva, and H. Hervonen (1978) Effects of aging on the histochemically demonstrable catecholamines and acetylcholinesterase of human sympathetic ganglia. *J. Neurocytol.* **7,** 11–23.

49. Hill, C. E., G. D. S. Hirst, and D. F. Van Helden (1983) Development of sympathetic innervation to proximal and distal arteries of the rat mesentery. *J. Physiol.* **338,** 129–147.

50. Hutchinson, E. A., C. J. Percival, and I. M. Young (1962) Development of cardiovascular response in the kitten. *Q. J. Exp. Physiol.* **47,** 201–210.

51. Jones, R., M. Dennison, and G. Burnstock (1983a) The effect of decentralization or chronic hypogastric nerve stimulation in vivo on the innervation and responses of the guinea pig vas deferens. *Cell Tissue Res.* **232,** 265–279.

52. Jones, R., R. Yokota, and G. Burnstock (1983b) The long-term influence of decentralization or preganglionic hypogastric nerve stimulation on the reinnervation of minced vas deferens in the guinea pig. *Cell Tissue Res.* **232,** 281–293.

53. Le Douarin, N. (1981) Plasticity in the Development of the Peripheral Nervous System, in *Development of the Autonomic Nervous Sys-*

tem (Ciba Foundation Symposium 83), Pitman Medical, London, pp. 19–50.

54. Le Douarin, N. and M. A. Teillet (1974) Experimental analysis of the migration and differentiation of neuroblasts of the autonomic nervous system and of neurectodermal mesenchymal derivatives, using a biological cell marking technique. *Develop. Biol.* **41,** 162–184.

55. Le Douarin, N. M., D. Renaud, M. A. Teillet, and G. H. Le Douarin (1975) Cholinergic differentiation of presumptive adrenergic neuroblasts in interspecific chimeras after heterotopic transplantations. *Proc. Natl. Acad. Sci. USA* **72,** 728–732.

56. Ljung, B. and D. Stage (McMurphy) (1975) Postnatal ontogenetic development of neurogenic and myogenic control in the rat portal vein. *Acta Physiol. Scand.* **94,** 112–127.

57. Lorez, H. P., H. Kuhn, and J. P. Tranzer (1973) The adrenergic innervation of the renal artery and vein of the rat. *Z. Zellforsch.* **138,** 261–272.

58. Lundberg, J., B. Ljung, D. Stage, and A. Dahlström (1976) Postnatal ontogenetic development of the adrenergic innervation pattern in rat portal vein: a histochemical study. *Cell Tissue Res.* **172,** 15–27.

59. Mark, G., J. Chamley, and G. Burnstock (1973) Interactions between autonomic nerves and smooth and cardiac muscle cells in tissue culture. *Develop. Biol.* **32,** 194–200.

60. McGregor, G. P., P. L. Woodhams, D. J. O'Shaughnessy, M. A. Ghatei, J. M. Polak, and S. R. Bloom (1982) Developmental changes in bombesin, substance P, somatostatin, and vasoactive intestinal polypeptide in the rat brain. *Neurosci. Lett.* **28,** 21–27.

61. Olivetti, G., P. Anversa, M. Melissari, and A. V. Loud (1980) Morphometric study of early postnatal development of the thoracic aorta in the rat. *Circ. Res.* **47,** 417–424.

62. Olson, L., H. Björklund, T. Ebendal, K-O. Hedlund, and B. Hoffer (1981) Factors Regulating Growth of Catecholamine-Containing Nerves, as Revealed by Transplantation and Explantation Studies, in *Development of the Autonomic Nervous System* (Ciba Foundation Symposium 83), Pitman Medical, London, pp. 213–231.

63. Potter, D. D., S. C. Landis, and E. F. Furshpan (1980) Dual function during development of rat sympathetic neurons in culture. *J. Exp. Biol.* **89,** 57–72.

64. Purves, D. (1980) Neuronal competition. *Nature* **287,** 585–586.

65. Read, J. B. and G. Burnstock (1970) Development of the adrenergic innervation and chromaffin cells in the human fetal gut. *Develop. Biol.* **22,** 513–534.

66. Robinson, D. S., A. Nies, J. M. Davis, W. E. Bunney, R. W. Coburn, H. R. Bowrne, D. M. Shaw, and A. J. Coppen (1972)

Aging, monoamines, and monoamine oxidase levels. *Lancet* **1**, 290–291.

67. Saba, H., T. Cowen, A. J. Haven, and G. Burnstock (1984) Reduction in noradrenergic perivascular nerve density in the left and right cerebral arteries of old rabbits. *J. Cerebral Blood Flow* **4**, 284–289.

68. Sakanaka, M., S. Inagaki, S. Shiosaka, E. Senba, H. Takagi, K. Takatsuki, Y. Kawai, H. Iida, Y. Hara, and M. Tohyama (1982) Ontogeny of substance P-containing neuron system of the rat: immunohistochemical analysis II. Lower brain stem. *Neuroscience* **7**, 1097–1126.

69. Santer, R. M. (1979) Fluorescence histochemical evidence for decreased noradrenaline synthesis in sympathetic neurons of aged rats. *Neurosci. Lett.* **15**, 177–180.

70. Schweiller, G. H., J. S. Douglas, and A. Bouhuys (1970) Postnatal development of autonomic efferent innervation in the rabbit. *Am. J. Physiol.* **219**, 391–397.

71. Shibata, S., K. Hattori, I. Sakurai, J. Mori, and M. Fujiwara (1971) Adrenergic innervation and cocaine-induced potentiation of adrenergic responses of aortic strips from young and old rabbits. *J. Pharmacol. Exp. Ther.* **177**, 621–632.

72. Sporrong, B., P. Alm, C. Owman, N-O. Sjöberg, and G. Thorbert (1981) Pregnancy is associated with extensive adrenergic nerve degeneration in the uterus. An electronmicroscopic study in the guinea pig. *Neuroscience* **6**, 1119–1126.

73. Stage, D. and B. Ljung (1978) Neuroeffector maturity of portal veins from newborn rats, rabbits, cats, and guinea pigs. *Acta Physiol. Scand.* **102**, 218–223.

74. Su, C. (1981) Purinergic Receptors in Blood Vessels, in *Purinergic Receptors, Receptors and Recognition,* Series B; Vol. 12 (G. Burnstock, ed.) Chapman and Hall, London and New York, pp. 93–117.

75. Su, C., J. A. Bevan, N. S. Assali, and C. R. Brinkman (1977) Development of neuroeffector mechanisms in the carotid artery cf the fetal lamb. *Blood Vessels* **14**, 12–24.

76. Toda, T., N. Tsuda, I. Nishimori, D. E. Leszczynski, and F. A. Kummerow (1980) Morphometrical analysis of the aging process in human arteries and aorta. *Acta Anat.* **106**, 35–44.

77. Todd, M. E. (1980) Development of adrenergic innervation in rat peripheral vessels: a fluorescence microscopic study. *J. Anat.* **131**, 121–133.

78. Todd, M. E. and M. K. Tokito (1981) An ultrastructural investigation of developing vasomotor innervation in rat peripheral vessels. *Am. J. Anat.* **160**, 195–212.

79. Turner, C. J. (1981) Growth of sympathetic nerve terminals in the adult rat. *Brain Res.* **207**, 449–452.

80. Vanhoutte, P. M. (1978) Heterogeneity in Vascular Smooth

Muscle, in *Microcirculation*, Vol. II, (G. Kaley and B. M. Altura, eds.) University Park Press, Baltimore, pp. 181–309.

81. Waterson, J. G., D. B. Frewin, and J. S. Soltys. Age-related differences in catecholamine fluorescence of human vascular tissue. *Blood Vessels* **11**, 79–85.

82. Wexler, B. C. and C. W. True (1963) Carotid and cerebral arteriosclerosis in the rat. *Circ. Res.* **12**, 659–666.

83. Wexler, B. C. and J. Saroff (1970) Metabolic changes in response to acute cerebral ischaemia following unilateral carotid artery ligation in arteriosclerotic versus nonarteriosclerotic rats. *Stroke* **1**, 38–51.

84. Wyse, D. G., G. R. Van Petten, and W. H. Harris (1977) Responses to electrical stimulation, noradrenaline, serotonin, and vasopressin in the isolated ear artery of the developing lamb and ewe. *Can. J. Physiol. Pharmacol.* **55**, 1001–1006.

Chapter 9

Regulation of Regional Vascular Beds by the Developing Autonomic Nervous System

Nancy M. Buckley

1. Introduction

Recent review articles have been devoted to sequential events in the formation and maintenance of a functioning vascular neuroeffector junction (15) and to possible stages of neurotransmitter development in autonomic neurons (5). Evidence is accumulating that concurrent development of the innervation and vasculature not only is determined by the embryonic program, but that innervation itself determines the degree of vessel reactivity that characterizes each vessel type (4). Although these provocative approaches are being pursued, this chapter is presented to the reader as a further discussion of phenomenological observations on the development of some of the regional circulations. The femoral, renal, and mesenteric circulations have been selected for emphasis because they supply blood to organs with very different physiologic functions and requirements for blood flow regulation. The focus of this review chapter is the synchrony of maturation of these three vascular beds. Rather than provide an overview of concurrent development of vasoconstrictor and vasodilator properties (7), this chapter covers the maturation of regional circulatory control by autonomic neural stimuli supplied directly (10) or reflexively (22). Information is presented on the development of innervation, responses to nerve fiber stimulation in vitro and in vivo, and participation in autonomic reflexes.

2. Innervation of Blood Vessels

The maturation of vascular innervation in mammals has been studied with the aid of electron microscopy and histochemistry (2,16,19–21,25,30,32,35,39,43,44). Adrenergic fibers are identified by fluorescence and cholinergic fibers by cholinesterase methodologies (14) (*see* chapters by Slotkin and Giacobini). Although a single vessel type has been examined in many of these studies (19–21,32,39,43), some investigators have compared the developing innervation of vessels from different circulations of the same animal (16,35), or arteries and veins from the same vascular bed (2,44).

Femoral blood vessels become innervated at different ages during the growth of different mammals. Human fetal femoral arteries show specific adrenergic fluorescence by the 16th gestational wk, and veins after the 19th wk (2). The nerve fibers are visible in the adventitial layer of the vessels before smooth muscle cells of the media are completely differentiated. In rabbits, femoral artery innervation is also detectable in late gestation (16); it increases during six postnatal weeks and then the adrenergic fiber density diminishes. In the rat, innervation of the femoral and saphenous arteries does not begin until at least 2 d after birth (43,44); the appearance and increase of fluorescence parallels the appearance of dense-core storage vesicles. The pattern of innervation of these two arteries develops at a different rate over a 30-d period and the femoral vein eventually develops more transmitter stores than does the femoral artery (43). A delay in the innervation of the canine femoral artery also has been reported (21); fluorescence appears only at 1 wk after birth, at which time the adrenergic fibers can accumulate tritiated norepinephrine. As in the rabbit, the fiber density decreases late in postnatal development. In none of these studies were cholinergic fibers identified.

Renal arteries in developing rats exhibit specific adrenergic fluorescence by the 22nd day of gestation (35). The density of the perivascular fibers in the renal cortex increases progressively to term and attains a fully developed pattern by the second postnatal week. In rabbits and guinea pigs, there is considerable fluorescence of periarterial fibers at birth (16,19), at which time the vascular smooth muscle in rabbit renal arteries is well developed (16). The fiber density and the number of varicosities continue to increase in rabbit vessels during the subsequent 6 mo (16, 19), but decrease in guinea pig vessels by the 6th postnatal week (19). In puppies, a postnatal increase in adrenergic fiber density occurs as renal catecholamine content increases (22). However, there is an important intrarenal difference in that the fibers are present in the cortex at birth, but appear in the medulla only 21 d later.

The intestinal circulation receives much of its blood from the mesenteric arteries and drains into the portal vein. Both types of vessel have been examined with respect to development of innervation. Rat intestinal vessels generally exhibit fluorescence by the 20th gestational day (35) and mature toward an adult pattern thereafter. However, there are some differences between the maturation rates for innervation of the ileal and jejunal arteries. There are no adrenergic fibers at birth in rat ileal arteries (25), but the innervation pattern resembles that of the adult by 2 wk after birth, and is attained faster in the distal branches than in the proximal vessel. At 2 wk after birth, the rat jejunal artery is already well innervated with granulated nonmyelinated fibers (39); fiber density increases throughout 12 wk after birth. The rat portal vein contains a plexus of adrenergic fiibers by the first postnatal week (30) and exhibits an adult pattern of innervation by the third week. In the rabbit, the mesenteric artery branch to the ileum contains fluorescing fibers at birth (16); maximum fiber density and number of varicosities is reached by 6 wk after birth. Human infant postmortem specimens as young as 2 wk after birth include mesenteric arteries containing fluorescing fibers and varicosities (20).

The model study comparing the development of perivascular adrenergic fibers in femoral, renal, mesenteric, and other arteries employed a quantitative fluorescence histochemical method with image analysis (16). Three stages in the general sequence of innervation were identified: (a) outgrowth of new axons along the vessel wall; (b) appearance of intensely fluorescent areas; and (c) rapid longitudinal growth with differentiation of terminal varicosities. The main differences among the arteries studied (from rabbits) were: (a) innervation appeared earlier in femoral and renal arteries than in mesenteric; (b) differentiation of vascular smooth muscle was earliest in renal arteries; (c) fiber density and number of varicosities were greatest in the renal artery within 6 mo after birth; and (d) innervation density actually decreased in the femoral artery some time after the sixth postnatal week. In spite of the species differences indicated in the preceding paragraphs of this section, such a model study has useful implications for the interpretation of the functional responses of the three regional vascular beds.

3. Responses of Isolated Blood Vessels to Autonomic Neurotransmitters

A commonly employed test of functional capacity of developing innervation is the in vitro response of isolated blood vessels to transmural electrical stimulation (3). This procedure elicits neurotransmitter release from nerve fiber terminals in the vessel wall. Unfortunately, only a few studies of this type have been carried out in developing vessels (19,28,41,42,48). One approach has been to select a major vessel as a prototype for investigation of both structural and functional aspects of its developing innervation (19). Another approach has been to compare responses of a particular vessel type from various mammalian species (19,28,41).

Since transmural stimulation of an isolated vessel produces a response that depends not only on functional capacity of the nerve fibers, but also on responsiveness of the vascular smooth muscle layer, exogenous administration of neurotransmitters has been utilized to test the reactivity of the muscle component of the response (7). In a few cases, both methods were applied to the same vessel (19,28,41,42) to provide differentiation between the two steps in the response.

A model study comparing the development of structure and functional responses of a prototype vessel was carried out on isolated renal arteries from newborn and adult rabbits and guinea pigs (28). At this juncture, only the rabbit artery experiments will be presented since species differences will be discussed below. Perivascular adrenergic nerve fibers with varicosities were found at birth, and both the density of fibers and number of varicosities increased until adulthood. In newborn arteries, the threshold frequency for transmural stimulation was ten times greater, and the slope of the frequency/response relationship was less steep, than in adult arteries. At low concentrations of exogenous norepinephrine, the newborn arteries were significantly less sensitive to norepinephrine than were the adult arteries. Thus, both neural release and vascular smooth muscle response were present at birth in an immature form.

Another model study, although less in vitro, related structural and functional aspects of the developing innervation in a mesenteric arterial tree attached to an isolated segment of rat ileum (25). Transmembrane potentials recorded from vascular smooth muscle of proximal and distal vessels in the tree showed no significant difference in resting membrane potential after the second postnatal day. Vasoconstriction was observed under the microscope from postnatal day 1 through 21 whenever perivascular nerves were stimulated with 300-ms pulses or the preparations were superfused with norepinephrine solution. Single neural stimuli of 0.5 ms duration and varying strengths initiated slow membrane depolarization potentials in preparations from 1–8-d-old animals, and excitatory junction potentials in preparations from animals older than 10 d. Fluorescence histochemical examination of the preparations revealed the presence of structures normally associated with noradrenergic transmission by the third postnatal day, but a mature pattern of innervation was present only after the ninth day.

One of the problems with extrapolating the observations from these model experiments to other mammalian species is the convincing evidence that functional innervation is present at different stages during postnatal development of regional circulations in different mammals. This has been reported in connection with studies on the renal artery (19) and the portal vein (28,41).

4. Responses of Regional Circulations to Autonomic Stimuli

4.1. Exogenous Administration of Neurotransmitters

Although neurotransmitters are delivered to blood vessels at nerve terminals within the adventitial layer, their intravenous or intra-arterial administration is a test of the in vivo responsiveness of the vascular smooth muscle layer. Single dose-response studies have been designed to elicit differences in threshold for vascular effects of circulating catecholamines (7). Infusion studies have been carried out to evaluate vascular effects of a continuous sympathomimetic influence (6,12,24,29,49).

The hindlimb circulation in the premature lamb constricts not only in response to norepinephrine, but also to tyramine (49), indicating that the norepinephrine release mechanism is functional at a very early age. In a variety of neonatal mammals, systemic norepinephrine infusion leads to a decrease in femoral blood flow and a significant increase in femoral vascular resistance (22). Controlled perfusion of the hindlimb *in situ* allows direct observation of constrictor responses in terms of increased perfusion pressure at constant flow. Applying this procedure in puppies (6) or swine (10) revealed vasoconstriction to intra-arterial norepinephrine on the day of birth. In the swine, the femoral arterial pressure/flow relationship was determined at different ages before and during intra-arterial infusion of norepinephrine (0.2 μg/kg/min) or acetylcholine (4–12 μg/min). At all ages from day of birth to 1 mo, neurotransmitter infusions in the denervated hindlimb *in situ* shifted the pressure/flow relationship in the direction of increased femoral vascular resistance (norepinephrine experiments) or decreased resistance (acetylcholine experiments).

The renal circulation in premature lamb fetuses constricts in response to norepinephrine, but not to tyramine (49), implying that the norepinephrine release mechanism is nonfunctional. In neonatal mammals, systemic norepinephrine infusion leads to a decrease in renal blood flow and a significant increase in renal vascular resistance (22). Furthermore, postnatal development in swine leads to maturation of mechanisms for autoregulatory escape from this vasoconstrictor influence (12). Controlled perfusion of the left kidney *in situ* in swine (10) revealed vasoconstriction to intra-arterial norepinephrine on the day of birth. When the renal arterial pressure/flow relationship was determined in swine of different ages, before and during intra-arterial infusion of norepinephrine (0.1 μg/kg/min) in the denervated kidney *in situ*, the relationship was shifted in the direction of increased renal vascular resistance.

The intestinal circulation of term fetal lambs is responsive to norepinephrine infusion, leading to redistribution of cardiac output to the coronary circulation in particular (29); the decrease in mesenteric blood flow was maintained during at least 20 min of infusion. Few such studies have been conducted in neonatal mammals except for our ongoing experiments in developing swine. Figure 1 illustrates that the mesenteric vasculature can escape the influence of continuous systemic infusion of norepinephrine in this species at birth. Similar results are obtained when the neurotransmitter is infused directly into the superior mesenteric artery.

At present, our ongoing experiments may be a model for evaluating this type of autoregulatory escape of blood flow from vasoconstrictor effect of norepinephrine. The preliminary results suggest that it occurs in the mesenteric circulation at an earlier age than in the renal or femoral circulations. The fact that the slope of the pressure/flow relationship in the perfused femoral or renal circulation was retained unaltered during norepinephrine infusion

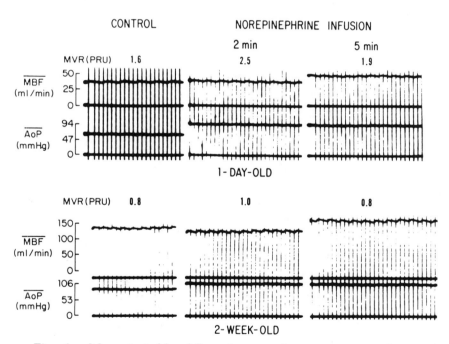

Fig. 1. Mesenteric blood flow decrease then return toward control level during continuous systemic infusion of norepinephrine (0.5 μg/kg/min) in a 1-d-old and 2-wk-old swine. Note maintenance of elevated aortic pressure throughout 2 and 5 min of infusion period in each animal. AoP, mean aortic pressure; MBF, mean blood flow in superior mesenteric artery; MVR, mesenteric vascular resistance. Original oscilloscopic recordings of mean pressure and mean flow at paper speed of 25 mm/s.

(*10*) indicates that blood flow autoregulation to changing perfusion pressure in these two beds had not been affected by the excess neurotransmitter. The possible role of local metabolic factors in these escape mechanisms will not be discussed in this chapter because there are so few studies on the development of such mechanisms.

4.2. Direct Stimulation of the Efferent Nerve Supply

With the exception of a few studies on the hindlimb circulation (*6, 21*), direct stimulation of the efferent nerve supply to a regional circulation has been carried out only in our laboratory (*8–11*). The model experiment is designed to determine the effects of nerve section and stimulation at various frequencies on regional vascular resistance in swine from birth through 1 mo of age. All animals were anesthetized with sodium pentobarbital (10–30 mg/kg, depending on age), ventilated artificially, and monitored with respect to body temperature, arterial blood gas composition, and pH. Aortic pressure was registered by a strain gage transducer attached to a catheter introduced through a femoral artery. The blood flow to the region under study was registered by an electromagnetic transducer placed around the exposed major artery. Vascular resistance was calculated as the ratio, mean arterial pressure/mean flow. After obtaining control measurements, the efferent nerve supply was doubly ligated and severed so that the distal trunk could be placed on an electrode for stimulation with different impulse frequencies at constant current and pulse duration.

Lumbar nerve control of the femoral circulation in developing swine (*9,10*) is summarized in Table 1. The left lumbar sympathetic chain exposed retroperitoneally terminated in fibers coursing toward the left hindlimb. Sectioning this bundle decreased femoral resistance in all animals age 1 wk, and in a few of the younger ones, indicating that neural vasomotor tone is minimal at birth in swine. This contrasts with findings in lambs (*17*) and calves (*33*). There were age-related differences in stimulus frequency threshold during the first postnatal month in the developing swine, indicating progressive maturation of the adrenergic innervation. At 1 mo, an atropine-blockable vasodilator response was obtained in response to low frequencies of stimulation, although femoral vascular resistance decreased at every age when

Table 1
Maturation of Lumbar Nerve Control of Femoral Blood Flow in Swine[a]

| Age[b] | Change in femoral vascular resistance[c] | | Frequency threshold for a decrease in blood flow,[e] *Hz* |
	After lumbar nerve section	During LNS[d] at 10 Hz	
≤ 1 d	+0.6 ± 0.4 (none in 5)	+8.7 ± 1.7 (+220%)	1
2–4 d	−1.6 ± 1.2 (none in 2)	+5.4 ± 1.5 (+106%)	1
1 wk	−1.0 ± 0.3 (−18%)	+3.2 ± 0.7 (+100%)	1
2 wk	−1.6 ± 0.5 (−28%)	+5.3 ± 1.3 (+120%)	0.5
1 mo	−0.7 ± 0.1 (−33%)	+2.3 ± 0.5 (+123%)	>1; 0.2–1.0 caused vasodilation[f]

[a]Adapted from Tables 2 and 3 and Fig. 5 of ref. *9,* and Table 5 of ref. *10.*
[b]Sample number (*n*) varied in age groups: 6 at ≤1 d, 8 at 2–4 d and 1 wk, 10 at 2 wk, and 9 at 1 mo.
[c]Resistance is the ratio of mean aortic pressure to mean femoral flow; mean change from preceding control, and SEM, in each age group.
[d]LNS, lumbar nerve stimulation.
[e]The lowest frequency eliciting a blood flow decrease with stimulation at 0.5 mA and 1.2 ms duration for 20 s.
[f]With stimulation at 0.2 mA and 1.2 ms duration for 20 s.

animals were given acetylcholine (2 μg) into the femoral artery. A cholinergic vasodilator component of the response to lumbar nerve stimulation had been found at birth in one of the studies on young dogs (*48*), but not in the other (*21*). However, the evidence is clear that swine exhibit delayed functional maturation of the cholinergic innervation of the hindlimb vasculature. We did not observe the escape of femoral blood flow during a 20-s train of low frequency stimuli at any age (*see* Fig. 2B). Flow or resistance responses to increasing frequency of lumbar nerve stimulation provided a frequency–response graph that was steeper in younger animals (*10*). The femoral vasoconstrictor effect of stimulation at 10 Hz was large at all ages.

Renal nerve control of the renal circulation in developing swine (*10,11*) is summarized in Table 2. The right renal nerve, exposed retroperitoneally, comprised a single bundle terminating on the right renal artery. Sectioning this nerve decreased renal vascular resistance in all animals, indicating that neural vasomotor tone was present at birth. There was an age-related difference in stimulus frequency threshold during the first postnatal month, indicating progressive maturation of the adrenergic innervation. We occasionally observed escape of renal blood flow during a 20-s

A)

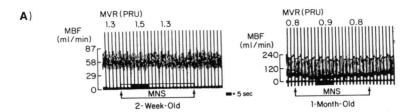

B)

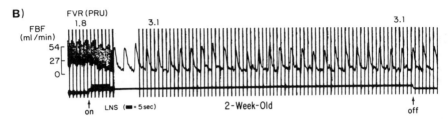

Fig. 2. Blood flow "escape" from stimulation of efferent vasoconstrictor fibers in the superior mesenteric artery circulation (A) of a 2-wk-
old and a 1-mo-old swine; but not in the femoral circulation (B) of a
2-wk-old swine. MNS, mesenteric nerve stimulation at 5 Hz, 1.0 mA and
1.2 ms for 20 s; LNS, lumbar nerve stimulation at 1 Hz, 0.5 mA, 1.2 ms
for 20 s; FBF and MBF, femoral and mesenteric blood flow; FVR and
MVR, femoral and mesenteric vascular resistance. Original oscilloscopic
recordings of phasic flow at paper speed of 2.5 mm/s (and increased to
25 mm/s during LNS).

Table 2
Maturation of Renal Nerve Control of Renal Blood Flow in Swine[a]

	Change in renal vascular resistance[c]		Frequency threshold for a decrease in blood flow,[e] Hz
Age[b]	After renal nerve section	During RNS[d] at 10 Hz	
≤ 1 d	−1.0 ± 0.4 (−12%)	+4.7 ± 1.1 (+60%)	3–5
2–4 d	−1.8 ± 0.3 (−24%)	+4.1 ± 0.9 (+54%)	3–5
1 wk	−2.0 ± 0.6 (−24%)	+3.3 ± 1.1 (+37%)	2
2 wk	−1.3 ± 0.3 (−19%)	+3.0 ± 0.5 (+43%)	2
1 mo	−0.1 ± 0.04 (−10%)	+1.3 ± 0.1 (+120%)	1

[a]Adapted from Table 2 and Fig. 1 of ref. *11*, and Table 4 of ref. *10*.
[b]For young swine, $n = 6$ in each age group; for the 1-mo age group, $n = 5$.
[c]Resistance is the ratio of mean aortic pressure to mean renal blood flow; mean
change from preceding control, and SEM, in each age group.
[d]RNS, renal nerve stimulation.
[e]The lowest frequency eliciting a blood flow decrease with stimulation at 1.0 mA and
1.2 ms duration for 20 s.

train of low frequency stimuli in the oldest animals. Flow or resistance responses to increasing frequency of stimulation provided a frequency–response graph that was steeper in the younger animals (*10*). The renal vasoconstrictor effect of stimulation at 10 Hz was large at all ages.

Splanchnic nerve control of the superior mesenteric artery blood flow to the small intestine was also studied in developing swine (*8*). The left splanchnic nerve exposed retroperitoneally was a large trunk supplying fibers to kidney, adrenal gland, and celiac ganglion. Sectioning this trunk decreased mesenteric arterial resistance in all animals, indicating that neural vasomotor tone is present at birth. Flow or resistance responses to increasing frequency of splanchnic nerve stimulation provided a frequency–response graph that was similar at all ages. However, an increasing magnitude of the pressor effect could be attributed to maturation of the adrenal medulla in swine (*22*), since the pressor responses in the older animals were attenuated after adrenalectomy. Postganglionic nerve control of superior mesenteric artery blood flow in developing swine is summarized in Table 3. The frequency threshold for stimulation (3 Hz) was similar up to 1 mo of age. The intestinal vasoconstrictor effect of 10 or 15 Hz tended to be greatest in the oldest animals. By age 2 wk, the mesenteric blood flow could escape from the influence of stimulation for 20 s

Table 3
Maturation of Postganglionic Nerve Control of Mesenteric Artery Blood Flow in Swine[a]

| | | Change in mesenteric vascular resistance[c] | |
| | | During MNS[e] at 10 Hz | During MNS[e] at 15 Hz |
Age[b]	After nerve section[d]		
2–7 d	-0.4 ± 0.1 (-13%)	$+0.1 \pm 0.02$ ($+8\%$)	$+0.2 \pm 0.06$ ($+29\%$)
2 wk	-0.3 ± 0.06 (-19%)	$+0.2 \pm 0.07$ ($+21\%$)	$+0.4 \pm 0.1$ ($+44\%$)
1 mo	-0.3 ± 0.1 (-25%)	$+0.05 \pm 0.015$ ($+11\%$)	$+0.2 \pm 0.09$ ($+43\%$)

[a]Unpublished observations by the author.
[b]Sample $n = 7$ for 2–7-d-old swine, and 5 for other age groups.
[c]Resistance is the ratio of mean aortic pressure to mean flow in the superior mesenteric artery; mean change from preceding control, and SEM, in each age group.
[d]Section of preganglionic splanchnic nerve and postganglionic fiber bundle coursing along the superior mesenteric artery.
[e]MNS, mesenteric nerve stimulation (1.5–2.5 mA and 0.5 ms duration) for 20 s.

with 5 Hz (*see* Fig. 2A), a response similar to that in the 1-mo-old swine.

The nerve stimulation experiments described above include another model of regional vascular escape from a vasoconstrictor stimulus, although only 20-s trains of stimuli were used to elicit the vasoconstriction. The results indicate that the mesenteric vasculature of the small intestine can escape effects of postganglionic fiber stimulation at an earlier age than can either the renal or femoral vasculatures. The possible role of putative modulators of adrenergic neurotransmission will not be discussed in this chapter because there are few studies on the development of such mechanisms.

4.3. Reflex Stimulation (Baroreceptor and Chemoreceptor)

Baroreceptor reflexes controlling arterial blood pressure may be less active in developing mammals (22,23), and the fetal circulatory system may be controlled predominately by chemoreceptor reflexes rather than by baroreceptor reflexes (27,34,40). There is as yet no model study in which both reflexes have been tested systematically in the same animal preparation as a function of maturational age.

Most studies on baroreceptor reflexes in developing animals focus on the blood pressure and heart rate responses. For example, the reflex increase in systemic arterial pressure during bilateral occlusion of the common carotid arteries in conscious puppies is depressed in comparison to the adult response (45). A unique study on isolated segments of rat aortic arch containing baroreceptors (1) revealed continuing development of a constant relationship between baroreceptor discharge rate and mechanical stress on the segment. One might conclude that the receptor maturity may not be the crucial factor in the apparent immaturity of the overall reflex response. There have been a few measurements of regional bloodflow changes in individual vascular beds during alterations in baroreceptor reflex activity. Forearm bloodflow in term neonates (36) and lower-leg bloodflow in premature infants (46) both decreased during head-up tilting, resulting in increased limb vascular resistance. Carotid sinus stimulation by local injection of saline under pressure led to decreased femoral and renal bloodflow in halothane-anesthetized piglets by the end of the first week after birth (13). A retrospective summary of the results of testing the carotid sinus reflex routinely in pentobarbital-anesthetized swine used in our nerve stimulation studies (8,9,11) is presented in Table 4 (at least two flows were recorded in each animal). From this table, it is apparent that both the femoral and the mesenteric vascular beds engage in the reflex response in newborns, whereas the renal circulation's participation in the reflex response is delayed until the end of the first week after birth.

Interpretation of the studies on chemoreceptor reflexes in developing mammals is complicated by the different modes of stimulation employed by the investigators (22). Although our laboratory has not been involved in such studies, this chapter will cover a few of the interesting issues being examined by others. The active chemoreceptor reflex responses noted in fetal mammals (27,34,40) are a consequence of early maturation of receptor cells and their innervation, e.g., the carotid bodies (26,37). The common clinical problem of asphyxia has been approached by experimentally clamping the umbilical cord (17,18,27,40) or the trachea (33). Fetal cord clamping in lambs produced increases in femoral (17) and renal (18) vascular resistance that could be attenuated by vagotomy. Tracheal clamping in neonatal calves (5–47 h old) produced an increase in vascular resistance of hindlimbs perfused *in*

Table 4
Vascular Effects of Bilateral Occlusion of the Common Carotid Arteries in Swine Anesthetized With Pentobarbital[a]

	Age[b]		
	≤ 1 d	1 wk	2 wk
Change in AoP[c] (mm Hg)	+9.6 ± 0.9	+13.0 ± 1.6	+13.2 ± 2.0
Percent change in:			
AoP[c]	+19.2 ± 1.7	+18.3 ± 2.0	+16.9 ± 1.8
FVR[d]	+16.8 ± 4.8	+10.7 ± 4.6	+20.2 ± 7.8
RVR[d]	(+8.0 ± 9.1)	+26.4 ± 8.4	+16.2 ± 4.9
MVR[d]	+12.3 ± 2.2	+9.0 ± 2.3	+15.8 ± 6.1

[a]Carotid arteries occluded for 30 s by plastic clamp, as in ref. *13;* changes are statistically different from zero change ($p < 0.01$, by the paired-sample *t*-test) except for the RVR values in parentheses.
[b]Sample $n = 10$ at ≤1 d, 12 at 1 wk, and 9 at 2 wk of age.
[c]AoP, mean aortic pressure.
[d]FVR, RVR, and MVR: femoral, renal, and mesenteric resistances, respectively, calculated as in footnote *c* of Tables 1–3.

situ (*33*), a change that was abolished by sectioning the sciatic nerve. In addition to the obvious problem of asphyxia providing two different chemical stimuli, experiments with fetal cord clamping at controlled high levels of arterial oxygen tension (*40*) or after selective autonomic blockade (*27*) emphasize that baroreceptors also become engaged in the total reflex response. Pure hypoxemia has been produced experimentally by localized injections of cyanide (*17,18*), maternal breathing of low oxygen mixtures (*31,38*), or newborn breathing of a low oxygen mixture (*47*), in a few studies on bloodflow responses. In experiments localizing the response to cyanide as a predominantly aortic chemoreceptor reflex (*17*), femoral vascular resistance in fetal lambs increased but was unchanged after deafferentation. A similar finding was reported for the renal vascular resistance in fetal lambs (*18*). Early or late gestation fetal lambs of mothers breathing 11% oxygen exhibited an increase in renal vascular resistance (*38*), as did neonatal lambs (2–38 d of age) breathing 12% oxygen (*47*), without an age-related difference in response magnitude. Furthermore, fetal rabbits of mothers breathing 8% oxygen exhibit a decrease in mesenteric bloodflow and vessel diameter (*31*).

Taken altogether, these observations lead to the impression that the peripheral chemoreceptors are very active in fetal life and that the chemoreceptor reflexes involve the femoral, renal, and mesenteric arterial circulations at an early age. On the other hand, the baroreceptor reflexes begin to play an important role at birth and develop toward an adult pattern of involvement of the regional circulations at rates that may differ among various mammalian species.

5. Conclusions

Vascular innervation develops at different rates in different regional circulations and in different mammals. Some vascular beds are under neural vasoconstrictor tone at birth and others appear not to be. Adrenergic innervation displays functional immaturity in terms of high threshold for norepinephrine release by transmural stimulation of nerve terminals in isolated vessels. Vascular smooth muscle displays its functional immaturity in terms of low magnitude of contractile response to neurotransmitters. There are ultrastructural correlates for both aspects of development of the neuroeffector junction, and both contribute to an age-dependent response of a regional circulation to direct stimulation of its efferent nerve supply. Maturation of these circulatory responses is not synchronous among the regional vascular beds. With postnatal development, some become responsive to mechanisms that decrease vasomotor tone during the continuing influence of vasoconstrictor stimuli. Peripheral chemoreceptor and baroreceptor reflexes are active with respect to control of blood pressure in the fetus and neonate, but the specific regional circulations participating in the reflexes vary with age and species.

Acknowledgments

Research reported in the tables and figures of this chapter was supported by grants from the National Heart, Lung, and Blood Institute (HL-15444 and HL-21865) and was conducted in collaboration with Dr. Paul Brazeau and Mr. Isaac Frasier.

References

1. Andresen, M. C., J. M. Krauks, and A. M. Brown (1978) Relation of aortic wall and baroreceptor properties during development in normotensive and spontaneously hypertensive rats. *Circ. Res.* **43,** 728–738.
2. Armati-Gulson, P. and G. Burnstock (1983) The development of adrenergic innervation in some human foetal blood vessels. *J. Autonom. Nerv. Sys.* **7,** 111–118.
3. Bevan, J. A. (1978) Response of Blood Vessels to Sympathetic Nerve Stimulation, in *Symposium on Molecular and Cellular Aspects of Vascular Smooth Muscle in Health and Disease. Blood Vessels* **15,** 1–3.
4. Bevan, J. A., and R. D. Bevan (1981) Developmental influences on vascular structure and function, in, CIBA Symposium #83, *Development of the Autonomic Nervous System* (G. Burnstock, ed.) Pitman Medical, London, pp. 94–101.
5. Black, I. B. (1982) Stages of neurotransmitter development in the autonomic nervous system. *Science* **215,** 1198–1204.
6. Boatman, D. L., R. A. Shaffer, R. L. Dixon, and M. J. Brody (1965) Function of vascular smooth muscle and its sympathetic innervation in the newborn dog. *J. Clin. Invest.* **44,** 241–246.

7. Buckley, N. M. (1983) Regional circulatory function in the perinatal period, in, *Perinatal Cardiovascular Function* (P. G. Gootman and N. Gootman, eds.) Dekker, NY, pp. 227–263.

8. Buckley, N. M., P. Brazeau, and I. D. Frasier (1985) Circulatory effects of splanchnic nerve stimulation in developing swine. *Am. J. Physiol* **248**, H69–H74.

9. Buckley, N. M., P. Brazeau, I. D. Frasier, and P. M. Gootman (1981) Femoral circulatory responses to lumbar nerve stimulation in developing swine. *Am. J. Physiol.* **240**, H505–H510.

10. Buckley, N. M., P. Brazeau, and P. M. Gootman (1983) Maturation of Circulatory Responses to Adrenergic Stimuli, in, Symposium on Development of the Autonomic Nervous System, *Fed. Proc.* **42**, 1643–1647.

11. Buckley, N. M., P. Brazeau, P. M. Gootman, and I. D. Frasier (1979) Renal circulatory effects of adrenergic stimuli in anesthetized piglets and mature swine. *Am. J. Physiol.* **237**, H690–H695.

12. Buckley, N. M., A. N. Charney, P. Brazeau, S. Cabili, and I. D. Frasier (1981) Changes in cardiovascular and renal function during catecholamine infusions in developing swine. *Am. J. Physiol.* **240**, F276–F281.

13. Buckley, N. M., P. M. Gootman, N. Gootman, G. D. Reddy, L. C. Weaver, and L. A. Crane (1976) Age-dependent cardiovascular effects of afferent stimulation in neonatal pigs. *Biol. Neonate* **30**, 268–279.

14. Burnstock, G. (1975) Innervation of vascular smooth muscle: histochemistry and electron microscopy. *Clin. Exp. Pharmacol. Physiol.* Suppl. **2**, 7–20.

15. Burnstock, G. (1981) Current Approaches to Development of the Autonomic Nervous System: Clues to Clinical Problems, in, CIBA Symposium #83, *Development of the Autonomic Nervous System* (G. Burnstock, ed.) Pitman Medical, London, pp. 1–14.

16. Cowen, T., A. J. Haven, C. Wen-Qin, D. D. Gallen, F. Franc, and G. Burnstock (1982) Development and aging of perivascular adrenergic nerves in the rabbit: A quantitative fluorescence histochemical study using image analysis. *J. Autonom. Nerv. Sys.* **5**, 317–336.

17. Dawes, G. S., B. V. Lewis, J. E. Milligan, M. D. Roach, and N. S. Talner (1968) Vasomotor responses in the hindlimb of foetal and new-born lambs to asphyxia and aortic chemoreceptor stimulation. *J. Physiol. (London)* **195**, 55–81.

18. Dunne, J. T., J. E. Milligan, and B. W. Thomas (1972) Control of the renal circulation in the fetus. *Am. J. Obst. Gynec.* **112**, 323–329.

19. Gallen, D. D., T. Cowen, S. G. Griffith, A. J. Haven, and G. Burnstock (1982) Functional and nonfunctional nerve-smooth muscle transmission in renal arteries of newborn and adult rabbit and guinea pig. *Blood Vessels* **19**, 237–246.

20. Gerke, D. C., D. B. Frewin, and S. S. Soltys (1975) Adrenergic innervation of human mesenteric blood vessels. *Austral. J. Exp. Biol. Med. Sci.* **53,** 241–243.

21. Gerová, M., J. Gero, S. Doležel, and M. Konečný (1974) Postnatal development of sympathetic control in canine femoral artery. *Physiol. Bohemoslov.* **23,** 289–296.

22. Gootman, P. M., N. M. Buckley, and N. Gootman (1979) Postnatal Maturation of Neural Control of the Circulation, in, *Reviews in Perinatal Medicine* Vol. 3 (E. M. Scarpelli and E. V. Cosmi, eds.) Raven Press, NY, pp. 1–72.

23. Gootman, P. M., N. Gootman, and B. J. Buckley (1983) Maturation of Central Autonomic Control of the Circulation, in, Symposium on Development of the Autonomic Nervous System *Fed. Proc.* **42,** 1648–1655.

24. Heymann, M. A., H. S. Iwamoto, and A. M. Rudolph (1981) Factors affecting changes in the neonatal systemic circulation. *Ann. Rev. Physiol.* **43,** 371–383.

25. Hill, C. E., G. D. S. Hirst, and D. F. van Helden (1983) Development of sympathetic innervation to proximal and distal arteries of the rat mesentery. *J. Physiol. (London)* **338,** 129–148.

26. Kanerva, L., A. Hervonen, and H. Hervonen (1974) Morphological characteristics of the ontogenesis of the mammalian peripheral adrenergic nervous system, with special remarks on the human fetus. *Med. Biol.* **52,** 144–153.

27. Lewis, A. B., M. Donovan, and A. C. G. Platzker (1980) Cardiovascular responses to autonomic blockade in hypoxemic fetal lambs. *Biol. Neonate* **37,** 233–242.

28. Ljung, B., and D. Stage (1975) Postnatal ontogenetic development of neurogenic and myogenic control in the rat portal vein. *Acta Physiol. Scand.* **94,** 112–127.

29. Lorijn, R. H., and L. D. Longo (1980) Norepinephrine elevation in the fetal lamb: oxygen consumption and cardiac output. *Am. J. Physiol.* **239,** R115–R122.

30. Lundberg, J., B. Ljung, D. Stage, and A. Dahlstrom (1976) Postnatal ontogenetic development of the adrenergic innervation pattern in rat portal vein: A histochemical study. *Cell Tiss. Res.* **172,** 15–27.

31. McCuskey, R. S., S. G. McClugage, Jr., T. J. Moore, and M. L. Miller (1969) Response of the fetal mesenteric microvascular system to maternal hypoxia. *Proc. Soc. Exp. Biol. Med.* **132,** 636–639.

32. McKenna, O. C. and E. T. Angelakos (1970) Development of innervation in the puppy kidney. *Anat. Rec.* **167,** 115–125.

33. Milligan, J. E., M. R. Roach, and N. S. Talner (1967) Vasomotor responses in the hindlimb of newborn calves. *Circ. Res.* **21,** 237–244.

34. Mott, J. C. (1982) Control of the foetal circulation (review). *J. Exp. Biol.* **100,** 129–146.

35. Owman, Ch., N.-O. Sjöberg, and G. Swedin (1971) Histochemical and chemical studies on pre- and postnatal development of the different systems of "short" and "long" adrenergic neurons in peripheral organs of the rat. *Z. Zellforsch.* **116,** 319–341.

36. Picton-Warlow, G. G., and F. E. Mayer (1970) Cardiovascular responses to postural changes in neonates. *Arch. Dis. Child.* **45,** 354–359.

37. Purves, M. J. (1981) The Neural Control of Respiration Before and After Birth, in, *Reviews in Perinatal Medicine* Vol. 4 (E. M. Scarpelli and E. V. Cosmi, eds.) Raven Press, NY, pp. 299–336.

38. Robillard, J. E., R. E. Weitzman, L. Burmeister, and F. G. Smith, Jr. (1981) Developmental aspects of the renal response to hypoxemia in the lamb fetus. *Circ. Res.* **48,** 128–137.

39. Scott, T. M., and S. C. Pang (1983) Correlation between development of sympathetic innervation and development of medial hypertrophy in jejunal arteries in normotensive and spontaneously hypertensive rats. *J. Autonom. Nerv. Sys.* **8,** 25–32.

40. Siassi, B., P. Y. K. Wu, C. Blanco, and C. B. Martin (1979) Baroreceptor and chemoreceptor responses to umbilical cord occlusion in fetal lambs. *Biol. Neonate* **35,** 66–73.

41. Stage, D. and B. Ljung (1978) Neuroeffector maturity of portal vein from newborn rats, rabbits, cats, and guinea pigs. *Acta Physiol. Scand.* **102,** 218–223.

42. Su, C., J. A. Bevan, N. S. Assali, and C. R. Brinkman III (1977) Development of neuroeffector mechanisms in the carotid artery of the fetal lamb. *Blood Vessels* **14,** 12–24.

43. Todd, M. E. (1980) Development of adrenergic innervation in rat peripheral vessels; a fluorescence microscopic study. *J. Anat.* **131,** 121–133.

44. Todd, M. E. and M. K. Tokito (1981) An ultrastructural investigation of developing vasomotor innervation in rat peripheral vessels. *Am. J. Anat.* **160,** 195–212.

45. Vatner, S. F. and W. T. Manders (1979) Depressed responsiveness of the carotid sinus reflex in conscious newborn animals. *Am. J. Physiol.* **237,** H40–H43.

46. Waldman, S., A. N. Krauss, and P. A. M. Auld (1979) Baroreceptors in preterm infants: their relationship to maturity and disease. *Devel. Med. Child Neurol.* **21,** 714–722.

47. Weismann, D. N. and W. R. Clarke (1981) Postnatal age-related renal responses to hypoxemia in lambs. *Circ. Res.* **49,** 1332–1338.

48. Wysse, D. G., G. R. van Petten, and W. H. Harris (1977) Responses to electrical stimulation, noradrenaline, serotonin, and vasopressin in the isolated ear artery of the developing lamb and ewe. *Can. J. Physiol. Pharmacol.* **55,** 1001–1006.

49. Zink, J. and G. R. van Petten (1981) Noradrenergic control of blood vessels in the premature lamb fetus. *Biol. Neonate* **39,** 61–69.

Chapter 10

Relationships Between the Sympathetic Nervous System and Functional Development of Smooth Muscle End Organs

Peter G. Smith

1. Introduction

The sympathetic nervous system (SNS) is an essential link between the central nervous system and effector cells that mediate adaptive homeostatic responses. Appropriate internal or external stimuli modulate central outflow of sympathetic nerve impulses that ultimately produce changes in effector cell activity. A question arises concerning SNS control of end organ activity; What is the manner in which SNS activity is matched with end organ performance to produce physiologically appropriate responses? Effector organs innervated by the SNS apparently lack well-defined feedback systems. For example, although activation of SNS vasomotor pathways resulting in vasoconstriction and elevated blood pressure does result in baroreceptor-mediated inhibition of SNS outflow, this feedback system does not provide information concerning resistance of specific vascular beds and can adapt rapidly to changes in pressure (25). The apparent absence of discrete feedback loops to end organs innervated by the SNS suggests that other factors must operate to assure that the effector unit output is appropriate to the level of neural stimulation.

Developmental factors acting to regulate maturation of SNS effector cells represent obvious mechanisms through which end organ responsiveness may be modulated, and several lines of evidence suggest that the SNS itself plays an important role in regulation end organ growth. In adult animals, administration of adrenergic agonists or electrical stimulation of sympathetic nerves enhances cell division and growth in salivary glands (28,31,34), whereas chemical sympathectomy and norepinephrine depletion antagonize salivary gland growth (6). Similarly, guanethidine-induced sympathectomy suppresses intestinal cell proliferation (18). In the neonatal rat, adrenergic agonists inhibit thymidine incorporation in cardiac muscle cells (7). Studies in immature rabbits provide evidence that sympathetic decentralization and denervation cause structural and functional deficits in vascular smooth muscle cell maturation (1–3,30). Furthermore, it has been demonstrated that adrenergic agonists stimulate growth of vascular smooth muscle cells in the tissue culture (5). Together, these studies suggest that the SNS plays an important role in regulating growth of end organs though adrenergic receptor-coupled mechanisms.

The hypothesis that the SNS plays a major role in regulating end organ development is especially pertinent to current concepts concerning the etiology of essential hypertension. A widely accepted theory proposes that sustained hypertension is a consequence of hypertrophy of vascular smooth muscle, occurring secondary to hyperactivity of the SNS (12). In this context, enhanced influence of the SNS on vascular smooth muscle would lead to abnormal end organ growth. The net result of this would be the establishment of physiologically inappropriate levels of end organ activity, expressed as hypertension. A recent ultrastructural study of the development of sympathetic innervation to jejunal arteries provides circumstantial support for this theory (32). In this study, Scott and Pang report that the jejunal vasculature is hyperinnervated in 2-wk-old spontaneously hypertensive rats (SHR) relative to normotensive Wistar-Kyoto controls (WKY), and that this occurs prior to the hypertrophy of vascular smooth muscle cells and the establishment of hypertension. The authors conclude that " . . . the medial hypertrophy which occurs in SHR during development is neither the cause nor the result of elevated blood pressure, but is more closely related to the development of hyperinnervation" (32). If end organ cellular growth occurs secondary to advances in innervation, then indices of end organ function should reflect temporally the developmental status of the SNS. The present report examines interactions in the functional development of the SNS and vascular and nonvascular smooth muscle end organs that it innervates.

2. Ontogeny of SNS Control of End Organ Activity in Sprague-Dawley Rats

The neonatal rat is an ideal animal in which to study the processes associated with SNS neuroeffector maturation, since both postganglionic neurons and end organ cells are highly immature in newborn rats relative to other mammals (13,22,23). Studies by Wekstein showed that the SNS exerts little influence on heart rate in rats prior to the sixth postnatal day (48). This observation has been confirmed and extended. It is now known that centrally acting stimuli that activate the cardiac sympathetic pathway of the adult rat do not alter end organ activity in neonatal rats. Thus, hypoglycemia produces SNS-mediated increases in cardiac ornithine decarboxylase activity in mature rats, but has little effect on enzyme activity in neonates prior to about 1 wk of age, although exogenously administered noradrenergic agonists are effective at this time (37).

The inception of end organ responses to SNS stimulation in neonates may be attendant upon the maturation of one or more specific components of this system, including (a) afferent pathways responsible for detecting stressful stimuli, (b) transmission in effector pathways central to the preganglionic sympathetic neuron, (c) ganglionic neurotransmission, and/or (d) postganglionic neural transmission. Biochemical studies have shown that indices of postganglionic noradrenergic synaptic transmission are immature in the neonate, and that significant maturational advances occur during the period when the SNS becomes functional (37). However, additional deficits in the SNS pathway may contribute to the absence of SNS function in the neonate.

We approached this problem by recording impulse activity in pre- and postganglionic cervical sympathetic nerves in developing rats anesthetized with urethane (44). The cervical sympathetic pathway (*see* Fig. 1) is a readily accessible branch of the SNS in which both pre- and postganglionic nerve impulse activity can be evaluated using classical electrophysiological techniques scaled down to a size appropriate to the neonate (approximately 10 g). Urethane anesthesia is routinely employed in all physiological studies because biochemical studies indicate that urethane, in the doses employed, does not alter resting levels or SNS-mediated increases in end organ ornithine decarboxylase activity, whereas other anesthetic agents can attentuate end organ responses (6B, and T. A. Slotkin, personal communication). Pre- and postganglionic nerve impulse activity was recorded and quantitated under several conditions: resting state, moderate stimulation by 2-deoxyglucose-induced hypoglycemia, and intense stimulation by prolonged asphyxia. Recordings from preganglionic neurons revealed that impulse activity in the resting and stimulated states in rats aged 2–11 d was comparable to that of young adults, and was transiently elevated to approximately twice the adult level at 17–19 d. In contrast, postganglionic impulse activity in 2–5-d-old rats was not significantly different from zero in the resting state or after 2-deoxyglucose, and was

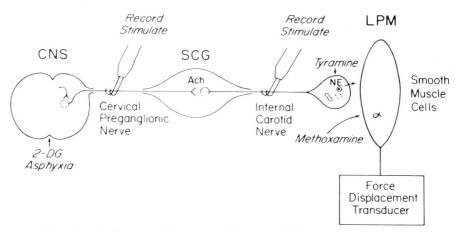

Fig. 1.   Schematic diagram of the cervical sympathetic pathway to Müller's smooth muscle, and the experimental interventions employed in these studies. Abbreviations: CNS, central nervous system; 2-DG, 2-deoxyglucose; SCG, superior cervical ganglion; Ach, acetylcholine; NE, norepinephrine; LPM, Müller's smooth muscle component of the levator palpebrae muscle.

barely detectable during asphyxia prior to death (Fig. 2). By 10–12 d of age, impulse activity in all conditions was comparable to adult levels. The absence of impulse activity in postganglionic nerves at 2–5 d is not caused by an inability of the axons to conduct action potentials, since direct electrical stimulation of the postganglionic nerve results in compound action potentials with comparable amplitudes and durations in 4–5- and 10–13-d-old rats.

From these experiments it is concluded that functional sympathetic ganglionic neurotransmission in the cervical sympathetic pathway is established in neonatal Sprague-Dawley rats between the 5th and 10th postnatal days. Physiological and ultrastructural studies provide additional support for this hypothesis. In a subsequent physiological study we found that electrical stimulation of preganglionic cervical sympathetic axons did not elicit end organ responses at 1 or 4 d, but was effective at 8–9 d of age (*38*) (*see* section 3.1 in this chapter). Quantitative ultrastructural studies of synaptogenesis in the superior cervical ganglion of developing rats indicate that a rapid increase in the number of ganglionic synapses occurs in the first postnatal week (*45*).

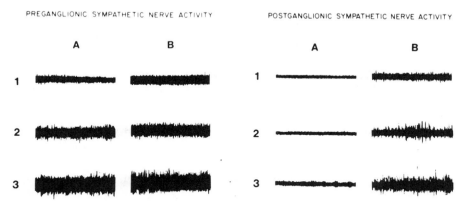

Fig. 2. Recordings from pre- and postganglionic cervical sympathetic nerves in urethane-anesthetized rats. Left panel: Recordings from the preganglionic cervical nerve in rats aged 2 d (A) and 10 d (B). Right panel: Recordings from the postganglionic internal carotid nerve in rats aged 5 d (A) and 11 d (B). Row 1, Resting activity; Row 2, Activity 40 min after intraperitoneal injection of 2-deoxyglucose (1 g/kg); Row 3, Activity during asphyxia. Vertical bars = 80 mV. Horizontal bars = 2 s (from ref. *44*).

There is evidence to suggest that ganglionic neurotransmission is established in other sympathetic pathwways also at the end of the first postnatal week. It is at this time that SNS control of cardiac activity is established (*37,48*), and this event is associated with a five-fold increase in synapses counted in the inferior cervical sympathetic ganglion (which provides the major cardiac sympathetic innervation) between 2 and 8 d of age (*41*). Furthermore, functional splanchnic-adrenomedullary innervation is established by the end of the first postnatal week (*37*). This is also a time during which cholinergic nerve terminals are observed proliferating within adrenomedullary tissue (*26*).

It is concluded that the development of SNS control of end organ activity in the superior cervical, and probably other, sympathetic pathways occurs in several distinct phases. Initially, the SNS exerts no physiological influence on the end organ because of the virtual absence of ganglionic neurotransmission during the first postnatal week. Ganglionic neurotransmission apparently is established rapidly, so that impulse activity in the postganglionic neuron increases from zero to adult levels at the beginning of the second postnatal week. Subsequently, sympathetic tone is elevated transiently in the third postnatal week before returning to the mature level.

3. Ontogeny of Nonvascular Smooth Muscle Activity in Sprague-Dawley Rats

3.1. Normal Development

Studies of SNS-end organ interactions during development require that a suitable preparation be available for monitoring maturational changes in end-organ function. For this purpose, we have utilized Müller's muscle (superior palpebral muscle, tarsal muscle) (*see* Fig. 1). which has been used by others to assess efficacy of sympathectomy (*11*). Müller's muscle is present in the eyelids of higher vertebrates and acts, in association with the striated levator palpebrae muscle, to retract the eyelid. In the rat, Müller's muscle is an extension of the levator palpebrae muscle composed entirely of smooth muscle cells (Fig. 3) receiving a dense noradrenergic innervation from the superior cervical ganglion (*29*). We have shown that Müller's muscle contraction in response to electrical stimulation of the preganglionic cervical nerve

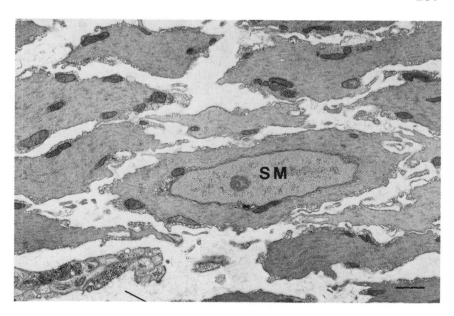

Fig. 3.   Electron micrograph of smooth muscle (SM) cells in Müller's muscle from an 45-d-old SHR. Note nerve varicosities containing dense core vesicles in lower left (arrow). Bar, 1 μm.

is attributable specifically to release of norepinephrine from postganglionic sympathetic nerve terminals, since administration of drugs that prevent norepinephrine release or that block alpha-noradrenergic receptors eliminate >90% of the contractile response (38). In this end organ preparation, functional development can be ascertained directly in vivo by recording smooth muscle contractile force induced by sympathetic stimulation in rats of different ages.

End organ responses in Müller's muscle were studied by selectively activating the sympathetic neuroeffector system at different levels (38) (Fig. 1). Methoxamine is a direct-acting, alpha-noradrenergic agonist that is neither taken up by sympathetic nerve terminals nor degraded by enzyme pathways that metabolize catecholamines (46,47). Contractile responses to methoxamine require the presence of functional postjunctional alpha$_1$-noradrenergic receptors and postreceptor contractile apparatus, and do not reflect directly the status of the SNS innervation. Tyramine is an indirectly acting noradrenergic agonist that is actively transported into sympathetic nerve terminals, resulting in the displacement of endogenous norepinephrine stores (46). Contractile responses to tyramine require functional end organ cells, as well as the presence of sympathetic innervation containing transmitter stores sufficient to activate the postjunctional receptors. Thus, tyramine is used to test the functional status of the postganglionic innervation, as well as end-organ performance. Electrical stimulation of preganglionic axons will elicit a contractile response only if ganglionic neurotransmission, nerve terminal release of noradrenalin, and receptor-mediated end-organ responses are functional.

The pattern for development of contractile force in response to direct alpha-noradrenergic receptor stimulation by methoxamine indicates that end organ development in this system is a discontinuous process, marked by periods of rapid maturational advance interposed with periods of latency (Fig. 4). Between 1 and 4 d of age, the methoxamine response increases dramatically from 3–21% of the adult (70–75 d) value. This is followed by a period from 4–9 d in which there is no increase in the force of contraction. A substantial increase of approximately 20% of the adult value occurs between 9 and 12 d, again followed by a period of no advancement through 19 d of age. In the following 4 wk (19–50 d of age) contractile force increases by 20% of the adult value; the remaining 35% is attained over the next 4 wk.

Contractions elicited by tyramine were comparable to methoxamine responses at all ages. These responses are caused by release of endogenous neurotransmitter and not by direct effects of tyramine on smooth muscle cells; administration of reserpine, which depletes nerve terminal norepinephrine, or desmethylimiprimamine, which blocks uptake of tyramine, abolishes tyramine-induced smooth muscle contractions in this system. It is concluded that the sympathetic innervation to Müller's muscle contains quantities of neurotransmitter sufficient to evoke maximum contractions at all ages throughout this developmental period.

Electrical stimulation of preganglionic axons at 1 and 4 d of age did not elicit contractions, but was as effective as tyramine or methoxamine at all other times. Since tyramine responses are comparable to those of methoxamine at 1 and 4 d, the absence of responses to electrical stimulation is not caused by insufficient stores of norepinephrine or numbers of sympathetic nerve terminals innervating Müller's muscle. As discussed above, these

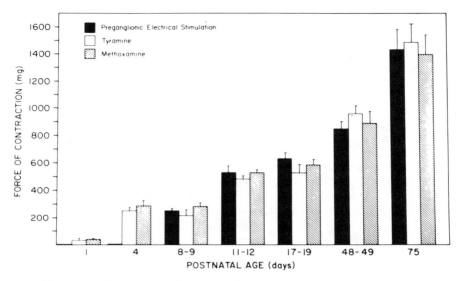

Fig. 4.   Ontogeny of contractile responses of Müller's muscle in Sprague-Dawley rats (from ref. *38*).

findings are consistent with the hypothesis that functional ganglionic neurotransmission is absent in the neonate and is not established until the end of the first postnatal week.

The relationship between maturational events occurring in the SNS and nonvascular smooth muscle in the cervical sympathetic Müller's muscle neuroeffector pathway of SD rats can be summarized as follows. End-organ function is extremely immature at birth, but increases eightfold between 1 and 4 d of age. This increase occurs during a period when the end organ is innervated by postganglionic sympathetic neurons containing norepinephrine, but is not stimulated by SNS activity because ganglionic neurotransmission is not functional. Ganglionic neurotransmission is established at 6–7 d of age and the end organ now can be influenced by SNS impulse activity that is at a level comparable to that of the adult. However, onset of SNS function is not accompanied by any immediate increase in end-organ performance, since force of contraction remains unchanged between 4 and 9 d of age. The contractile response doubles between 9 and 12 d, and remains at this level through 19 d of age. This is a period when SNS activity of central origin increases transiently to twice the adult level. Consequently, it appears that this elevation in SNS activity does not induce any immediate response in end-organ functional development. An additional 50% increase in the contractile response is evident in the period from 19–50 d, and a 75% increase from 50–75 d of age.

3.2. Effects of Disturbances in Sympathetic Innervation

Previous investigations have shown that smooth muscle structure and function can be altered by interventions affecting the SNS (1–3,30). These studies have employed methods that have relatively drastic effects on innervation, such as decentralization of the postganglionic neuron or sympathetic denervation of the end organ. We have examined end organ development in the cervical sympathetic–Müller's muscle neuroeffector system under several conditions in which SNS function has been altered in order to answer two important questions. First, to what extent is end-organ maturation dependent on nerve activity and the presence of sympathetic innervation in this pathway? Second, can alterations in sympathetic innervation less severe than decentralization or denervation result in disturbed development in end organ function?

3.2.1. Decentralization

The role of SNS impulse activity in end-organ development was investigated by excising a segment of the preganglionic cervical sympathetic nerve at 1 d of age, resulting in permanent decentralization of the superior cervical ganglion (38). Responses to methoxamine in decentralized preparations studied at 18–19 and 45–50 d of age were not significantly different from those of sham-operated animals. However, there was a significant deficit in the methoxamine response in decentralized rats at 70–75 d of age. On the basis of these experiments we conclude that sympathetic nerve activity in postganglionic neurons does not play a major role in the functional development of this smooth muscle end organ system through 50 d of age, but does contribute to the increase in contractile force that occurs between 50 and 75 d of age.

3.2.2. Surgical Sympathectomy

The possibility that SNS may influence end organ development through mechanisms not associated with impulse activity must also be considered. Thus, the presence of sympathetic nerve terminals may exert a trophic influence on smooth muscle cells during development. We investigated this possibility by extirpating the superior cervical ganglion in rats aged 1–3 d (Smith, in preparation). To test for reestablishment of functional innervation caused by incomplete denervation, the ipsilateral preganglionic nerve was stimulated electrically. In no case did this elicit a response. Furthermore, gross examination of the region of the carotid bifurcation did not reveal the presence of any structures resembling the ganglion. However, tyramine was effective in eliciting contractions, and these responses were blocked by desmethylimipramine.

To determine whether the tyramine responses in these animals was caused by sprouting of sympathetic nerve terminals to denervated tissues, as is known to occur in the central nervous system (8), we stimulated the contralateral preganglionic cervical sympathetic nerve. Indeed, contralateral stimulation evoked a contractile response in these "denervated" animals. This response was abolished by ganglionic blockade with chlorisondamine and by cutting the contralateral postganglionic internal carotid nerve as it entered the skull. We have stimulated the contralateral preganglionic cervical sympathetic nerve in a large number of control and sham-operated animals to determine whether there is normally some bilateral distribution of the postganglionic innervation. Contralateral stimulation was never effective in eliciting contractions in these groups. In addition, animals denervated at 30 d, or decentralized at 1 d of age, did not develop contralateral innervation. In light of these unexpected findings, responses obtained in preparations denervated by this method cannot be considered as accurate reflections of development in the absence of sympathetic innervation. Preliminary findings using guanethidine to induce chemical sympathectomy tend to support the conclusions of Bevan and Tsuru (3) that the functional deficit observed following denervation is similar to that occurring after decentralization and, therefore, can be attributed primarily to the absence of impulse activity.

The functional reinnervation following surgical sympathectomy is consistent with the hypothesis that end-organ cells are capable of regulating the development of sympathetic innervation (4,15). It is known that end-organ removal prevents the normal development of postganglionic neurons (10). Furthermore, interventions resulting in end-organ hypertrophy can induce aug-

mented innervation (9). It would appear that neonatal sympathectomy effectively increased the size of the target organ field for
the contralateral superior cervical ganglion. It is possible that the
functional reinnervation of the denervated end organ is associated with enhanced survival of contralateral postganglionic neurons. Thus, denervation at 30 d, a time when normal cell death
that occurs during development is complete (16), did not induce
functional reinnervation.

3.2.3. Neonatal Hyper- and Hypothyroidism

Development of the SNS is perturbed markedly by alterations in
neonatal thyroid status (36, 37). Triiodothyronine (T3) injections
in newborn rats result in precocious onset of SNS control of end-
organ activity (19) associated with accelerated ganglionic
synaptogenesis (41) and enhanced development of sympathetic
nerve terminals and postsynaptic receptor mechanisms
(19,20,40). Conversely, suppression of thyroid function by propylthiouracil (PTU) delays development of sympathetic nerve terminals and postsynaptic receptors (21) and onset of centrally mediated responses in the adrenal medulla (14). We performed
longitudinal studies on developing animals in which SNS function had been disturbed by alterations in thyroid status to determine if end organ function was affected. Experiments were performed on rats receiving T3 (pups injected on postnatal days 0–6),
PTU (dams injected gestational day 19 through postpartum day 9;
pups injected postnatal days 0–9), vehicle injections using the
same schedules as for drug treatments, and on control rats
receiving no treatments. The Müller's muscle preparation was
used to assess end-organ function in rats aged 9–86 d. In all cases,
contractile responses to electrical stimulation, tyramine, and
methoxamine were comparable in drug-treated and vehicle-
injected animals. Therefore, perturbations in sympathetic innervation induced by neonatal hyper- or hypothyroidism do not
have significant effects on end organ development in this cervical
sympathetic neuroeffector pathway.

4. Ontogeny of Nonvascular Smooth Muscle Function in SHR and WKY

The development of smooth muscle function in spontaneously
hypertensive rats (SHR) is of particular interest in view of the postulated role of the SNS in vascular smooth muscle hypertrophy

and hypertension (12,32). Experiments in which SNS activity has been recorded show that impulse activity of central origin is elevated in developing SHR relative to normotensive Wistar-Kyoto (WKY) rats (17,27). Studies of the cervical sympathetic–Müller's muscle neuroeffector system were conducted to determine whether smooth muscle function in SHR compared with WKY might reflect this enhanced level of sympathetic tone (42 and unpublished observations). Contractile responses to electrical stimulation of preganglionic cervical sympathetic nerves in developing SHR and WKY are presented in Table 1. The values for electrical stimulation are similar to those obtained with tyramine and methoxamine. Responses in SHR are depressed significantly from 9–20 d relative to WKY, whereas responses at other times are comparable. It should be noted that in a previous study (42) we found that contractile responses to tyramine were suppressed in SHR at 8–19 d, but significantly elevated at 41–46 d. Those experiments were conducted in rats obtained from a different breeder (Taconic Farms) than those used in the present study (Charles River). In the former group, the weights of SHR at 41–46 d were significantly greater than those of WKY, whereas the weights of both strains used in the present study were comparable throughout development. Therefore, the increased response observed previously at 41–46 d appears to be a function of increased body weight, whereas the depressed response at 8–20 d is not associated with body weight and is a consistent feature in animals obtained from different breeders. Decreased contractile response

Table 1
Müller's-Muscle Contractile
Responses to Electrical Stimulation
of Preganglionic Axons in
SHR and WKY (mg)

Age, d	SHR	WKY
2–3	77 ± 41	77 ± 7
5	106 ± 17	126 ± 17
9	219 ± 28[a]	290 ± 22
20	380 ± 22[a]	495 ± 34
42	820 ± 45	808 ± 20
82–86	1140 ± 72	1171 ± 125

[a]$p < 0.05$ by one-way analysis of variance.

has been reported also for vascular smooth muscle in young SHR (*35*).

5. Ontogeny of Vascular Smooth Muscle Function

Studies of vascular smooth muscle function present special problems since direct measurements of end-organ performance are not easily obtained. In addition, observations from a single vessel or vascular bed may not reflect the functional status of the vasomotor system in general. We have assessed vascular smooth muscle function in developing rats by recording systemic blood pressure under conditions designed to provide indices of the neural contribution to resting blood pressure and the contractile capacity of vascular smooth muscle (*43*). Blood pressure was recorded directly by arterial cannulation in urethane-anesthetized rats at different ages (Fig. 5). Stable pressure values following surgery were taken as the resting blood pressure. Neurogenic contributions to resting blood pressure were eliminated by ganglionic blockade with chlorisondamine. The dose of chlorisondamine used in these experiments is supramaximum for blocking ganglionic neurotransmission in the cardiac autonomic pathway (*33*) and cervical sympathetic pathway (Smith and Mills, unpublished) in Sprague-Dawley rats, and in the cervical sympathetic pathway of SHR and WKY (*43*). Stable blood pressure values following chlorisondamine were taken as the basal blood pressure. The neurogenic contribution to resting blood pressure was estimated by subtracting basal blood pressure from resting blood pressure (i.e., the blood pressure decrease after chlorisondamine). A cardioselective β-blocking agent (atenolol) was then administered to eliminate cardiac chronotropic and inotropic effects of noradrenergic stimulation. Pressor responses obtained under these conditions are considered to be representative of vascular smooth muscle response. Contractile function was tested by injecting graded doses of methoxamine and tyramine until a maximum pressure increase was obtained. The ED_{50} for tyramine and methoxamine was determined, and the methoxamine and tyramine pressor responses were calculated by subtracting the basal blood pressure from the maximum pressure attained during noradrenergic stimulation (Δ mm Hg) (*43*).

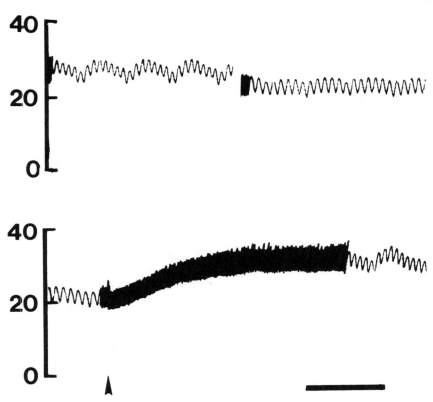

Fig. 5.   Arterial blood pressure recording from the carotid artery of a 1-d-old SHR. Left top: resting blood pressure. Right top: blood pressure after ganglionic blockade with chlorisondamine. Lower panel: pressor response to intraperitoneal administration of tyramine (30 mg/kg) at arrow (no pretreatment with atenolol). Bar = 2 min or 2 s.

5.1. Neurogenic Control of Vascular Smooth Muscle Function in Sprague-Dawley Rats

The neurogenic contribution to resting blood pressure (pressure decrease after chlorisondamine) and vascular smooth muscle contractile function (methoxamine pressor response) for Sprague-Dawley rats from age 20–86 d are presented in Fig. 6 (Smith and Mills, unpublished). It is apparent that both indicies develop in parallel. There are two likely explanations for the parallel associa-

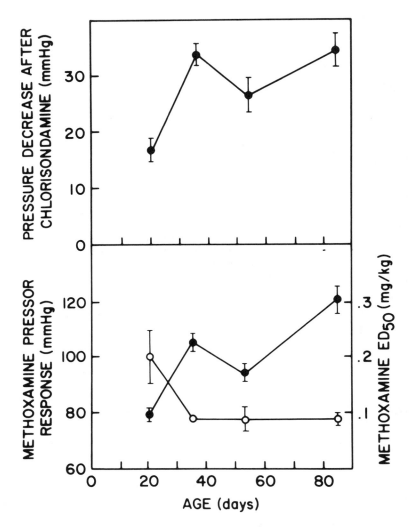

Fig. 6. Ontogeny of neurogenic contribution to resting blood pressure (pressure decrease after chlorisondamine, top panel), vascular smooth muscle reactivity (methoxamine pressor response, bottom panel, closed circles), and methoxamine ED_{50} (open circles) in Sprague-Dawley rats. Pressure decrease after chlorisondamine = resting pressure minus pressure after ganglionic blockade. Methoxamine response = maximum pressor response minus blood pressure after ganglionic blockade.

tion between the neurogenic component and the pressor response to methoxamine. First, variations in the neurogenic component reflect changes in sympathetic nerve activity that directly regulate the development of vascular smooth muscle function. Second, developmental changes in the neurogenic component are linked to variations in the functional capacity of the smooth muscle cells in the presence of relatively constant sympathetic nerve influences. The first hypothesis does not appear to be consistent with data from nerve recording experiments (44). Thus, the increase in the neurogenic component to resting blood pressure and in the methoxamine pressor response between 20 and 36 d occurs at a time when sympathetic impulse activity is decreasing. Additionally, the methoxamine ED_{50} is decreased at 20 d, so the effects of enhanced SNS activity would be attentuated at the level of the end organ receptor. Consequently, the developmental pattern for blood pressure decreases after ganglionic blockade probably pertains directly to the functional capacity of the vascular smooth muscle and not to specific variations in sympathetic tone.

5.2. Neurogenic Control of Vascular Smooth Muscle Function in SHR and WKY

Comparisons of the blood pressure decrease after chlorisondamine (*see* Fig. 7) and the methoxamine pressor response (*see* Fig. 8) reveal significantly different patterns of development in SHR and WKY (43). Analysis of strain differences in the neurogenic component by ANOVA indicates that the contribution to blood pressure in SHR is significantly greater throughout development. The difference between strains is not consistent at all ages (age × strain interaction) because of disproportionately greater decreases in SHR blood pressure after ganglionic blockade at 20 and 42 d of age. Furthermore, the rate of increase in the chlorisondamine-sensitive component of blood pressure is greater in SHR than WKY, as determined by regression analysis.

These differences between SHR and WKY in the neurogenic component of blood pressure are not evident in the development of the methoxamine pressor response. Although responses to methoxamine are significantly greater in SHR than in WKY, the differences remain constant with age, and slopes for rates of increase are comparable (*see* Fig. 8). Values obtained for methoxamine ED_{50} in SHR and WKY (Fig. 9) are not consistent with the

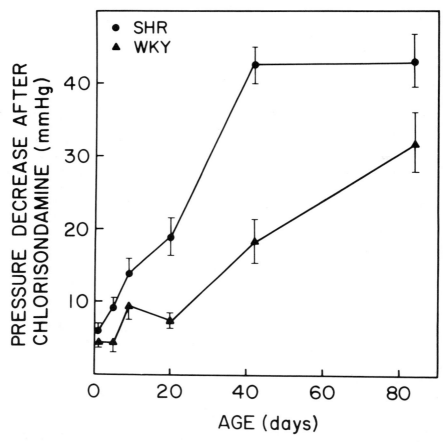

Fig. 7. Ontogeny of neurogenic contribution to resting blood pressure (pressure decrease after chlorisondamine = resting blood pressure minus blood pressure after ganglionic blockade) in SHR and WKY (from ref. 43).

idea that differences in receptor sensitivity alone can account for the elevated neurogenic component at 20 through 42 d in SHR, since the ED_{50} is not disproportionately elevated in this period. Indeed, experiments in which preganglionic impulses were recorded directly indicated that central outflow of SNS activity in SHR is elevated at this time (*17,27*). However, the decreased sensitivity to methoxamine in SHR at 82–86 d may be associated with the sustained values for the neural contribution to blood pressure between 42 and 86 d of age. We consider these results to be com-

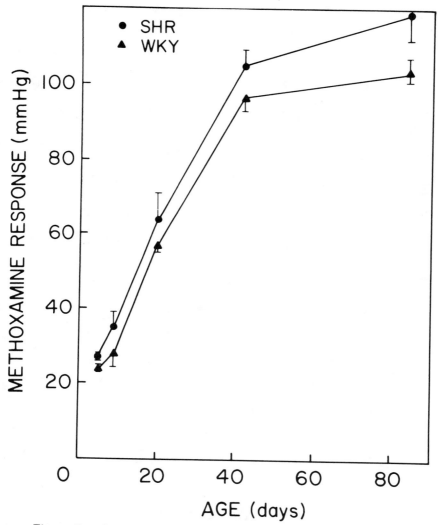

Fig. 8. Ontogeny of vascular smooth muscle reactivity (Methoxamine Response) in SHR and WKY. Methoxamine response = maximum pressor response minus blood pressure after chlorisondamine (from ref. 43).

patible with the idea that both vascular smooth muscle function and sympathetic nerve activity are augmented in SHR during development, but that these defects occur in parallel and are not inextricably linked.

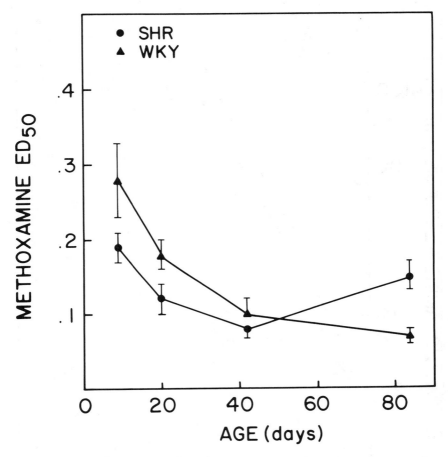

Fig. 9. Ontogeny of methoxamine ED_{50} in SHR and WKY.

5.3. *Postnatal Maternal Stress*

It is accepted that hereditary factors play a major role in the pre-disposition toward essential hypertension. However, environmental factors, including stress, also are believed to be important in the etiology of blood pressure abnormalities. In the course of conducting experiments on rats subjected to the stresses of perinatal injections and handling, we observed profound effects on blood pressure regulation during maturation (39). Disturbances in blood pressure were observed in Sprague-Dawley rats receiving subcutaneous injections of vehicles (alkaline saline) used to solubilize T_3 and PTU. As noted above, drug treatments were ineffective in altering end organ functional development in the cervical sympathetic–Müller's muscle pathway (see Section 3.2.3 in this chapter) and had few specific effects different from vehicle injections on vascular function through 86 d of age. In contrast, a regimen consisting of injections to gravid dams on gestational d 19 through d 9 postparturition and to pups on postnatal d 0–9 had dramatic effects on cardiovascular function in the developing animals. Rats in this group were hypertensive at 19–23 d and subsequently hypotensive at 82–86 d of age, compared with uninjected controls (see Fig. 10).

The developmental disturbances in cardiovascular function induced by this injection regimen are attributable to maternal stress rather than to direct effects on the pups, since injections to pups alone did not result in alterations in blood pressure (Fig. 10). Preliminary observations suggest that the functional defects in development are a consequence of the injections administered to the nursing mother and not to the gravid dam. Offspring of rats receiving prenatal saline injections during the last 2 wk of pregnancy appear to develop normally, whereas saline injected during the suckling period results in developmental irregularities in blood pressure regulation accordant with our previous findings (Smith and Mills, unpublished). It thus appears that postnatal maternal stress associated with handling and injections of nursing dams can induce developmental alterations in cardiovascular function of offspring, probably by transference of some factor(s) during suckling. Studies in which hypertension was induced in normotensive pups fostered by hypertensive dams (24) provide support for the hypothesis that factors affecting cardiovascular development can be transferred from mother to pup during suckling.

Postnatal maternal stress produced alterations in both the neural contribution to resting blood pressure and methoxamine pressor response (see Fig. 11), compared with normally developing Sprague-Dawley rats (see Fig. 6). The blood pressure elevation at 19–23 d of age was caused, in part, by a significantly enhanced chlorisondamine-sensitive component to resting blood pressure. The elevation in the neural component occurred despite a marked decrease in vascular smooth muscle receptor sensitivity, as indicated by an increased ED_{50} at this time. This suggests that the source of enhanced neural component was increased impulse ac-

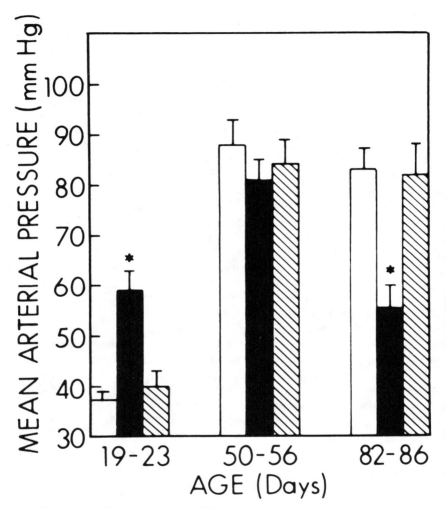

Fig. 10. Development of blood pressure after perinatal injections of alkaline saline. Open bars: uninjected controls. Solid bars: pre- and postnatal injections to dams, postnatal injections to pups. Hatched bars: postnatal injections to pups. Mean ±SEM. Asterisks: greater than control at 19–23 d ($p < 0.001$) and lower at 82–86 d of age ($p < 0.01$) by one-way anova (from ref *39*).

tivity of central origin. The increase in sympathetic tone at 19–23 d was not reflected in vascular smooth muscle function, since the pressor response to methoxamine was at the control level at 19–23

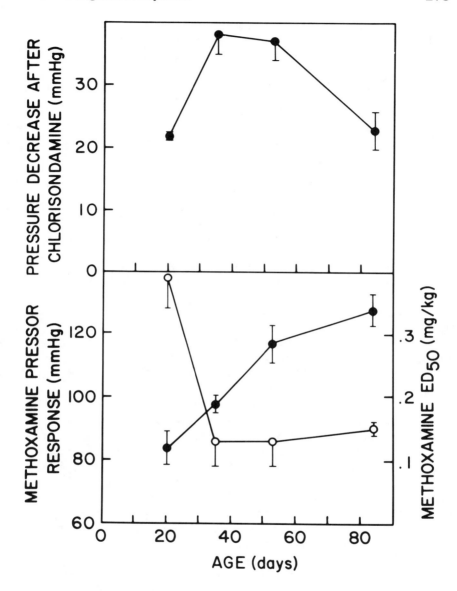

Fig. 11.   Ontogeny of neural contribution to resting blood pressure (pressure decrease after chlorisondamine, upper panel), vascular smooth reactivity (methoxamine pressor response, lower panel, closed circles), and methoxamine ED_{50} (open circles) in offspring of rats subjected to postnatal maternal injection stress.

d and significantly depressed at 33–36 d of age. Hypotension at 82–86 d was associated with a decrease in the neural contribution to resting blood pressure, whereas the methoxamine response was comparable to controls at this age. It appears that postnatal maternal stress affects both the neural contribution to resting blood pressure and vascular smooth muscle function, but not in a manner that implies a direct association.

6. Conclusions

There is sufficient evidence to conclude that the SNS plays a limited role in the early developmental phases of smooth muscle end-organ function. In the cervical sympathetic–Müller's muscle neuroeffector system of the Sprague-Dawley rat, maturational events in neonatal contractile function do not appear to be linked to ontogeny of the SNS; this is also the case for development of vascular smooth muscle responses in SHR, WKY and Sprague-Dawley rats developing normally or subjected to postnatal maternal stress. In SHR, periods in which impulse activity of central origin is enhanced are, at times, accompanied by subresponsiveness of smooth muscle. In addition, perturbations in thyroid status affecting the timing of establishment of neurotransmission in the sympathetic pathway do not produce demonstrable alterations in end-organ function of nonvascular and vascular smooth muscle. Finally, decentralization of the postganglionic neuron in newborn Sprague-Dawley rats is without effect on contractile performance through 50 d of age. Consequently, end-organ development in the neonate appears to occur to a large extent independently of variations in sympathetic innervation. Factors other than the SNS are likely to be the major determinants of development of smooth muscle function in early postnatal life. These factors may include hormones, since pups nursed by stressed dams display abnormalities in vascular smooth muscle response. Impulse activity in the postganglionic neuron is important in the later stages of development, since decentralization produces a significant deficit in the maturation of smooth muscle contractile response in Sprague-Dawley rats between 50 and 75 d of age.

Studies on the ontogeny of blood pressure regulation in normotensive Sprague-Dawley rats indicate that the neural contribution to resting blood pressure is linked to the functional status of the vascular smooth muscle. In SHR, both the neurogenic component to blood pressure and vascular smooth muscle response are elevated, but they do not appear to be directly linked. Similarly, postnatal maternal stress produces disturbances in blood pressure by affecting both neurogenic and vascular smooth muscle function, but not in a parallel fashion. It is postulated that disturbances in blood pressure regulation can occur when neural activity is matched inappropriately to end organ performance.

Acknowledgments

The author is an Established Investigator of the American Heart Association. Research was supported, in part, by grant HL-29403. Studies discussed in this chapter were completed prior to October, 1983.

References

1. Bevan, R. D. (1975) Effect of sympathetic denervation on smooth muscle cell proliferation in the growing rabbit ear artery. *Circ. Res.* **37,** 14–19.
2. Bevan, R. D. and H. Tsuru (1979) Long-term denervation of vascular smooth muscle causes not only functional but structural change. *Blood Vessels* **16,** 109–112.
3. Beven, R. D. and H. Tsuru (1981) Long-Term Influence of the Sympathetic Nervous System on Arterial Structure and Reactivity: Possible Factor in Hypertension, in *Disturbances in Neurogenic Control of the Circulation* (F. M. Aboud, H. A. Fozzard, J. P. Gilmore, and D. J. Reis, eds.) Am. Physiol Soc., Bethesda, pp. 153–160.
4. Black, I. B. (1978) Regulation of autonomic development. *Ann. Rev. Neurosci.* **1,** 183–214.
5. Blaes, N. and J. P. Boissel (1983) Growth-stimulating effect of catecholamines on rat aortic smooth muscle cells in culture. *J. Cell. Physiol.* **116,** 167–172.
6A. Brenner, G. M. and H. C. Stanton (1970) Adrenergic mechanisms responsible for submandibular salivary gland hypertrophy in the rat. *J. Pharmacol. Exp. Ther.* **173,** 166–175.
6B. Butler, S. R., M. R. Suskind, and S. M. Schanberg (1978) Maternal behavior as a regulator of polyamine biosynthesis in brain and heart of the developing rat pup. *Science,* **199,** 445–447.
7. Claycomb, W. C. (1976) Biochemical aspects of cardiac muscle differentiation: Possible control of deoxyribonucleic acid synthesis

and cell differentiation by adrenergic innervation and cyclic adenosine 3'5'-monosphosphate. *J. Biol. Chem.* **251**, 6082–6089.

8. Crutcher, K. A. (1982) Histochemical studies of sympathetic sprouting: Fluorescence morphology of noradrenergic axons. *Brain Res. Bull.* **9**, 501–508.

9. Dibner, M. D. and I. B. Black (1978) Biochemical and morphological effects of testosterone treatment on developing sympathetic neurons. *J. Neurochem.* **30**, 1479–1483.

10. Dibner, M. D., C. Mytilineou, and I. B. Black (1977) Target organ regulation of sympathetic neuron development. *Brain Res.* **123**, 301–310.

11. Finch, L., G. Haeusler, and H. TThoenen (1973) A comparison of the effects of chemical sympathectomy by 6-hydroxydopamine in newborn and adult rats. *Br. J. Pharmac.* **47**, 249–260.

12. Folkow, B., M. Hallback, Y. Lundgren, R. Sivertsson, and R. Weiss (1973) Importance of adaptive changes in vascular design for establishment of primary hypertension, studied in man and in spontaneously hypertensive rats. *Circulation Res.* **32**, (suppl. 1) 2–16.

13. Gootman, P. M., N. M. Buckley, and N. Gootman (1979) Postnatal maturation of neural control of the circulation. *Rev. Perinatal Med.* **3**, 1–72.

14. Gripois, D. and A. Diarra (1983) Adrenal epinephrine depletion after insulin-induced hypoglycaemia in young hypothyroid rats. *Molec. Cell. Endocrinol.* **30**, 241–245.

15. Hendry, I. A. (1976) Control in the development of the vertebrate sympathetic nervous system. *Rev. Neurosci.* **2**, 149–194.

16. Hendry, I. A. and J. Campbell (1976) Morphometric analysis of rat superior cervical ganglion after axotomy and nerve growth factor treatment. *J. Neurocytol.* **5**, 351–360.

17. Judy, W. V. and S. K. Farrell (1979) Arterial baroreceptor reflex control of sympathetic nerve activity in the Spontaneously Hypertensive Rat. *Hypertension* **1**, 605–614.

18. Klein, R. M. and J. Torres (1978) Analysis of intestinal cell proliferation after guanethidine-induced sympathectomy. *Cell. Tissue Res.* **195**, 239–250.

19. Lau, C. and T. A. Slotkin (1979) Accelerated development of rat sympathetic neurotransmission caused by neonatal triiodothyronine administration. *J. Pharmacol. Exp. Ther.* **208**, 485–490.

20. Lau, C. and T. A. Slotkin (1980) Maturation of sympathetic neurotransmission in the rat heart. II. Enhanced development of presynaptic and postsynaptic components of noradrenergic synapses as a result of neonatal hyperthyroidism. *J. Pharmacol. Exp. Ther.* **212**, 126–130.

21. Lau, C. and T. A. Slotkin (1982) Maturation of sympathetic neurotransmission in the rat heart. VIII. Slowed development of noradrenergic synapses resulting from hypothyroidism. *J. Pharmacol. Exp. Ther.* **220**, 629–636.

22. Lipp, J. A. M. and A. M. Rudolph (1972) Sympathetic nerve development in the rat and guinea-pig heart. *Biol. Neonate* **21**, 76–82.

23. McMurphy, D. S. and B. Ljung (1978) Neuroeffector maturity of portal veins from newborn rats, rabbits, cats, and guinea pigs. *Acta Physiol. Scand.* **102**, 218–223.

24. McMurty, J. P., G. L. Wright, and B. C. Wexler (1981) Spontaneous hypertension in cross-suckled rats. *Science* **211**, 1173–1175.

25. Mifflin, S. W. and D. L. Kunze (1982) Rapid resetting of low pressure vagal receptors in the superior vena cava of the rat. *Circ. Res.* **51**, 241–249.

26. Mikhail, Y. and Z. Mahran (1965) Innervation of the cortical and medullary portion of the adrenal gland of the rat during postnatal life. *Anat. Rec.* **152**, 431–437.

27. Mueller, S. M. and E. J. Ertel (1983) Association between sympathetic nerve activity and cerebrovascular protection in young spontaneously hypertensive rats. *Stroke* **14**, 88–92.

28. Muir, T. C., D. Pollock, and C. J. Turner (1975) The effects of electrical stimulation of the autonomic nerves and of drugs on the size of salivary glands and their rate of cell division. *J. Pharmacol. Exp. Ther.* **195**, 372–381.

29. Page, R. E. (1973) The distribution and innervation of the extraocular smooth muscle in the orbit of the rat. *Acta Anat.* **85**, 10–18.

30. Rusterholz, D. B. and S. M. Mueller (1982) Sympathetic nerves exert a chronic influence on the intact vasculature that is age related. *Ann. Neurol.* **11**, 365–371.

31. Schneyer, C. A. (1972) Regulation of Salivary Gland Size, in *Regulation of Organ and Tissue Growth* (R. J. Gross, ed.) Academic Press, NY, pp. 211–232.

32. Scott, T. M. and S. C. Pang (1983) The correlation between the development of sympathetic innervation and the development of medial hypertrophy in jejunal arteries in normotensive and spontaneously hypertensive rats. *J. Autonomic Nervous System* **8**, 25–32.

33. Seidler, F. J. and T. A. Slotkin (1979) Presynaptic and postsynaptic contributions to ontogeny of sympathetic control of heart rate in the preweanling rat. *Br. J. Pharmac.* **65**, 431–434.

34. Selye, H., R. Veilleux, and M. Cantin (1961) Excessive stimulation of salivary gland growth by isoproterenol. *Science* **133**, 44–45.

35. Shibata, S., K. Kurahashi, and M. Kuchii (1973) A possible etiology of contractility impairment of vascular smooth muscle from spontaneously hypertensive rats. *J. Pharmacol. Exp. Ther.* **185**, 406–417.

36. Slotkin, T. A. (1984) Endocrine Control of Synaptic Development in the Sympathetic Nervous System: The Cardiac-Sympathetic Axis, in this publication.

37. Slotkin, T. A., P. G. Smith, C. Lau, and D. L. Bareis (1980) Functional Aspects of Development of Catecholamine Biosynthesis and Release in the Sympathetic Nervous System, in *Biogenic Amines in Development* (H. Parvez and A. Parvez, eds.) Elsevier, Amsterdam, pp. 29–48.

38. Smith, P. G., G. Evoniuk, C. W. Poston, and E. Mills (1983) Relation between functional maturation of cervical sympathetic innervation and ontogeny of alpha-noradrenergic smooth muscle contraction in the rat. *Neuroscience* **8**, 609–616.

39. Smith, P. G. and E. Mills (1983) Abnormal development of blood pressure and growth in rats exposed to perinatal injection stress. *Life Sci* **32**, 2497–2501.

40. Smith, P. G., E. Mills, and T. A. Slotkin (1979) Effects of T3 on neonatal rat heart innervation: Ultrastructure. *Trans. Am. Soc. Neurochem.* **10**, 196.

41. Smith, P. G., E. Mills, and T. A. Slotkin (1981) Maturation of sympathetic neurotransmission in the efferent pathway to the rat heart: Ultrastructural analysis of ganglionic synaptogenesis in euthyroid and hyperthyroid neonates. *Neuroscience* **6**, 911–918.

42. Smith, P. G., C. W. Poston, and E. Mills (1982) Abnormal functional development of sympathetic nerve terminal-smooth muscle complex in neonatal spontaneously hypertensive rats. *Life Sci.* **31**, 889–892.

43. Smith, P. G., C. W. Poston, and E. Mills (1984) Ontogeny of neural and nonneural contributions to arterial blood pressure in spontaneously hypertensive rats. *Hypertension,* **6**, 54–60.

44. Smith, P. G., T. A. Slotkin, and E. Mills (1982) Development of sympathetic ganglionic neurotransmission in the neonatal rat. Pre- and postganglionic nerve response to asphyxia and 2-deoxyglucose. *Neuroscience* **7**, 501–507.

45. Smolen, A. and G. Raisman (1980) Synapse formation in the rat superior cervical ganglion during normal development and after deafferentation. *Brain Res.* **181**, 315–323.

46. Trendelenberg, U. (1972) Classification of Sympathomimetic Amines, in *Handbuch der Experimentellen Pharmakologie* (H. Blaschko and E. Muschool, eds.) Springer-Verlag, Berlin, pp. 336–362.

47. Trendelenberg, U., R. A. Maxwell, and S. Pluchino (1970) Methoxamine as a tool to assess the importance of intraneuronal uptake of 1-norepinephrine in the cat's nictitating membrane. *J. Pharmac. Exp. Ther.* **172**, 91–99.

48. Wekstein, D. R. (1965) Heart rate of the preweanling rat and its autonomic control. *Am. J. Physiol.* **208**, 1259–1262.

Chapter 11

Development of Central Autonomic Regulation of Cardiovascular Function

Phyllis M. Gootman

1. Introduction

Although investigators have been interested in autonomic regulation of cardiovascular function for well over 100 yr, regulation in the perinatal period has been investigated systematically for only the last 25 yr. Whereas considerable progress has been made concerning neural regulation of cardiovascular function in the adult, the amount of quantitative information concerning the perinatal period is quite limited (5,32,35,38,52). It is only recently that the neonatal cardiovascular regulatory system and its postnatal maturation, particularly with reference to asynchronous development of autonomic regulation, has begun to interest investigators and information is becoming available. In collaboration with a number of different investigators, I have been examining postnatal development of central neural regulation of cardiovascular function. This chapter will review the information that we have gathered over the past 18 yr, and attempt to place it in the framework of ongoing studies of postnatal autonomic regulation of cardiovascular function by other investigators.

We have employed four primary methods of investigation: (a) changing inputs from visceral and/or somatic afferents (11,

30,31,34,36,48,50,53,54,72,73); (b) stimulating central vasoactive sites (30,33,34,50–53); (c) subjecting animals to stresses, e.g., hemorrhage (4,50,53), hypercapnia (50,53,72), and hypoxia (28,29,50,51,53); and (d) recording from efferent sympathetic nerves (32,37,38,50,53).

Our work has shown that, depending upon the afferent stimulated, the neonatal pig has almost the adult pattern of response to afferent stimulation [e.g., J-receptor reflexes (50,73)], moderately slow postnatal maturation [e.g., baroreceptor reflexes (11,30,34, 35,50)], or a prolonged period of postnatal maturation [e.g., Bezold–Jarisch reflex (48,50,54)]. Because of the age-related differences in adrenergic mechanisms observed in different vascular beds (5–12,34,50), a limited study was carried out on isolated vessels from different vascular beds (50,81,82). Since afferent somatic and visceral stimulation elicited marked reflex responses in the renal and splanchnic vasculatures, we decided to examine spontaneous activity of their major preganglionic sympathetic nerve supply (32,37,38,53), i.e., the greater splanchnic nerve. These recordings indicate whether the discharge pattern of spontaneous splanchnic activity in newborn pigs is similar to that observed in adult mammals (14,15,40–47,49).

One important area of investigation involves the ability of the neonate's cardiovascular regulatory mechanisms to respond to the effects of adverse stresses. The vulnerability to adverse stresses is higher in the newborn period than at any other stage of life in the child, i.e., infant mortality is highest during the first month. It is also during this period that the infant must cope with the changes of extrauterine existence in addition to other possible adverse stresses. Therefore, one of the areas of research that we have been pursuing is the response of the piglet, from birth to 2 mo, to a variety of stresses including hypercapnia, hypoxia, and hemorrhage. A subset of experiments has involved simulation of gastroesophageal reflux (62), a mechanism that has been implicated by a number of different investigators in the etiology of sudden infant death syndrome (56,65). Because the gastrointestinal tract is a frequent site of pathological changes in stressed neonates, we examined regional circulatory changes (particularly superior mesenteric) in piglets during feeding under both control and stress conditions (53,84–87).

1.1. General Methodology

1.1.1. The Animal Preparations

Animals ranging in age from birth to 2-mo-old and sexually mature miniature swine were anesthetized with either halothane (0.25–0.5%, depending on age) or pentobarbital sodium (10–25 mg/kg, depending on age). Animals were grouped according to age: <1d, 2–4 d, 1 wk, 1 mo, 2 mo. Details of general methodologies as well as explanations of specific protocols can be obtained from earlier publications (*16,17,26,27,30,34,53,72*). Femoral, carotid, renal, and mesenteric arterial blood flows were registered with Biotronix or Statham flow transducers. At least three flows were recorded in each experiment.

A control period of at least 1 h was allowed for stabilization of cardiovascular function before proceeding with a particular experimental protocol. The protocols included: (1) effects of stress, i.e., hypoxia, hypercapnia, hemorrhage; (2) baroreceptor stimulation and inhibition; (3) stimulation of sites in the central nervous system; and (4) combined interactions of afferent stimulation.

In another series of experiments, the left greater splanchnic nerve was located retroperitoneally and subdiaphragmatically, cut just proximal to the celiac ganglion, and desheathed in pigs more than 1 wk of age. The left phrenic nerve was located in the dorsolateral region of the neck, tied as far caudally as possible, and desheathed in piglets 1 wk of age and older. Recordings from each nerve were obtained from bipolar hook electrodes, with the nerve immersed in a pool of mineral oil through which 5% CO_2/95% O_2 had been bubbled (*38*). Efferent splanchnic activity (monophasic recordings, bandpass 0.1–1000 Hz) and efferent phrenic discharge (bandpass 10–10,000 Hz) were recorded on analog tape. The splanchnic bandpass allowed faithful reproduction of slow potential changes (*14*). Phrenic activity was used as the monitor of the respiratory rhythm generator (*13*). Relative changes of splanchnic and phrenic activity were examined by integration procedures. Integrated neuronal activity was obtained by passing the signals through an integrator with a time constant of 100 ms. Periodicities in spontaneous splanchnic activity were studied by auto- and crosscorrelation techniques, and by power spectral analyses using a DEC PDP 11/45 computer (*38*).

2. Results and Discussion

2.1. Effects of Stress on Cardiovascular Function

Stressing animals acutely by hypercapnia (34,50,72), hemorrhage (4,33,34,50,72), or hypoxia (28,50,51,53), common stresses of the perinatal period, provides one method of testing the functional capacity of the cardiovascular regulatory system.

2.2.1. Hemorrhage

The cardiovascular effects of graded arterial or venous hemorrhage were evaluated in developing swine at ages of <1 d, 2–5 d, 1 wk, and 2 wk (4). Serial aliquots of 5 mL/kg arterial or venous blood were removed at 3–4-min intervals to a cumulative total of 20 mL/kg. For each age group, a statistically significant decrease in mean aortic pressure was produced by 5, 10, 15, and 20 mL/kg arterial or venous hemorrhage. The volume of blood withdrawn had a significant effect on the decrease in mean aortic pressure, to both arterial and venous hemorrhage: The decrease was greater the larger the volume withdrawn at all ages. Aortic pressure decreases to arterial hemorrhage were significantly larger in the younger animals than in the older. However, aortic pressure responses to venous hemorrhage were not age-dependent (4). Tachycardia occurred in most animals. The increases in heart rate were not different between arterial and venous hemorrhage and did not differ with age.

2.1.2. Compensation to Hemorrhage

Compensation was defined in this study (4) as the ability of the neonate's cardiovascular regulatory system to restore aortic pressure to within 20% of the prehemorrhage level before the next withdrawal of blood. Table 1 presents the maximum amount of blood withdrawn at which compensation still occurred after either arterial or venous hemorrhage in each age group. Compensation to arterial hemorrhage (left column) was not retained in the 1–5 d group when the volume of blood loss was greater than 10 mL/kg, in contrast to the effect in the older animals. However, compensation to venous hemorrhage (right column) did not differ with age. In the 1–5-d-old animals, compensation was demonstrable after a larger volume of blood loss during venous hemorrhage

Table 1
Largest Volume of Blood Withdrawn
(mL/kg)[a] Following Which
Compensation[b] Was Retained[c]

Age	Arterial hemorrhage	Venous hemorrhage
1 d	8.5 ± 1.2 (n = 9)	14.6 ± 1.5 (n = 6)
2–5 d	8.8 ± 2.1 (n = 7)	16.0 ± 2.1 (n = 9)
1 wk	15.0 ± 2.0 (n = 7)	14.8 ± 2.1 (n = 7)
2 wk	18.2 ± 1.5 (n = 8)	15.9 ± 1.2 (n = 7)

[a]Mean change ±SEM; $p \leq 0.01$ for all groups.
[b]Mean aortic pressure ≤ 20% of pre-hemorrhage. Level by 3–4 min after blood withdrawal.
[c]From Buckley et al. (4), with permission of the publisher.

than during arterial hemorrhage, a difference that was not observed in older animals.

These results indicated maturational differences in the ability of the central nervous cardiovascular regulatory system to compensate reflexively for the stress of hemorrhage (4,33,50,53,72).

A recent paper by Le Gal (67) supported our general findings and indicated marked age-related differences in longer-term responses to hemorrhage, e.g., plasma proteins, increase in erythropoietic activity.

2.1.3. Hypoxia

The effects of moderate and severe hypoxia were studied in three age groups of developing swine (*see* Table 2) (50,53). Under control conditions, arterial blood gases and pH were within the normal ranges. When the percent of inspired O_2 was decreased, thereby decreasing arterial PO_2, aortic pressure and carotid resistance decreased significantly in piglets 2 wk old or younger. However, in 2-mo-old swine, aortic pressure increased during severe hypoxia, without change in carotid resistance to either level of

Table 2
Cardiovascular Effects of Hypoxia in Neonatal Swine[a]

PO_2 (torr)	59.5 ± 2.1	31.6 ± 0.6
pH	7.40 ± 0.06	7.33 ± 0.01
2–4 d		
(n = 12)		
HR (bpm)	+18.3 ± 9.3	+15.3 ± 10.5
AoP (mm Hg)	−8.8 ± 2.7[b]	−20.2 ± 4.7[b]
RenF (mL/min)	−0.9 ± 0.8	−5.0 ± 1.2[b]
RenR (PRU)	−0.7 ± 0.6	+0.5 ± 1.1
FemF (mL/min)	+1.3 ± 1.2	−2.8 ± 1.6
FemR (PRU)	−1.7 ± 0.5[b]	−0.9 ± 1.0
CarF (mL/min)	+13.4 ± 5.1[b]	+27.3 ± 5.7[b]
CarR (PRU)	−1.0 ± 0.2[b]	−1.9 ± 0.5[b]
MesF (mL/min)	−3.6 ± 3.9	−19.3 ± 6.2[b]
MesR (PRU)	−0.2 ± 0.1	−0.2 ± 0.2
2 wk		
(n = 14)		
HR (bpm)	+7.3 ± 4.0	+4.3 ± 11.1
AoP (mm Hg)	−6.4 ± 1.7[b]	−36.4 ± 6.2[b]
RenF (mL/min)	+0.5 ± 1.1	−8.5 ± 1.7[b]
RenR (PRU)	−0.3 ± 0.3	+2.1 ± 1.1
FemF (mL/min)	+1.2 ± 0.7	−2.1 ± 2.0
FemR (PRU)	−1.3 ± 0.6	−2.2 ± 0.7[b]
CarF (mL/min)	+8.4 ± 2.5[b]	+24.6 ± 5.5[b]
CarR (PRU)	−0.5 ± 0.2[b]	−1.5 ± 0.02[b]
MesF (mL/min)	+0.6 ± 5.0	−11.4 ± 6.9
MesR (PRU)	−0.3 ± 0.06[b]	−0.4 ± 0.1[b]
2 mo		
(n = 7)		
HR (bpm)	+31.8 ± 6.7[b]	+29.4 ± 14.2
AoP (mm Hg)	+10.4 ± 3.3[b]	+2.5 ± 5.4
RenF (mL/min)	+5.0 ± 4.5	−34.7 ± 7.7[b]
RenR (PRU)	+0.09 ± 0.02[b]	+0.61 ± 0.13[b]
FemF (mL/min)	+12.0 ± 3.9[b]	−1.6 ± 8.7
FemR (PRU)	−0.03 ± 0.11	−0.08 ± 0.35
CarF (mL/min)	+23.5 ± 6.9[b]	+79.9 ± 10.1[b]

Table 2 (*continued*)

CarR (PRU)	-0.03 ± 0.02	-0.24 ± 0.09
MesF (mL/min)	-3.4 ± 22.8	-237 ± 49.8^b
MesR (PRU)	$+0.02 \pm 0.008^b$	$+0.29 \pm 0.06^b$

[a]Mean change $\pm$ SEM.
[b]$p \leq 0.05$.
Abbreviations: HR, heart rate; AoP, aortic pressure; RenF, renal arterial flow; RenR, renal resistance; FemF, femoral arterial flow; FemR, femoral resistance; CarF, carotid arterial flow; CarR, carotid resistance; MesF, mesenteric arterial flow; MesR, mesenteric resistance; PRU, peripheral resistance units; n, number of animals. Portions of this table were adapted from Table 2 of ref. 53 and published in ref. 50, with permision of the publishers.

hypoxia. Mesenteric resistance decreased during severe hypoxia in the 2-wk-old piglets, whereas there was an increase in mesenteric resistance to both moderate and severe hypoxia in 2-mo-old swine. On the other hand, renal resistance increased only in swine 2 mo of age, and only during severe hypoxia (*see* Table 1). It appears, therefore, that there is an age-related development of the physiologic response to hypoxia.

2.1.4. Hypercapnia

During hypercapnia (*see* Fig. 1), produced by having animals breathe gas mixtures of CO_2 and O_2, arterial PO_2 was maintained above 100 torr as in the control period when CO_2 and pH were within normal ranges. Significant increases in heart rate and mean aortic pressure occurred during hypercapnia (50,72). Under moderate hypercapnic acidosis (Fig. 1, open bars), carotid and renal flows increased with little change in resistance, and with no consistent pattern in the femoral vascular bed. However, under severe hypercapnic acidosis (Fig. 1, shaded bars), carotid and renal flows decreased and resistance increased significantly in both of these beds (50,72).

Thus, the responses observed in newborn piglets to stresses occurring quite frequently in the perinatal period suggest the cardiovascular controlling system is not completely mature at birth, and is, therefore, not completely compensating for these stresses in the manner observed in adults of other species (24,40).

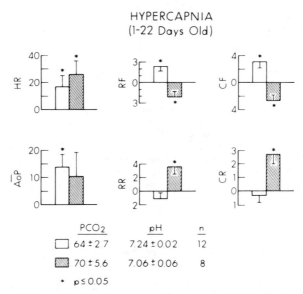

Fig. 1. Effect of two degrees of hypercapnia (indicated on figure) on mean aortic pressure (AoP, mm Hg), heart rate (HR, bpm), renal arterial flow (RF, mL/min) and resistance (RR, PRU), and carotid arterial flow (CF, mL/min) and resistance (CR, PRU) in swine.

2.2. Effects of Afferent Stimulation on Cardiovascular Function

Studies concerning age of onset and rate of maturation of cardiovascular reflexes in the fetus and newborn have been reviewed recently (32,35,39). The reported variations in response to the effects of afferent stimulation may reflect species differences, different methods of stimulation, and the absence, presence, type, and/or depth of anesthesia. Since our recent review (39), there have been some further reports of studies on the possible role of baroreceptors in the regulation of fetal circulation (19,60,88). The results of these investigations suggest that baroreceptors are probably not critically involved in maintenance of fetal mean aortic pressure and heart rate.

2.2.1. Baroreceptor Reflex

Baroreceptor stimulation has been found to reduce femoral and renal blood flows in developing swine (9,11,34,50). We have recently observed that baroreceptor stimulation (74) with equipressor doses of phenylephrine (30 mm Hg increase in blood pressure) in 2–4-d-old and 2-mo-old pigs (*see* Table 3) elicited a reflex bradycardia that was significant in both age groups, although the response was significantly greater in the older piglets (*see* Table 2).

A 45% increase in mean aortic pressure (Table 3) was required to elicit a 9% decrease in heart rate in piglets 2–4 d of age. On the other hand, a 26% increase in mean aortic pressure elicited the 25.6 beats/minute (bpm) decrease in heart rate in 2-mo-olds. Furthermore, immaturity of alpha adrenergic receptors in developing swine has been previously reported by us (9,12,53) and is additionally documented by the larger dosage requirement of phenylephrine in the younger animals (5 μg/kg) as compared to 2-mo-old swine (1.5 μg/kg).

It would appear, based on the results of our investigation of age-related changes in baroreceptor function, plus those of other investigators studying development in a number of different species (32,35,39), that in most species evidence exists for age-related development of baroreceptor reflexes. It is of particular interest to

Table 3
Cardiovascular Effects of Equipressor Doses (iv) of
Phenylephrine in Developing Swine[a]

Cardiovascular parameters	Age	
	2–4 d ($n = 23$)	2 mo ($n = 31$)
AoP (mm Hg)	$+29.9 \pm 1.0^b$	$+30.4 \pm 1.0$
HR (mm Hg)	-19.1 ± 2.3^c	-25.6 ± 2.2
MesF (mL/min)	-13.3 ± 3.0	-101.8 ± 8.2
MesR (PRU)	$+1.5 \pm 0.3$	$+0.21 \pm 0.02$
RenF (mL/min)	-3.0 ± 0.5	-86.8 ± 6.7
RenR (PRU)	$+5.8 \pm 1.2$	$+1.67 \pm 0.26$

[a]Values are mean change ±SEM; all responses are significantly different from 0 ($p \leq 0.05$).

[b]Comparison of HR responses; significantly different ($p \leq 0.01$).

Abbreviations: AoP, mean aortic pressure; HR, heart rate; MesF, mean superior mesenteric arterial flow; MesR, mesenteric resistance; RenF, mean renal arterial flow; RenR, renal resistance; n, number of animals.

this reviewer that we were finally able to elicit small but significant changes in heart rate with simultaneous stimulation of all baroreceptors. Furthermore, the magnitude of the rise in arterial pressure required for stimulation was considerably greater than would normally be observed in neonates (22,63,66, 70,75,83). Marked age-related maturation of baroreceptors can be seen in our studies (and those of other investigators) when one examines the reflex responses to alterations in activity from one set of baroreceptors, e.g., carotid sinus. A brief stimulus at one carotid sinus (*see* Fig. 2) did not elicit cardiac responses in piglets <1 w old (9,11).

Carotid sinus inhibition by bilateral common carotid occlusion in 66 piglets was accompanied by similar pressor responses at all ages. These responses were significantly smaller than those obtained in eight mature swine, regardless of anesthetic agents employed (9,11,50). There was no immediate heart rate response in most piglets, and no bradycardia at the peak of the pressor effect, except in three older piglets and in the mature swine (*see* Fig. 3). The lack of a cardiac response to bilateral common carotid occlu-

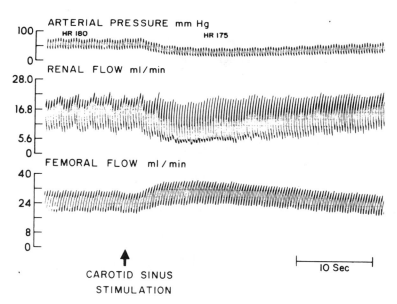

Fig. 2.   Reflex effects of unilateral carotid sinus stimulation on aortic pressure and renal and femoral arterial blood flows in a 3-d-old piglet with vagi intact (from ref. *11*, with permission of the publisher).

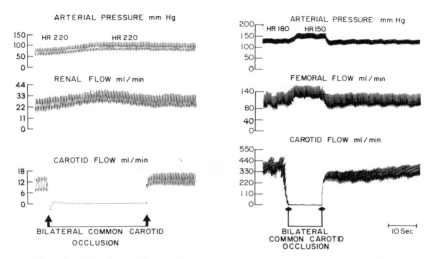

Fig. 3. Reflex effects of bilateral common carotid occlusion on aortic arterial pressure and regional blood flows in a 3-d-old piglet with vagi intact (left), and in a 5–6-mo-old Mini Swine with vagi intact (right) (from ref. *11*, with permission of the publisher).

sion was observed under light anesthesia as well as without general anesthesia (*see* Fig. 4).

Immaturity of the baroreceptor reflex has been reported in other species, including absence of heart rate changes during inhibition of the carotid sinus baroreceptors (*21,32,35,39*).

Recently, Tomomatsu and Nisha (*78*) obtained evidence, using single fiber recording techniques in newborn and adult rabbits, that the carotid sinus baroreceptors were not underdeveloped in newborns. Their results in combination with those from other laboratories (*32*), including our own (*9,11,34,50*), suggest that the sites of immaturity in neonates are within the central controlling system, its output (autonomic discharge), and/or the effectors (cardiac and vascular smooth muscle).

2.2.2. Somatosympathetic Reflexes

The circulatory responses of adult mammals to stimulation of the central end of a mixed nerve, such as the sciatic or median nerve of the brachial plexus, are known to be mediated by sympathetic activation or inhibition (*32*). In piglets, when the sciatic nerve or the median nerve of the brachial plexus was stimulated, several patterns of responses were obtained, depending on the combina-

WITH HALOTHANE

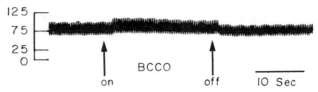

Fig. 4. Aortic pressure response to bilateral common carotid occlusion (BCCO) in a 12-h-old piglet under 0.25% halothane (upper panel) and local anesthesia with xylocaine (lower panel) (from ref. *35*, with permission of the publisher).

tion of stimulus parameters (*11,34,36,50,72*). High frequency or high intensity stimulation of somatic nerves (*see* Fig. 5, upper left) produced increases in blood pressure and in renal and femoral vascular resistances (*36,50,72*). Low frequency or low intensity stimulation produced decreases in pressure and flows with no accompanying change in pressure and no accompanying heart rate changes in piglets up to 2 wk old (*11,36*).

We also examined the effects of combined somatic afferent stimulation on cardiovascular responses in newborn piglets (*36,50*). Combined high frequency and intensity stimulation produced responses smaller than the estimated sum of the responses to each of the nerves alone (*36*) (*see* Table 4). Low-frequency or low-intensity somatic nerve stimulation resulted in depressor responses accompanied by significant decreases in mean femoral flow without change in heart rates; combined stimulation produced responses smaller than the estimated sums. When pressor (i.e., high frequency) and depressor (i.e., low frequency) patterns of stimulation were combined, pressor responses always dominated. The absence of facilitation to any combination of interactions of somatic afferent input implies that the cardiovascular regulatory system is not fully developed at birth, and undergoes postnatal maturation. The results of depressor afferent stimulation indicate that the cardiovascular regulatory system in neonatal

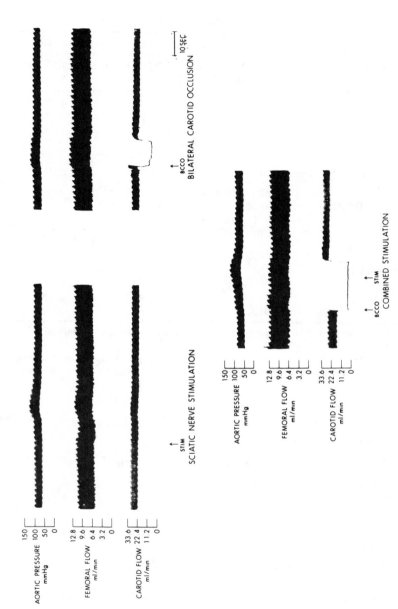

Fig. 5.   Responses obtained in a 2-4-d-old piglet to interactions between high frequency sciatic nerve stimulation (0.5 mA, 50 Hz for 10 s) and bilateral common carotid occlusion (BCCO). Top set of traces: Left; sciatic nerve stimulation (stim) alone. Right; bilateral common carotid occlusion (BCCO) alone. Bottom set of traces: combined stimulation with BCCO.

Table 4
Somatic and Visceral Afferent Interactions[a]

Stimulus	Aortic pressure	Femoral flow
High frequency sciatic (SNS) and brachial (BNS) nerve stimulation ($n = 6$; age 2–3 wk)		
SNS	22.0 ± 3.2	12.6 ± 4.7
BNS	22.8 ± 5.7	20.2 ± 6.1
SNS then BNS	32.8 ± 5.6	26.4 ± 9.1
BNS then SNS	34.6 ± 7.4	32.0 ± 7.6
Simultaneous	28.3 ± 3.9	25.2 ± 6.5
Albebraic sum of SNS and BNS	44.8 ± 7.9	32.8 ± 9.8
High frequency sciatic nerve stimulaton (SNS) and Bilateral common carotid occlusion (BCCO) ($n = 13$; age 1–22 d)		
SNS	22.6 ± 4.7	11.8 ± 2.2
BCCO	13.9 ± 4.1	13.7 ± 4.5
SNS then BCCO	30.9 ± 5.5	31.5 ± 17.1[b]
BCCO then SNS	31.4 ± 5.5	17.1 ± 4.8
Simultaneous	38.1 ± 6.2	25.4 ± 5.8
Algebraic sum of SNS and (BCCO)	40.5 ± 6.5	26.7 ± 6.0

[a]Percent increase above control expressed as arithmatic mean ± standard error $p \leq 0.05$.

[b]Not significant due to variations in magnitude of change. Adapted from Table II of ref. *36*, and Table I of ref. *31* and published in ref. *50*, with permission of the publishers.

swine is functioning at a low level of activity and, therefore, that little inhibition can occur.

2.2.3. Combinations of Visceral and Somatic Nerve Stimulation

Our results of combined baroreceptor stimulation and inhibition with somatic nerve stimulation have shown that the neonatal piglets' CNS regulatory system is capable of some integration of afferent inputs (*31,34,50*).

Combinations of sciatic nerve stimulation and carotid sinus stimulation by infusion or inhibition by bilateral common carotid occlusion were studied in piglets up to 2 wk of age. Combined stimulation of somatic and visceral afferents produced no heart rate changes. Combinations of pressor stimuli (i.e., bilateral com-

mon carotid occlusion and high frequency sciatic nerve stimulation) produced the clearest evidence for an additive effect in mean
aortic pressure and femoral flow (Fig. 5). However, combinations
of depressor stimuli (that is, carotid sinus stimulation) and low
frequency or intensity of somatic afferent stimulation resulted in
occlusive responses. In combinations of pressor with depressor
stimuli, the pressor pattern of responses dominated (31,34,50).

The observation of interactions between somatic and visceral
affferent inputs indicates that neonatal pigs are capable of some
central nervous system integration of afferent inputs. In contrast
with our findings concerning interactions of two somatic afferent
inputs, facilitation was observed with some combinations of visceral and somatic afferents. It is of interest that two similar types
of excitatory inputs (34,36,50) never led to summation or facilitation, whereas two dissimilar inputs (somatic and visceral) did
lead to summation or facilitation (31,50). This contrast suggests
that although peripheral vascular effectors are capable of larger
responses to combined stimulation than to either stimulus alone,
the neonatal integrating system appears to be better equipped to
handle different, rather than similar, inputs. This implies that
various substations of the cardiovascular controlling system involved in the total response may be maturing at different rates.
Our results supply further support for the impression that there is
a low level of basal excitatory activity in the neonate's
cardiovascular controlling system, since only occlusion was observed in the responses to combined stimulation of inhibitory inputs. Our results from these interactions studies suggest that the
low blood pressure observed at birth in most species may be a
consequence of low basal efferent control of vascular smooth
muscle (31,34,36,50). Certainly there is now considerable evidence of maturation of the adrenergic receptors (5,8), as well as of
efferent autonomic outflow (32,37,38,50,53,69).

2.2.4. Cardiopulmonary Receptors

The use of mixtures of Veratrum viride alkaloids (VVA) (2,64,71)
has shown that the Bezold-Jarisch reflex (48,50,54) can be elicited
in developing swine (see Fig. 3 in 50). Only one other laboratory,
to the best of my knowledge, has examined the Bezold-Jarisch
reflex in the perinatal period. After administration of veratridine,
heart rate and aortic blood pressure decreased in term fetal lambs
but increased in immature lambs (1). In 1–2-d-old swine, the heart

rate component of the Bezold-Jarisch reflex is present in response
to even the lowest dose of VVA. However, the hypotension and
peripheral vasodilation observed in all animals >1 wk of age were
not present in the youngest animals, even when the dose of VVA
was increased to toxic levels. In contrast to the baroreceptor heart
rate reflex, in which the cardiac component is not present in very
young animals, the cardiac component of cardiopulmonary receptor reflexes was present at birth (48,50). Mesenteric, renal, and
femoral circulations dilated in piglets >1 wk of age in response to
stimulation of cardiopulmonary receptors with VVA (*see* Fig. 6);
this was not observed in piglets <2 d old. Vagotomy abolished
the cardiovascular responses to VVA, indicating that the afferent
pathway is carried in the vagus trunk (18,71,77). In order to elicit
responses to veratrum veride (i.e., the Bezold-Jarisch reflex) considerably greater doses were required as compared to those reported for adults of other species (18,20,71). However, once the
cardiopulmonary receptors were stimulated, at least in piglets >2
wk of age, the pattern of responses was similar to that observed in
other mammalian adult species, i.e., decreases in blood pressure,
heart rate, and peripheral vascular resistance in the renal and
femoral beds (20,71,77). Furthermore, we have shown that in
swine the mesenteric vascular bed also dilated (Fig. 6) (48,54).

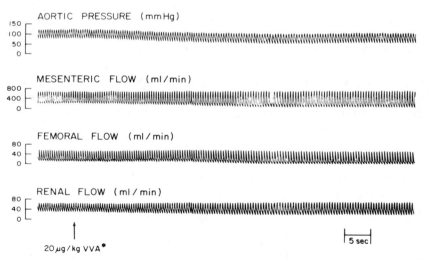

Fig. 6. Circulatory responses to left ventricular injection of 20
µg/kg of a mixture of veratrum veride alkaloids (VVA).

2.2.5. Juxtacapillary (J) Receptors

We have also shown that neonatal swine have mature cardiorespiratory responses to stimulation of J receptors with phenyl diguanide (50,73) similar to those reported for adult mammals (*see* refs. in *32*). J-receptor stimulation (*see* Fig. 7), even in the youngest piglets studied (2 d), elicited a fall in aortic pressure, bradycardia, and transient cardiac arrhythmias consisting of second and third degree heart block (*see* Fig. 8 and Fig. 4 in *50*). Cardiac arrhythmias had not previously been reported to the best knowledge of the author. Apnea followed by rapid shallow breathing was also observed in most animals. This respiratory effect was more pronounced in the younger animals. After atropine, phenyl diguanide elicited increased aortic pressure; the cardiac rhythm and arrhythmias were abolished. However, the apneic episodes were still present. Bilateral vagotomy abolished the hypotensive effect of phenyl diguanide. Thus, the pattern of response to J-receptor stimulation seen in adults of other species, i.e., hypotension, bradycardia, and apnea followed by tachypnea (*see* refs. in *32*), was observed in piglets by 2 d of age. This is unlike reports in kittens in whom the J-receptor reflex could not be elicited before 1 wk of age (*32,61*).

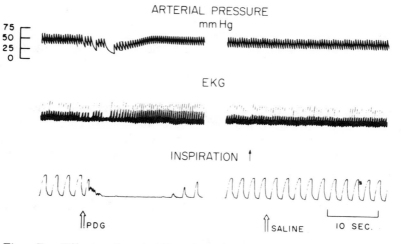

Fig. 7.   Effects of a single intraarterial dose of 77 μg/kg phenyl diguanide (PDG) in a 7-d-old piglet (left). Effect of control injection of saline (right).

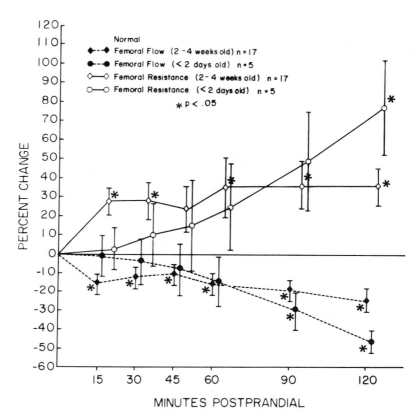

Fig. 8. Postprandial changes in femoral blood flow and femoral vascular resistance in two age groups of piglets. Values are mean ±SEM, in percent change from control measurements.

The results of different types of afferent stimulation have shown age-related postnatal maturation of these different reflexes, as well as species differences. The results also suggest immaturity of the vasomotor controlling system when combinations of afferent stimulation were employed.

2.3. Cardiovascular Responses to Alterations in Afferent Discharge From the Upper Gastrointestinal Tract

2.3.1. Feeding

Food intake is a stimulus to mesenteric blood flow in adult mammals, including humans. Within 20 min after a meal, superior mesenteric vascular resistance begins to decrease (55) to a level that represents an increase of 28–132% in superior mesenteric flow. Because the gastrointestinal tract is a frequent site of pathological changes (23,25,79) in stressed neonates, we decided to determine whether age-related changes in superior mesenteric blood flow occurred with feeding in piglets <2 d old and 2–4 wk old under control and stressed (hemorrhage) conditions (53, 84–87a). Gavage feeding was carried out with commercially formulated cow's milk. Feeding induced significant increases in mesenteric flow (13–23%) and decreases in mesenteric vascular resistance within 15 min of the 2-h postprandial observation period in piglets <2 d old. Piglets 2–4 wk of age also had significant postprandial increase in mesenteric flow (19–29%) that started 30 min following feeding and lasted for the remainder of the 2-h observation period (see Table 5). In addition, unlike the hemodynamic changes observed in younger piglets, femoral blood flow decreased and femoral resistance increased significantly (Fig. 8) in 2–4-wk-old pigs, suggesting that regional vascular redistribution of the cardiac output favoring the gastrointestinal circulation occurred in the older pig. Heart rate and mean aortic pressure did not change significantly during feeding. Following a hemorrhage of 15% of the estimated blood volume (76), feeding induced relatively transient changes in the mesenteric circulation in <2-d-old piglets. On the other hand, significant and sustained increases in mesenteric flow and decreases in mesenteric resistance occurred in the 2–4-wk-old piglets (Table 5). In these older

Table 5
Postprandial Percent Changes[a] in Regional
Circulations From 2–4-w-Old Swine

Cardiovascular parameters	Feeding (n = 20)	Feeding following hemorrhage (n = 10)
MesF (mL/min)	+29.9 ± 10.7	+44.2 ± 14.3
MesR (PRU)	−15.4 ± 7.8 (NS)	−31.5 ± 8.9
FemF (mL/min)	−24.8 ± 7.2	+43.7 ± 15.6
FemR (PRU)	+39.3 ± 17.9	−28.6 ± 6.9

[a]mean ±SEM; all responses statistically different from 0 ($P \leq 0.005$) (except NS, not significant).
See Table 2 for explanation of abbreviatons.

piglets, femoral flow increased and femoral resistance decreased, accompanied by increased pulse pressure (87a). It would appear, therefore, that feeding can induce age-related regional circulatory responses. The stress of hemorrhage resulted in both increased and altered responses to feeding (84–87).

2.3.2. Simulated Gastroesophageal Reflux

We have recently reported our findings of simulated gastroesophageal reflux on cardiovascular function in 2–4-d-old piglets (62). As previously stated, gastroesophageal reflux has been implicated as a possible cause in the etiology of Sudden Infant Death Syndrome (56,65). We decided to determine whether bradycardia could occur during reflux experimentally induced in newborn animals independent of changes in respiration. Simulated gastroesophageal reflux was carried out by infusion of saline solutions of increasing acidity through a catheter that was introduced into the distal esophagus. Almost all piglets responded to the simulated reflux at a pH of 1.5 (see Fig. 1 in 62) with a decrease in heart rate (Fig. 9, right). Atropine eliminated the bradycardia (Fig. 9, left), indicating that excitation of the vagal drive to the heart was elicited by simulated gastroesophageal reflux. Since these experiments were carried out under our general methodology, i.e., animals immobilized with decamethonium bromide and artificially ventilated, the marked cardiac response to simulated reflux was independent of peripheral changes in ventilation. Thus bradycardia can be elicited in neonates independent of re-

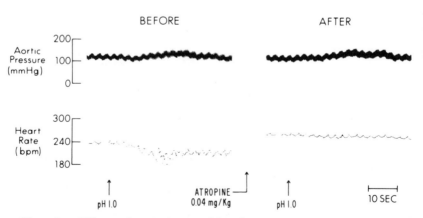

Fig. 9. Effect of atropine on blood pressure response and heart rate responses to simulated gastroesophageal reflux (pH 1.0 at arrows) (from ref. 62, with permission of the publisher).

spiratory changes. The bradycardia response may explain why gastroesophageal reflux has clinical relevance.

2.4. Age-Related Responses of Isolated Blood Vessels

The findings of the age-related differences in maturation of adrenergic mechanisms in different vascular beds (*see* refs. in *5,6*) prompted a brief in vitro study of isolated blood vessels. This series of experiments examined the age-related responsiveness of various vessels to a variety of vasoactive agents. The vessels studied included the superior mesenteric artery, iliac artery, portal vein, and thoracic aorta (*53,81,82*). These in vitro studies included cummulative dose-response curves, as well as stimulation by single doses (ED50–ED60) of epinephrine, norepinephrine, isoproterenol, serotonin, vasopressin, and angiotensin II. The methodology involved placing helically cut arterial strips and longitudinally cut portal vein segments in an oxygenated muscle bath containing normal Krebs–Ringer bicarbonate solution (*80*). After determining the optimal resting tension, dose–response curves were obtained in the absence and presence of appropriate antagonist (*53,81,82*). It was found that contractile tension elicited by the various agonists (particularly epinephrine and norepinephrine) progressively increased with increasing postnatal age (*see* Figs. 10,11). Thus age-related responses were obtained to almost all of

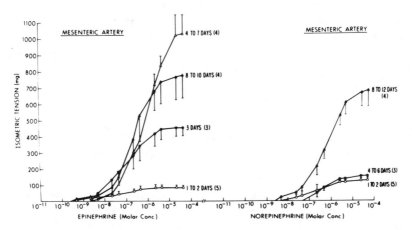

Fig. 10. Age-dependent dose–response curves of piglet mesenteric vascular smooth muscle strips to epinephrine (left) and norepinephrine (right). Number in parentheses ( ) indicates number of vessels used.

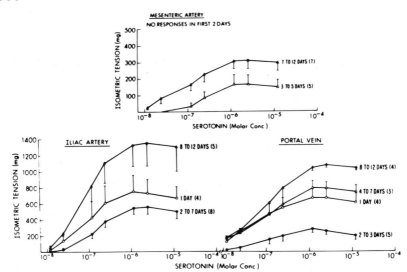

Fig. 11.   Age-dependent dose-response curves in piglet vascular smooth muscle to serotonin. Number in parentheses ( ) indicates number of vessels used.

the vasoactive agents (53,81,82), e.g., from the superior mesenteric artery (*see* Fig. 11). In 1-d-old piglet, norepinephrine and epinephrine could only elicit weak contractile response. By day 2–3, postnatally, dose-dependent contractile responses were observed to angiotensin II and serotonin in addition to norepinephrine and epinephrine. By 1 wk of age, vasopressin elicited contractile responses. Finally, at 8–12 d of age, significantly greater maximum tensions were obtained to all the agonists (Figs. 10 and 11). The results obtained from this in vitro study were supportive of the age-related responses observed while examining regional circulatory changes in vivo (*see* refs. in 5,6) indicating the varying rates of vascular smooth muscle receptor maturation.

2.5. Cardiovascular Responses to Stimulation of Vasoactive Sites in the Central Nervous System

There have been a limited number of attempts to localize central neural vasoactive sites in the fetus and the neonate (*see* refs. in 32,35) (30,34,50). Vasoactive sites in developing swine have been located throughout the brainstem within the reticular formation of the medulla, pons, and mesencephalon (*see* Figs. 12,13) (30,33–35,50–53). Vasoactive sites in the thalamus and diencepha-

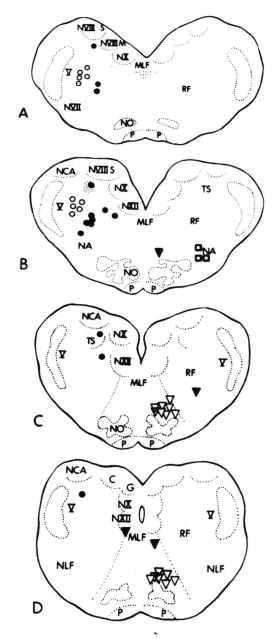

Fig. 12. Diagrammatic coronal sections of piglet brain stem
(drawn from histological material) showing regions in medulla from

(continued on next page)

lon have also been located (*see* Fig. 12) (*33–35,50–53*). Age-related aortic pressure (Fig. 14) and femoral flow responses (Fig. 15) have been obtained from a number of different sites (*30,33–35,50–53*) within the brain stem of neonatal pigs. An example of the responses obtained to stimulation of vasoactive sites in the thalamus (*52*) is shown in Fig. 16. Age-related differences in magnitude of responses to stimulation within the lateral hypothalmus have also been obtained (Figs. 13, 14). In the medulla, stimulation of the vasopressor area (Fig. 12, open circles) in the dorsolateral reticular formation resulted in increases in mean aortic pressure and femoral, renal, mesenteric, and carotid resistances without accompanying heart rate changes (*34,50,53*). Stimulation of the classical depressor area (Fig. 12, open triangles) in the ventromedial reticular formation elicited short latency decreases in mean aortic pressure and flows. Stimulation of sites in the nucleus ambiguus (Fig. 12, squares) produced marked decreases in heart rate (*30,34*). We have also recently reported that cardiac vagal motoneurons are located in the ventralateral nucleus ambiguus (*58,59*). Evoked increases in blood pressure to stimulation of the medullary pressor area were not significantly altered in the presence of hypoxia (*50,51*) or hemorrhage (Fig. 17a, b) (*34,35,53*). During hypoxia, however, stimulation of sites medial to this vasopressor area elicited markedly diminished magnitudes of evoked responses (*51*).

Active vasoactive sites have also been found in the rostral pons, and sites have been located from which increases in blood pressure are obtained. These sites were located within the dorsolateral

Fig. 12. (*cont.*) which changes in aortic pressure, regional arterial flows, and heart rate were obtained: ○ = marked changes in pressure and flows; • = small increases in arterial pressure sometimes accompanied by changes in heart rate; ▽ = marked decreases in pressures and flows ▼ = small decreases in pressure sometimes accompanied by alterations in heart rate and rhythm; □ = decreased heart rate only. Usually one site was intensively studied per animal. Anatomical designations (*90*): C, nucleus cuneatus; G, nucleus gracilis; MLF, fasciculus longitudinalis medialis; NA, nucleus ambigus; NCA, nucleus cuneatus accessarius; NLF, nucleus fasciculi lateralis; NO, nucleus olivaris; NVII, nucleus nervi facialis; NVIII S, nucleus vestibularis inferior; NVIII M, nucleus vestibulanis medialis; NX, nucleus dorsalis nervi vagi; NVIII, nucleus vestibularis mechalis; NXIII, nucleus nervi hypoglossi; P, pyramis; RF, formatio reticularis; TS, tractus solitarius; V, nucleus tractus spinalis nervi trigemeni (from ref. 34).

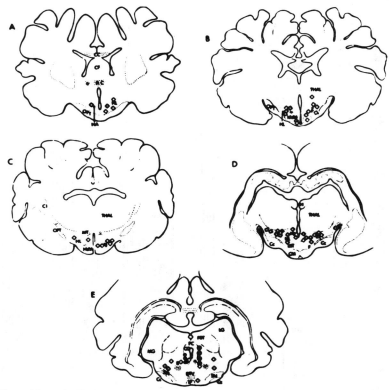

Fig. 13. A–D, Diagrammatic coronal sections of piglet diencephalon (drawn from histological material) showing sites from which marked alterations in aortic pressure, heart rate, and arterial flows were obtained with stimulation (◇). Results of explorations in planes 0.5 mm anterior and 0.5 mm posterior to each level of section are included. Sites shown are usually from more than one piglet. E, Diagram of coronal section of mesencephalon through the posterior commissure indicating locations of sites from which responses were obtained. ○ and ● = sites from which responses were obtained in mapping experiments in two different animals. ◇, sites from which responses were obtained in nine other piglets. Anatomical designations (90): AC, commissura anterior, CC, corpus callosum truncus; CF, corpus fornicus; CG, substantia grisea centralis; CI, capsula interna; CM, corpus mamillare; Cc, crus cerebri; DTV, decussatio tegmenti ventralis; F, columna fornicis; H, habenula; HA, nucleus hypothalamicus anterior; HL, nucleus hypothalamicus lateralis; HP, nucleus hypothalamicus posterior; HVM, nucleus hypthalamicus ventromedialis; IP, nucleus interpeduncularis; LG, corpus geniculatum laterale; MG, corpus geniculatum mediale; MT, fasciculus mamillothalamicus; OPT, tractus opticus; PC, commissura posterior; PRT, pretectum; R, nucleus ruber; SN, substantia nigra; THAL, thalamus; ZI, zona incerta (from ref. 34).

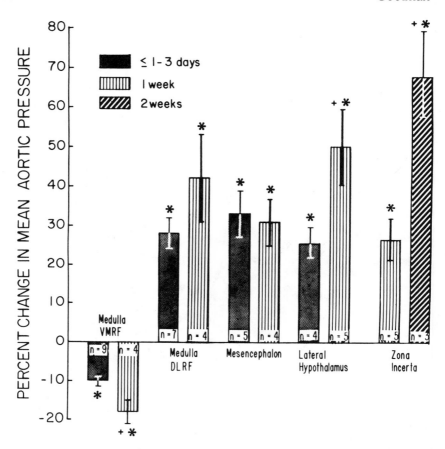

Fig. 14. Age-related percent changes (±SEM) in mean aortic pressure to stimulation of five vasoactive sites in the central nervous system of developing swine. DLRF, dorsolateral medullary reticular formation (Fig. 12, open circles); VMRF, ventromedial medullary reticular formation (Fig. 12, open triangles); lateral hypothalamus (Figure 13, B, C, D); Mesencephalon (Figure 13, E); Zona incerta: (Figure 13, D). n = number of animals in each group. (*) Indicates statistically significant responses; (+) indicates significantly greater responses in older group (from ref. 50, with permission of the publisher).

rostral pons, a vasoactive area previously described in adult cats and sheep (*see* refs. in *32*). An earlier paper by us (*30*) reported that, in the spontaneously breathing neonatal pig, changes in respiration could be elicited by stimulation of those regions that have been classically been described as the medullary inspiratory and

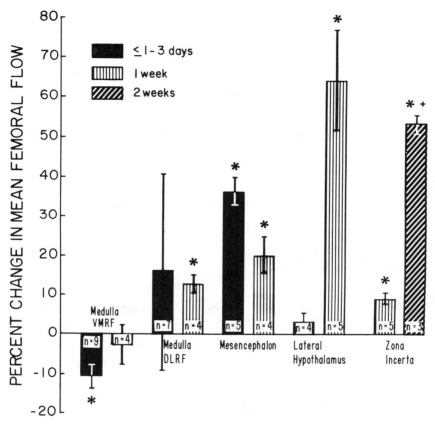

Fig. 15. Age-related changes (±SEM) in mean femoral flow to stimulation of five vasoactive sites in the central nervous system of developing swine. See Fig. 14 for details (from ref. *50,* with permission of the publisher).

expiratory areas (Fig. 18) and the pneumotaxic center in the rostral pons (*13*) (*see* refs. in *32*).

Vasoactive sites that evoked increases in mean aortic pressure and femoral flows were also located at 15 loci within the mesencephalon (Fig. 13E). Significant increases in heart rate were obtained only in animals at least 1 wk old (*34,50*). These response patterns were similar to those reported for adult mammals (*see* refs. in *32*). Vasoactive sites are found in the diencephalon from the level of the preoptic area to the mammillary bodies (*see* Fig. 13A–D). Age-related changes in cardiovascular responses were

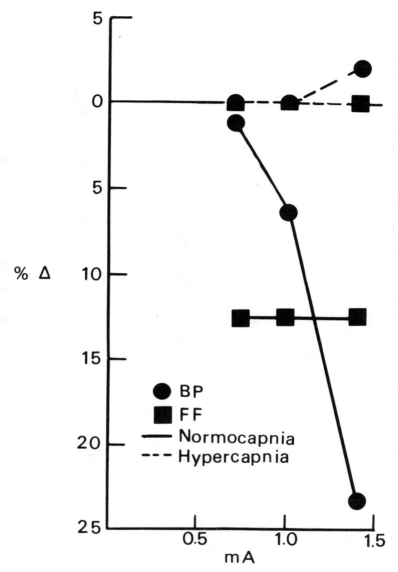

Fig. 16. Effects of hypercapnia on cardiovascular responses (BP, aortic pressure; FF, femoral arterial blood flow) to stimulation (0.5–1.5 mA at 50 Hz for 10 s, 0.1 ms pulses) of nucleus anteriomedialis thalami in a 5-d-old piglet.

evoked in several sites in the diencephalon, e.g., lateral hypothalamus and zona inserta (Figs. 14 and 15). The cardiovascular responses to diencephalic stimulation in neonatal pigs were similar
to those reported for adult mammals (*see* refs. in *32*).

2.5.1. Effects of Stresses on Responses to CNS Stimulation

Responses to diencephalic stimulation were either lost (in piglets
<2 wk of age) or diminished (in piglets >2 wk of age) in the pres-
ence of such stresses as hypercapnia (Figs. 16, 19) or hemorrhage
(Fig. 20) (*34,35,50*). In Fig. 20, the responses were lost with removal of 10 mL/kg blood from a 1-d-old piglet; reinfusion of blood
previously removed resulted in a return of the responses (*50*).
Hypoxia has also been reported to diminish aortic pressure and
cardiovascular responses to posterior hypothalamic stimulation in
puppies 4–6 wk of age (*3*). These investigators also reported that
posterior hypothalamic stimulation under normoxia elicited an increase in blood pressure and decrease in colonic blood flow with
an increase in resistance. Under hypoxic conditions, the rise in
blood pressure was markedly diminished, and no changes in flow
or resistance in colonic circulation occurred. We have reported
that aortic pressure and mesenteric resistance responses to medullary pressor area stimulation are still present under the stress of
hypoxia (*50,51,53*) or hemorrhage (Fig. 16) (*35,50*) in piglets 2 d to
3 wk of age. However, with stimulation of sites adjacent to the
classical medullary pressor area, responses were markedly diminished under conditions of hypoxia (*51*).

It would appear, therefore, that in the perinatal period, medullary vasoactive sites are involved in those reflex changes
occurring under conditions of stress, whereas the more rostral regions of the cardiovascular regulatory system appear to be more
sensitive to stress, and lose their ability to respond to electrical
stimulation in the young piglet (*34,35,53*). These findings would
suggest that the more rostral neuraxis becomes important later in
postnatal development, when more complex regulation appears
to mature (e.g., temperature regulation, changes in blood flow to
feeding and exercise). Our results show that cardiovascular responses to stimulation of sites in the central nervous system of
the neonatal pig were similar to those reported in adult mammals
(*see* refs. in *32*). Furthermore, these central neural vasoactive sites

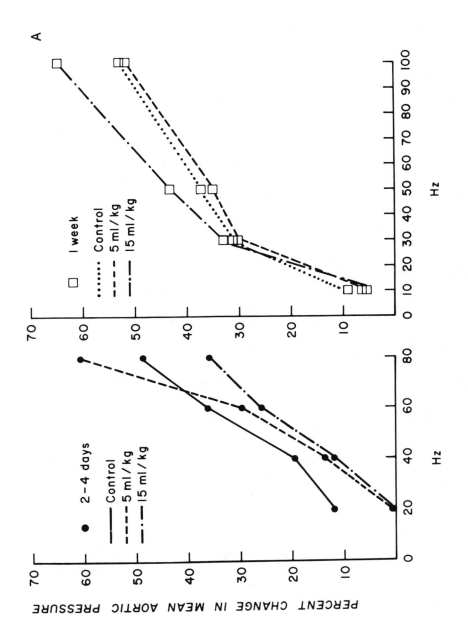

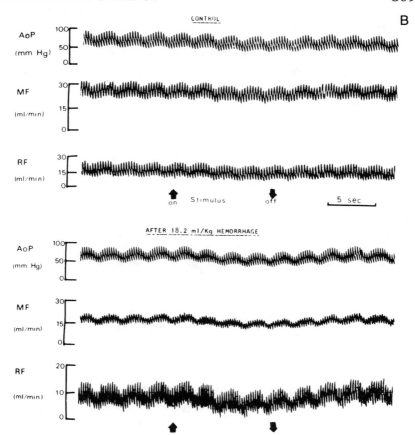

Fig. 17. A, Examples of aortic pressure responses from a 2–4-d-old (left) and 1-wk-old (right) piglet to stimulation of medullary pressor area. (Fig. 12, open circles) before and following withdrawal of 5 and 15 mL blood/kg. B, Original polygraph tracings showing effect of hemorrhage on responses to stimulation of medullary depressor area (Fig. 12, open triangles) at 50 Hz for 10 s 1.0 mA.

undergo postnatal maturation. Their immaturity is indicated by the age-related magnitudes of responses to stimulation under normal experimental conditions, by the loss of such responses in the presence of stresses, as well as by the failure to process two simultaneous afferent inputs of similar types.

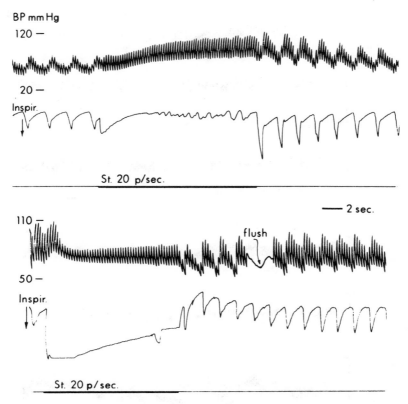

Fig. 18.   Effect on blood pressure (BP) and respiration (inspiration indicated by direction of arrow) of stimulation at two medullary sites in a 7.5-h-old piglet. Stimulus parameters: 1.0 mA, 20/s, for 10 s. Site of stimulation. *Upper trace;* 3 mm rostral to obex in dorsolateral reticular formation, 4 mm ventral to dorsal surface. *Lower trace:* 0.5 mm rostral to obex in ventromedial reticular formation, 1 mm dorsal to inferior olive (from ref. *30,* with permission of the publisher).

2.6. Location of Vagal Cardiac Motor Neurons

We have determined the distribution of the cells of origin of both the cervical vagus and cardiopulmonary nerves (58,59). These cell bodies were located by utilizing the horseradish peroxidase technique (57,68). Following injection of horseradish peroxidase into the cervical vagus nerve, retrogradely labeled neurons were pres-

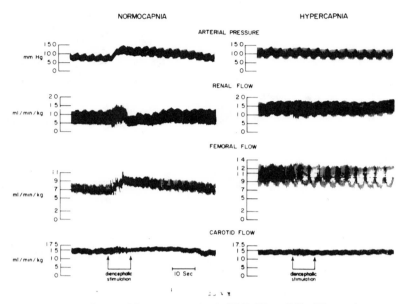

Fig. 19. Effect of hypercapnia (pH 7.27, pCO_2 53 torr) on mean aortic pressure, heart rate, and mean regional flow responses to stimulation of a site in ventromedial nucleus of the hypothalamus in a 10-d-old piglet (0.9 mA, 50 Hz, for 10 s). Decreases in responses under hypercapnia: mean aortic pressure, 51 to 7%; heart rate, 31 to -3%; RenF, 47 to 8%; FemF, 25 to 9%; CarF, 11 to 4% (from ref. *34*, with permission of the publisher).

ent in the dorsal motor nucleus of the vagus nerve (DMV), the nucleus of the solitary tract, the nucleus ambiguus (NA), ventrolateral to the NA, and in an intermediate zone between the DMV and the NA (*see* Fig. 21). Two unique clusters of neurons were observed: One group was located lateral to the most caudal levels of the DMV and extended as far caudally as the C1 spinal segment: The second distinctive group was located ventrolateral to the nucleus ambiguus in a cell column identified as the ventrolateral nucleus ambiguus (VLNA). After injections of HRP into cardiopulmonary nerves, the majority of neurons were found in the VLNA (Fig. 21).

The distribution of neurons in the medulla oblongata projecting the cardiopulmonary nerves in the piglet is similar to that de-

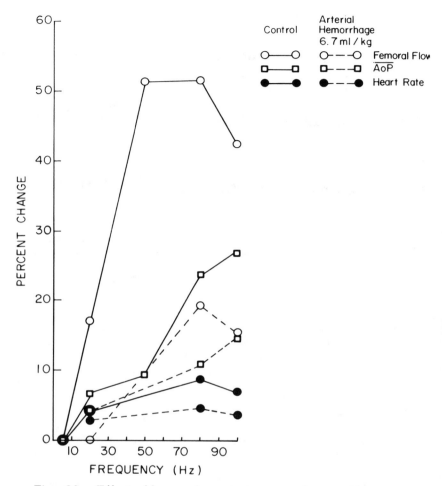

Fig. 20. Effect of hemorrhage on percent changes in mean aortic pressure (□) heart rate (•) and mean femoral flow (o) of responses to increasing frequency of stimulation at a site in the posterior hypothalamus (Fig. 13, C) in a 1-d-old piglet.

scribed in other species, i.e., the nucleus ambiguus, especially its ventrolateral cell column, is the primary site of cardiomotor neurons (*see* refs. in *58,59*). In addition, in the piglet there is a morphologically distinct cluster of cells related to the heart, and possibly the lungs, that does not appear to be present in other species.

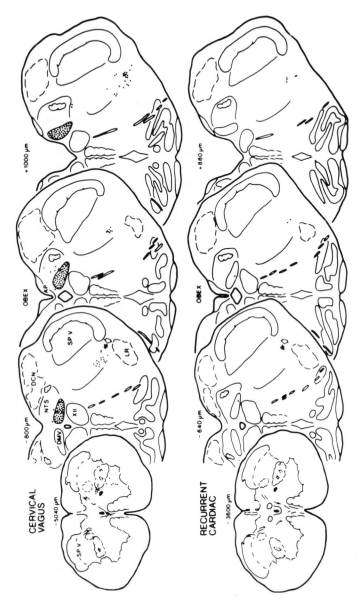

Fig. 21. Drawings of the medulla from 7-d-old piglets showing the distribution of retrogradely labeled neurons after injection of HRP into the cervical vagus nerve and the recurrent cardiac nerve. The solid arrows point to the location of cell clusters lateral to DMV cell column in the caudal medulla oblongata. Open arrows point to the location of compact ventrolateral clusters of labeled cells. AP, area postrema; DCN, dorsal column nuclei; DMV, dorsal motor nucleus of vagus nerve; LR, lateral reticular nucleus; NTS, nucleus of the solitary tract; SPV, spinal trigeminal complex (from ref. 59, with permission of the publisher).

2.7. *Spontaneous Efferent Sympathetic Discharge*

Spontaneous efferent splanchnic discharge, monophasically re-corded, consisted predominantly of synchronized action potentials in the form of slow waves (32,37,38,50,53). The predominant frequencies were about 19 and 29 Hz. An example, from a 2-d-old piglet, of periodicities recorded in the splanchnic nerve, is shown in the power spectra of Fig. 22. Two other major types of periodicities observed in neonatal sympathetic discharge were: (1) oscillations in phase with the cardiac cycle (Fig. 23) and, (2) oscillations in phase with the central respiratory cycle (Fig. 24). An example of the cardiac modulation of splanchnic activity is shown in Fig. 25 from a piglet 13 d old. The cross-correlation histograms of Fig. 25 indicated that minimal splanchnic activity occurred during systole. Oscillations of sympathetic discharge in phase of the central respiratory cycle were usually present with maximum splanchnic activity occurring during the inspiratory phase and minimum activity during the expiratory phase (*see* Fig. 26). These oscillations

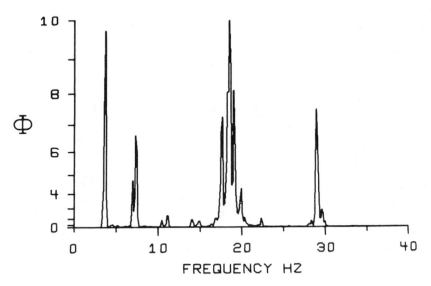

Fig. 22. Power spectra of efferent splanchnic activity from a 2-d-old piglet. Sampling rate 128 Hz with 20 epochs of 1024 data points. Signal filtered with a 4–40 Hz bandpass.

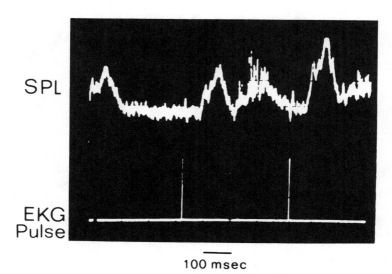

100 msec

Fig. 23. Oscilloscope traces of efferent splanchnic (SPL) activity (negativity upward) in a 1-d-old piglet. Sweeps were triggered from pulses derived from the R wave of the electrocardiogram (EKG). Top trace: monophasic recording of SPL activity; bottom trace: pulses derived from the R wave of the EKG (from ref. 53, with permission of the publisher).

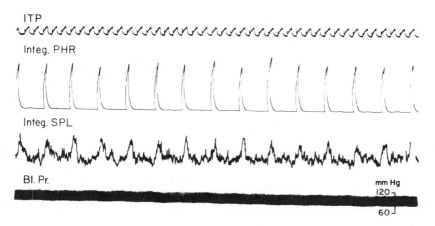

Fig. 24. Spontaneous efferent activity in the phrenic and the splanchnic nerves from a 1-wk-old piglet. ITP, intratracheal pressure; Integr. PHR, integrated phrenic activity (time constant 100 ms); Integr. SPL, integrated splanchnic activity (time constant 100 ms); BL. Pr., aortic blood pressure (from ref. 53, with permission of the publisher).

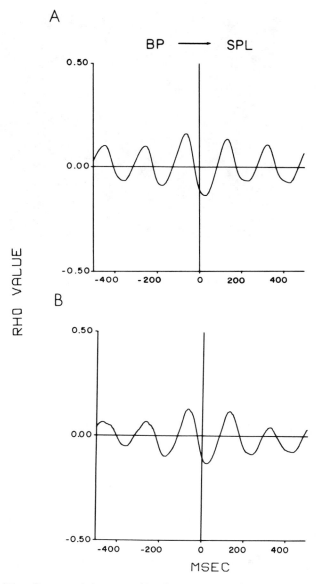

Fig. 25. Sequential normalized cross-correlation histograms between blood pressure (BP) and splanchnic (SPL) activity from a 13-d-old piglet. 500 2-ms bins, 20,000 data points. SPL signal filtered with a 2.5–40 Hz bandpass. Zero time represents peak systole during BP cycle (from ref. *38*, with permission of the publisher).

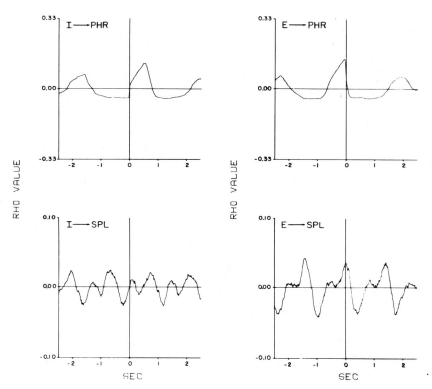

Fig. 26. Normalized cross-correlation histograms of simultaneously recorded phrenic (PHR) activity (after processing through an 80 Hz Highpass filter) and splanchnic (SPL) activity vs inspiration onset (I), pulses (left), or expiration onset (E) pulses (right) from a 13-d-old piglet. 500 10-ms bins; 20,000 data points (from ref. *38*, with permission of the publisher).

are similar to those observed in adult animals (*14,15,40–47,49*). Figure 24 shows integrated phrenic and splanchnic activity from a 1 wk-old piglet. Note the increase of splanchnic activity during inspiration and the decrease during the expiratory phase of the central respiratory cycle. Thus, spontaneous sympathetic discharge contained periodicities that were modulated by both the cardiac and central respiratory cycles in piglets at birth (*37,38,53*). By use of the cycle-triggered-pump (*13,15,49*), it was possible to show modulation of splanchnic activity by pulmonary afferents, as had been reported in adult mammals (*15,49*). The results of these experiments indicate that tonic sympathetic discharge to the celiac ganglion and adrenal medulla is present in piglets at birth (*38,50,53*). The presence of splanchnic activity could count for the marked changes in renal and superior mesenteric arterial blood flows observed during either afferent or central stimulation.

3. Conclusions

Considerable postnatal development of central neural regulation of cardiovascular function in mammals is suggested by the results of our studies and those of other investigators working with different species (*32,35,39*).

Acknowledgments

The work of the author and her colleagues, Drs. N. M. Buckley, N. Gootman, B. J. Buckley, as well as Drs. H. L. Cohen, A. C. Yao, P. D. M. V. Turlapaty, A. M. Steele, D. A. Hopkins, S. M. DiRusso, and K. Kenigsberg, summarized and reviewed in this chapter, was supported by grants from the National Institutes of Health (HL-20864) and the American Heart Association, Nassau Chapter. Unitensin® (cryptenamine) was kindly supplied by Wallace Laboratories.

References

1. Assali, N. S., C. P. Brinkman III, R. Woods, Jr., A. Dandavino, and B. Nuwayhid (1978) Ontogenesis of the autonomic control of cardiovascular functions in the sheep. in *Fetal and Newborn Cardiovascular Physiology*, Vol. 1, Developmental Aspects, Garland STPM Press, New York, pp. 47–91.
2. Benfordo, J. M., W. Flacke, C. R. Swaine and W. Mosimann (1960) Studies on veratrum alkaloids XXIX. The Action of some germine esters and of enratridine upon blood pressure, heart rate, and femoral blood flow in the dog. *J. Pharmacol. Exptl. Therap.* **130,** 311–320.
3. Broadie, T. A., M. Davedas, J. Rysavy, J. P. Delaney, and A. S. Leonars (1973) The effect of hypoxia and posterior hypothalamic stimulation on colonic blood flow in the weanling puppy. *J. Pediatri. Surg.* **8,** 747–756.

4. Buckley, B. J., N. Gootman, J. S. Nagelberg, P. G. Griswold, and P. M. Gootman (1984) Cardiovascular responses to arterial and venous hemorrhage in neonatal swine. *Am. J. Physiol.* **247**, R626–R633.

5. Buckley, N. M. (1983) Regional circulatory function in the perinatal period, in *Perinatal Cardiovascular Function.* (N. Gootman and P. M. Gootman, eds.) Marcel Dekker New York, pp. 227–264.

6. Buckley, N. M. (1984) Regulation of regional vascular beds by the developing autonomic nervous system, in *Developmental Neurobiology of the Autonomic Nervous System,* (P. M. Gootman, ed.) Humana Press, NJ, pp 00–00.

7. Buckley, N. M., P. Brazeau, and I. D. Frasier (1983) Circulatory effects of splanchnic nerve stimulation in developing swine. *Fed. Proc.* **42**, 1115.

8. Buckley, N. M., P. Brazeau, I. D. Frasier, and P. M. Gootman (1981) Femoral circulatory responses to lumbar nerve stimulation in developing swine under pentobarbital anesthesia. *Am. J. Physiol.* **240**, H505–H510.

9. Buckley, N. M., P. Brazeau, and P. M. Gootman (1983) Maturation of circulatory response to adrenergic stimuli. *Fed. Proc.* **42**, 1645–1649.

10. Buckley, N. M., P. Brazeau, P. M. Gootman, and I. D. Frasier (1979) Renal circulatory effects of adrenergic stimuli in anesthetized piglets and mature swine. *Am. J. Physiol.* **237**, H690–H695.

11. Buckley, N. M., P. M. Gootman, N. Gootman, G. D. Reddy, L. C. Weaver, and L. A. Crane (1976) Age-dependent cardiovascular effects of afferent stimulation in neonatal pigs. *Biol. Neonate,* **30**, 268–279.

12. Buckley, N. M., P. M. Gootman, E. L. Yellin, and P. Brazeau (1979) Age-related cardiovascular effects of catecholamines in anesthetized piglets. *Circ. Res.* **45**, 282–292.

13. Cohen, M. I. (1979) Neurogenesis of respiratory rhythm in the mammal. *Physiol. Rev.* **59**, 1105–1173.

14. Cohen, M. I. and P. M. Gootman (1970) Periodicities in efferent discharge of splanchnic nerve of the cat. *Am. J. Physiol.* **218**, 1092–1101.

15. Cohen, M. I., P. M. Gootman, and J. L. Feldman (1980) Inhibition of sympathetic discharge by lung inflation, in *Arterial Baroreceptors and Hypertension* (P. Sleight, ed.) Oxford, pp. 161–167.

16. Crane, L., N. Gootman, and P. M. Gootman (1974) The effects of decamethonium (C-10) on blood pressure and heart rate in newborn piglets. *Arch. Int. Pharmacodyn. Therap.* **208**, 52–60.

17. Crane, L., N. Gootman, and P. M. Gootman (1975) Age-dependent cardiovascular effects of halothane anesthesia in neonatal pigs. *Arch. Int. Pharmacodyn. Ther.* **214**, 180–187.

18. Dawes, G. S. and J. H. Comroe, Jr. (1954) Chemoreflexes from the heart and lungs. *Physiol. Rev.* **34**, 167–201.

19. Dawes, G. S., B. M. Johnston, and D. W. Walker (1980) Relationship of arterial pressure and heart rate in fetal, newborn and adult sheep. *J. Physiol.* **309**, 405–417.

20. Dawes, G. S., J. C. Mott, and J. G. Widdicombe (1951) Respiratory and cardiovascular reflexes from the heart and lungs. *J. Physiol.* **115**, 258–291.

21. Downing, S. E. (1960) Baroreceptor reflexes in newborn rabbits. *J. Physiol.* **150**, 201–213.

22. Earley, A. Ng. S. Fayers, E. A. Shinebourne, and M. de. Swiet (1980) Blood pressure in the first 6 weeks of life. *Arch Dis. Child.* **55**, 755–757.

23. Frantz, I. D. III, P. L'Heureux, R. R. Engel, and C. E. Hunt (1975) Nectrotizing enterocolitis. *J. Pediatr.* **86**, 259–263.

24. Gelhorn, E. (1939) Effect of hemorrhage, reinjection of blood and dextran on the activity of the sympathetic and parasympathetic systems. *Acta Neuroveg.* **22**, 291–299.

25. Goldman, H. I. (1980) Feeding and necrotizing enterocolitis. *Am J. Dis. Child.* **134**, 553–555.

26. Gootman, N., B. J. Buckley, P. M. Gootman, P. G. Griswold, J. D. Mele, and D. B. Nudel (1983) Maturation-related differences in regional circulatory effects of dopamine infusion in swine. *Dev. Pharmacol. Ther.* **6**, 6–22.

27. Gootman, N., B. J. Buckley, P. M. Gootman, and J. S. Nagelberg (1982) Age-related effects of single injections of dopamine on cardiovascular function in developing swine. *Dev. Pharmacol. Ther.* **4**, 139–150.

28. Gootman, N., P. M. Gootman, B. J. Buckley, P. G. Griswold, and B. J. Peterson (1982) Age-related alterations in circulatory function during hypoxia. *Pediatr. Res.* **16**, 100A.

29. Gootman, N., P. M. Gootman, B. J. Buckley, P. G. Griswold, and M. Sugarman (1981) Postnatal maturation of autonomic responses of the superior mesenteric artery. *Fed. Proc.* **40**, 559.

30. Gootman, N., P. M. Gootman, N. M. Buckley, M. I. Cohen, M. I. Levine, and R. Spielberg (1972) Central vasomotor regulation in the newborn piglet (*Sus scrofa*). *Am. J. Physiol.* **222**, 994–999.

31. Gootman, N., P. M. Gootman, L. A. Crane, and B. J. Buckley (1979) Integrated cardiovascular responses to combined somatic and visceral afferent stimulation in newborn piglets. *Biol. Neonate,* **36**, 70–77.

32. Gootman, P. M. (1983) Postnatal development of central nervous system regulation of cardiovascular function, in *Perinatal Cardiovascular Function* (N. Gootman and P. M. Gootman, eds.) Dekker, NY, pp. 265–328.

33. Gootman, P. M., B. J. Buckley, N. Gootman, and P. G. Griswold (1979) Effects of hemorrhage on cardiovascular responses to medullary pressor area stimulation in neonatal pigs. *Fed. Proc.* **138,** 1385.

33a. Gootman, P. M., B. J. Buckley, N. Gootman, and M. E. Salinas-Zeballos (1985) Localization of vasoactive sites in the thalamus of newborn swine, in *Physiological Development of Fetus and Newborn* (C. T. Jones and P. W. Nathanielsz, ed.) Academic Press, NY, pp. 599–604.

34. Gootman, P. M., N. M. Buckley, and N. Gootman (1978) Postnatal maturation of the central neural cardiovascular regulatory system, in *Fetal and Newborn Cardiovascular Physiology,* Vol. 1, Developmental Aspects (L. D. Longo and D. D. Reneau, eds.) Garland STPM Press, New York, pp. 93–152.

35. Gootman, P. M., N. M. Buckley, and N. Gootman (1979) Postnatal development of neural control of the circulation, in *Reviews in Perinatal Medicine,* Vol. 3 (E. M. Scarpelli and E. V. Cosmi, eds.), Raven Press, New York, pp. 1–72.

36. Gootman, P. M., N. M. Buckley, N. Gootman, L. A. Crane, and B. J. Buckley (1978) Integrated cardiovascular responses to combined somatic afferent stimulation in newborn piglets. *Biol. Neonate* **34,** 187–198.

37. Gootman, P. M., H. L. Cohen, S. M. DiRusso, A. P. Rudell, and L. Eberle (1982) Efferent splanchnic activity in neonatal swine. *Neurosci. Abst* **8,** 6.

38. Gootman, P. M., H. L. Cohen, S. M. DiRusso, A. P. Rudell, and L. P. Eberle (1984) Characteristics of spontaneous efferent splanchnic discharge in developing swine, in *Catecholamines* Part A: *Basic and Peripheral Mechanisms* (E. Usdin, A. Dahlstrom, and A. Carlsson, eds.) Liss, New York, pp. 369–374.

39. Gootman, P. M., H. L. Cohen, and N. Gootman (1986) Autonomic control of heart rate in the perinatal period, in *New Directions in Pediatric Electrocardiography* (J. Liebman, R. Plansey, Y. Rudy, M. Nijhoff, eds.) The Hague.

40. Gootman, P. M. and M. I. Cohen (1970) Efferent splanchnic activity and systemic arterial pressure. *Am. J. Physiol.* **219,** 897–903.

41. Gootman, P. M. and M. I. Cohen (1971) Evoked potentials produced by electrical stimulation of medullary vasomotor regions. *Exptl. Brain Res.* **13,** 1–14.

42. Gootman, P. M. and M. I. Cohen (1973) Periodic modulation (cardiac and respiratory) of spontaneous and evoked sympathetic discharge. "Symposium on Central and Peripheral Adrenergic Mechanisms," Warsaw, 1971 *Acta Physiol. Polon.* **24,** 99–109.

43. Gootman, P. M. and M. I. Cohen (1974) The interrelationships between sympathetic discharge and central respiratory drive, in

Central-Rhythmic and Regulation (Hippokrates-Verlag, S., W. Umbach, and H. P. Koepchen, eds.) pp. 195–209.

44. Gootman, P. M. and M. I. Cohen (1980) Origin of rhythms common to sympathetic outflows at different spinal levels, in *Arterial Baroreceptors and Hypertension* (P. Sleight, ed.) Oxford, pp. 154–160.

45. Gootman, P. M. and M. I. Cohen (1981) Sympathetic rhythms in spinal cats. *J. Autonomic Nerv. System* **3**, 379–387.

46. Gootman, P. M. and M. I. Cohen (1983) Inhibitory effects on fast sympathetic rhythms. *Brain Res.* **270,** 134–136.

47. Gootman, P. M., M. I. Cohen, M. P. Piercey, and P. Wolotsky (1975) A search for medullary neurons with activity patterns similar to those in sympathetic nerves. *Brain Res.* **87,** 395–406.

48. Gootman, P. M., S. M. DiRusso, B. J. Buckley, N. Gootman, A. C. Yao, P. E. Pierce, P. G. Griswold, M. P. Epstein, and H. L. Cohen. Age-related responses to stimulation of cardiopulmonary receptors in swine. Submitted to *Am. J. Physiol.*

49. Gootman, P. M., J. L. Feldman, and M. I. Cohen (1980) Pulmonary afferent influences on respiration modulation of sympathetic discharge, in *Central Interactions Between Respiratory and Cardiovascular Control System* (H. P. Koepchen, S. M. Hilton, and A. Trzebski, eds.) Springer-Verlag, New York, pp. 172–179.

50. Gootman, P. M., N. Gootman, and B. J. Buckley (1983) Maturation of central autonomic control of the circulation. *Fed. Proc.* **42,** 1648–1655.

51. Gootman, P. M., N. Gootman, B. J. Buckley, P. G. Griswold, and B. J. Peterson (1982) Effects of hypoxia on cardiovascular responses to medullary pressor area stimulation in neonatal piglets. *Fed. Proc.* **41,** 1520.

52. Gootman, P. M., N. Gootman, B. J. Buckley, and M. -E. Salinas-Zeballos (1983) Cardiovascular responses to stimulation of vasoactive thalamic sites in neonatal swine. *Fed. Proc.* **42,** 1120.

53. Gootman, P. M., N. Gootman, P. D. M. V. Turlapaty, A. C. Yao, B. J. Buckley, and B. M. Altura (1981) Autonomic regulation of cardiovascular function in neonates, in *Development of the Autonomic Nervous System* Ciba Foundation Symposium 83. (G. Burnstock, ed.) London, Pitman Medical, pp. 70–93.

54. Gootman, P. M., A. C. Yao, S. M. DiRusso, P. E. Pierce, B. J. Buckley, and N. Gootman (1981) Age-related responses to stimulation of cardiopulmonary receptors in swine. *Fed. Proc.* **40,** 523.

55. Granger, D. N., P. D. I. Richardson, P. R. Kvietys, and N. A. Mortillaro (1980) Intestinal blood flow. *Gastroenterology* **78,** 837–863.

56. Herbst, J. J., S. D. Minton, and I. S. Book (1979) Gastroesophageal reflux causing respiratory distress and apnea in newborn infants. *Pediatrics* **95,** 763–768.

57. Hopkins, D. A. and J. A. Armour (1982) Medullary cells of origin of physiologically identified cardiac nerves in the dog. *Brain Res. Bull.* **8**, 359–365.

58. Hopkins, D. A., P. M. Gootman, N. Gootman, S. M. DiRusso, and M. -E. Zeballos (1981) Localization of vagal cardiomotor neurons in neonatal pig. *Neurosci. Abstr.* **7**, 115.

59. Hopkins, D. A., P. M. Gootman, N. Gootman, S. M. DiRusso, and M. -E. Zeballos (1984) Brainstem cells of origin of the cervical vagus and cardiac nerves in the neonatal pig (*Sus scrofa*). *Brain Res.* **306**, 63–72.

60. Itskovitz, J., E. F. LaGamma, and A. M. Rudolph (1983) Baroreflex control of the circulation in chronically instrumented fetal lambs. *Circ. Res.* **52**, 589–596.

61. Kalia, M. (1976) Visceral and somatic reflexes produced by J pulmonary receptors in newborn kittens. *J. Appl. Physiol.* **41**, 1–6.

62. Kenigsberg, K., P. G. Griswold, B. J. Buckley, N. Gootman, and P. M. Gootman (1983) Cardiac effects of esophageal stimulation: Possible relationship between gastroesophageal reflux (GER) and sudden infant death syndrome (SIDS). *J. Pediatr. Surg.* **18**, 542–545.

63. Kitterman, J. A, R. H. Phibbs, and W. H. Tooley (1969) Aortic blood pressure in normal newborn infants during the first 12 hours of life. *Pediatrics* **44**, 959–968.

64. Krayer, O. (1961) The history of the Bezold–Jarisch effect. *Arch. Exp. Path. O. Pharmak.* **240**, 361–368.

65. Leape, I. L., T. M. Holder, J. D. Franklin, R. A. Armory, and K. W. Ashcraft (1977) Respiratory arrest in infants secondary to gastroesophageal reflux. *Pediatrics* **60**, 924–927.

66. Lees, B. J. and L. A. Cabal (1981) Increased blood pressure following pupillary dilation with 2.5% phenylephrine hydrochloride in preterm infants. *Pediatrics* **68**, 231–234.

67. LeGal, Y. M. (1983) Effect of acute hemorrhage on some physiological parameters of the cardiovascular system in newborn pigs. *Biol. Neonate.* **44**, 210–218.

68. Mesulam, M. -M. (1978) Tetramethylbenzidine for horseradish peroxidase neurohistochemistry: A noncarcinogenic blue reaction-product with superior sensitivity for visualizing neural afferents and efferents. *J. Histochem. Cytochem.* **26**, 106–117.

69. Mills, E. and P. G. Smith (1983) Functional development of the cervical sympathetic pathway in the neonatal rat. *Fed. Proc.* **42**, 1639–1642.

70. Modanlou, H., S. -Y. Yeh, B. Siassi, and E. H. Hon (1974) Direct monitoring of arterial blood pressure in depressed normal infants during the first hour of life. *J. Pediatrics* **85**, 553–559.

71. Paintal, A. S. (1973) Sensory mechanisms involved in the Bezold–Jarisch effect. *Aust. J. Biol. Med. Sci.* **51**, 3–15.

72. Reddy, G. D., N. Gootman, N. M. Buckley, P. M. Gootman, and L. Crane (1974) Regional blood flow changes in neonatal pigs in response to hypercapnia, hemorrhage, and sciatic nerve stimulation. *Biol. Neonate.* **25**, 249–262.

73. Schleman, M., N. Gootman, and P. M. Gootman (1979) Cardiovascular and respiratory responses to right atrial injections of phenyl diguanide in pentobarbital-anesthetized newborn piglets. *Pediatr. Res.* **13**, 1271–1274.

74. Smyth, H. S., P. Sleight, and G. W. Pickering (1969) Reflex regulation of arterial pressure during sleep in man. *Circ. Res.* **24**, 109–121.

75. Swiet, M. de, R. Fancourt, and J. Peto (1975) Systolic blood pressure variation during the first 6 days of life. *Clin. Sci. Mol. Med.* **49**, 557–561.

76. Talbot, R. B. and M. J. Swenson (1970) Blood volume of pigs from birth through 6 weeks of age. *Am. J. Physiol.* **218**, 1141–1144.

77. Thoren, P. (1979) Role of cardiac vagal C-fibers in cardiovascular control. *Rev. Physiol. Biochem. Pharmacol.* **86**, 1–94.

78. Tomomatsu, E. and K. Nishi (1982) Comparison of carotid sinus baroreceptor sensitivity in newborn and adult rabbits. *Am. J. Physiol.* **243**, H546–H550.

79. Toulowkian, R. J., J. N. Posch, and R. Spencer (1972) The pathogenesis of ischemic gastroenterocolitis of the neonate: selective gut mucosal ischemia in asphyxiated neonatal piglets. *J. Pediatr. Surg.* **7**, 194–205.

80. Turlapaty, P. D. M. V., B. T. Altura, and B. M. Altura (1978) Influence of TRIS on contractile responses of isolated rat aorta and portal vein. *Am. J. Physiol.* **235**, H208–H213.

81. Turlapaty, P., B. T. Altura, P. M. Gootman, and B. M. Altura (1979) Vascular reactivity in blood vessels of neonatal pigs. *Fed. Proc.* **38**, 437.

82. Turlapaty, P. D. M. V., B. T. Altura, P. M. Gootman, and B. M. Altura (1980) Do neonatal mammalian arteries and veins exhibit receptors for specific vasodilator hormones? *Blood Vessels* **17**, 165–166.

83. Versmold, H. T., J. A. Kitterman, R. H. Phibbs, G. A. Gregory, and W. H. Tooley (1981) Aortic blood pressure during the first 12 hours of life in infants with birth weight 610–4220 grams. *Pediatrics* **67**, 607–613.

84. Yao, A. C., P. M. Gootman, N. Gootman, P. E. Pierce, and S. M. DiRusso (1982) Effect of induced hypovolemia (HV) on superior mesenteric artery blood flow (MBF) response to feeding in piglets. *Pediatr. Res.* **16**, 119A.

85. Yao, A. C., P. M. Gootman P. E. Pierce, and S. M. DiRusso (1984) Regional circulatory responses in 2-4 weeks old piglets. Effects of feeding and hemorrhage (H). *Pediatric Res.* **18**, 218A.

86. Yao, A. C., P. M. Gootman, P. E. Pierce, and S. M. DiRusso. Effects of Feeding on Regional Circulations of Developing Swine, in *Swing in Biochemical Research* (M. E. Tumbleson, ed.), Plenum, New York, in press.

87. Yao, A. C., P. M. Gootman, P. E. Pierce, S. M. DiRusso, and N. Gootman (1981) Feeding and superior mesenteric artery flow in newborn piglets. *Pediatr. Res.* **15,** 688.

87a. Yao, A. C., P. M. Gootman, P. E. Pierce, and S. M. DiRusso (1985) Effects of feeding and hemorrhage on the femoral circulation of developing swine, in *Physiological Development of Fetus and Newborn* (C. T. Jones, and P. W. Nathanielsz, ed.) Academic Press, NY, pp. 445–449.

88. Yardley, R. W., G. Bowes, M. Wilkinson, J. P. Cannata, J. E. Maloney, B. C. Ritchie, and A. M. Walker (1983) Increased arterial pressure variability after arterial baroreceptor denervation in fetal lambs. *Circ. Res.* **52,** 580–588.

89. Yoshikawa, T. (1968) Chapter V. The Brain of the Pig (Yorkshire Breed). *Atlas of the Brains of Domestic Animals,* University of Tokyo Press, Tokyo and Pennsylvania State University Press, University Park, PA, plates P1–P33.

90. Salinas-Zeballos, M.-E., G. A. Zeballos, and P. M. Gootman (1986). A Stereotaxic Atlas of Developing Swing (*Sus serofa*) Forebrain, in *Swine in Biomedical Research* (M. E. Tumbleson, ed.) Plenum, New York, in press.

Chapter 12

Developmental Changes in Neural Control of Respiration

Andrew M. Steele

1. Introduction

For humans and other mammals, the act of breathing (respiration) has a simple purpose: to maintain metabolic homeostasis by acquiring oxygen (during inspiration) and eliminating carbon dioxide (during expiration). Achieving this aim, however, requires a complex series of neural and neuromuscular interactions. Perhaps because of the rhythmic contractions of skeletal muscles involved in respiration, some investigators have described this as an autonomic function (42). However, in contrast to cardiac and vascular smooth muscle, respiratory muscles have no intrinsic automaticity. The actions of inspiratory muscles of the thoracic cavity, such as diaphragm, intercostals, and airways (e.g., efferent vagally innervated skeletal muscles of the pharynx and larynx), depend upon the generation of a rhythmic discharge originating in the central nervous system (CNS) (Fig. 1). However, the rate and amplitude of the respiratory rhythm generator (RRG) may be facilitated and/or inhibited by a number of afferent inputs, such as chemoreceptors, baroreceptors, J receptors (Fig. 2), as well as central mechanisms.

Generally speaking, studies of the neural control of respiration have concentrated on two problems: (i) how and where the re-

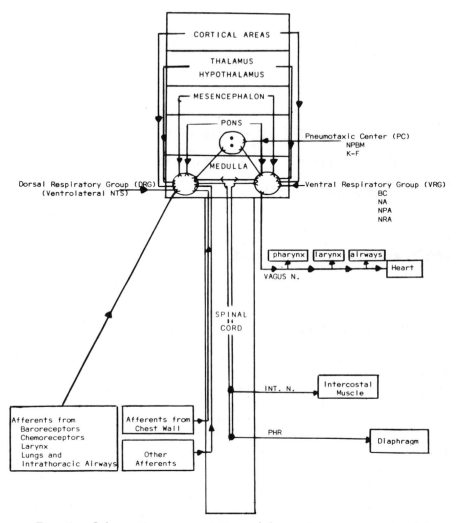

Fig. 1.   Schematic representation of the major components of the respiratory control system, some of the interconnections within the respiratory rhythm generator (RRG), and outputs to the respiratory musculature. Not shown are elaborate interconnections within the central neural controlling system and interneurons interspersed between inputs and outputs. Abbreviations: NTS, nucleus tractus solitarius; NPBM, nucleus parabrachialis medialis; K-F, Kolliker-Fuse nucleus; BC, Botzinger complex; NA, nucleus ambiguus; NPA, nucleus paraambigualis; NRA, nucleus retroambigualis; INT.N, intercostal nerve; PHR, phrenic nerve.

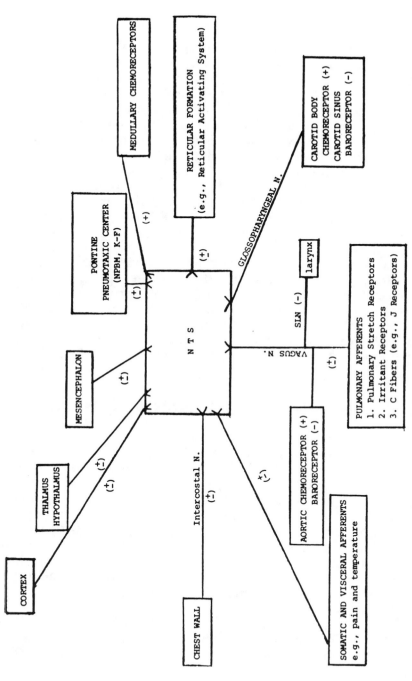

Fig. 2. Schematic of afferent inputs via nucleus tractus solitarius (NTS) to the respiratory rhythm generator (RRG). (+) indicates excitatory; (−) indicates inhibitory. Abbreviations: NPBM, nucleus parabrachialis medialis; K-F, Kölliker-Fuse nucleus; SLN, superior laryngeal nerve.

spiratory rhythm is generated (localization of the RRG), and (ii) what factors are involved in determining the rate and depth of breathing? The first problem has been studied primarily in adult animals, using (i) transections and lesions, (ii) electrical stimulation of various areas in the brain stem, (iii) antidromic stimulation, (iv) horseradish peroxidase staining, and (v) extracellular and intracellular recordings from respiratory-related neurons using microelectrodes. Data from these techniques have been correlated with the output of the RRG, e.g., phrenic nerve activity. The latter problem has been of great interest to investigators concerned with developmental changes in respiratory function. Studies in this regard have provided considerable information about maturational changes in modulation of the RRG by a number of afferent inputs.

The present review will be concerned with maturation of the respiratory rhythm generator (RRG). When possible, comparisons between newborn and adult respiratory functions will be presented. For the benefit of those readers more familiar with the autonomic nervous system (ANS) than the RRG, I will briefly review what is known, mainly from studies in the adult, about the anatomical and functional organization of the RRG.

2. Generation of the Respiratory Rhythm

Pacemaker cells (endogenously rhythmic cells) of respiration, which could be considered to be the primary on-switch, have not yet been identified in the CNS (282). At present, most investigators view the efferent discharge of the RRG as the result of several sequentially activated mechanisms that depend upon a number of synaptic interactions confined mainly to the lower brainstem (pons and medulla) (62,101,270). Such interactions have been observed to result in characteristic periodic patterns of activity of alpha motoneurons, e.g., phrenic (PHR), intercostals, hypoglossal (HYP), and recurrent laryngeal (RL), which innervate the respiratory musculature, e.g., diaphragm, intercostals, genioglossus, and posterior cricoarytenoid.

It is the pattern of discharge of the PHR (Fig. 3) that has been most frequently employed as a basis for reference to the phenomena or mechanisms that determine the output of the RRG and the various phases of the respiratory cycle (62,101,270). Figure 4 presents a schematic representation of an integrated PHR signal (INT

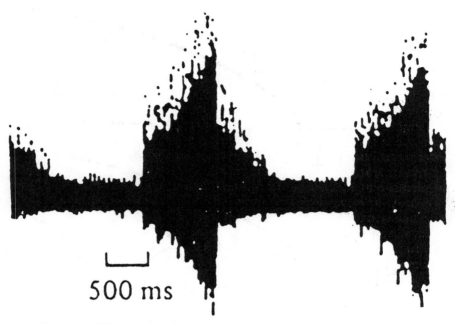

Fig. 3. Efferent discharge recorded from a phrenic nerve in a chloralose anesthetized cat (from ref. *270*, with permission of author and publisher).

PHR) in an adult cat: (1) a sudden generation of inspiratory neuronal activity following a period of silence (E to I or PHR on-switch), (2) a progressively augmenting "ramp-like" pattern that peaks at the end of inspiration, (3) an abrupt decrease in activity (I to E or PHR off-switch), (4) a brief period of silence followed by a postinspiration inspiratory activity (PIIA), with a decrementing pattern (early E), and (5) period of silence (late E). A similar, but not identical, pattern of PHR activity (PIIA was not observed) has been found in the newborn piglet (*125,307*), as can be seen in Fig. 5.

External intercostal neuronal activity in the cat has been found to be relatively similar to that of the PHR (*62*). Although the pattern of discharge of the external intercostal nerve(s) has not been reported in newborn animals, an augmenting pattern of activity has been observed in the EMG of intercostal muscle of kittens (*325*).

Whereas the augmenting discharge patterns of the PHR and intercostal nerves produce a gradual increase in the force of dia-

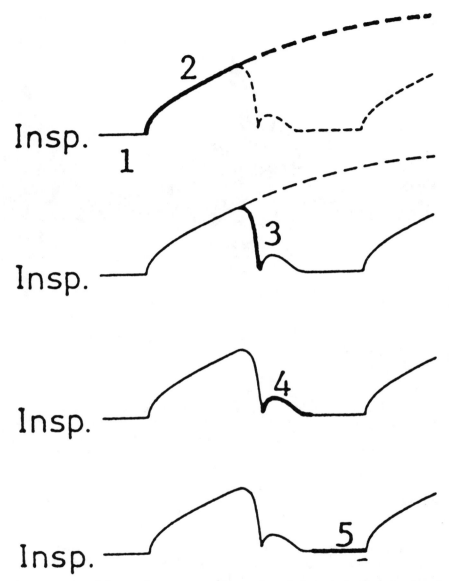

Fig. 4. Schematic representation of an integrated phrenic signal
showing various components of respiratory cycle. Insp., inspiration.
Numbers 1–5 correspond to various phases of respiratory cycle. For details, *see* text (from ref. *101*, with permission of author and publisher).

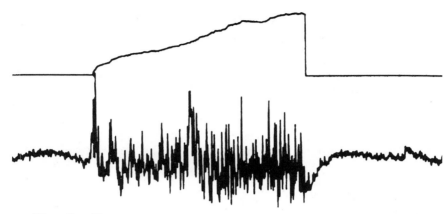

Fig. 5. Top: Computerized integration of phrenic nerve signal shown below. Note progressively augmenting ramp-like pattern peaking at end of inspiration. Bottom: Efferent discharge recorded from a phrenic nerve in an anesthetized piglet. Note burst of high frequency oscillations (HFO) during inspiration and absence of postinspiration inspiratory activity (PIIA) seen in adult cat in Fig. 3.

phragmatic and intercostal muscle contraction, HYP and RL discharge patterns appear adapted for maintenance of upper airway patency (299). In the cat, both the onset and maximal discharge of the HYP and RL have been observed to precede that of the PHR. Also, the HYP discharge was noted to be typically decrementing, whereas that of the RL is plateau-like. Cross-correlations of the activity of these neurons with PHR activity have not as yet been reported in newborn animals.

These patterns have raised several questions: (i) How is I initiated or generated? (ii) What factors determine the slope of the progressively augmenting PHR I "ramp"? (iii) What mechanisms control the duration of I (terminate I)? and (iv) What mechanisms are employed during E; specifically early vs late E?

The neural circuits that determine these aspects of the respiratory rhythm have not been fully delineated. However, several different populations of respiratory and respiratory-related neurons have been identified. These have been classified on the basis of anatomic location, discharge pattern, and response to various perturbations, e.g., lung inflation (21,60–63).

2.1. Anatomic and Functional Organization of Respiratory Neurons

Several regions have been shown to contain critical components for determining the basic respiratory pattern, although they have not been found to be independently capable of sustaining normal respiratory rhythmicity. These areas (Fig. 1) have included (1) the dorsal respiratory group (DRG), situated in the vicinity of the nucleus tractus solitarius (NTS), (2) the ventral respiratory group (VRG), in the ventrolateral medulla, and (3) several pontine structures.

2.1.1. Dorsal Respiratory Group

The DRG has been found to contain mainly inspiratory-related (I) neurons. These have been of considerable interest because (i) most project to the spinal cord, some making monosynaptic connections with the PHR alpha-motoneurons, and (ii) many receive projections from vagal afferents (pulmonary and nonpulmonary) (19,21,63) and afferents from the carotid bodies (chemoreceptor) and sinuses (baroreceptor) (62,63).

DRG neurons in ventrolateral NTS have been found to be heterogenous with respect to their basic discharge patterns and their responses to lung inflation (Table 1). It is believed that both the slope of the I ramp and off-switching of I (i.e., facilitation or inhibition) depend upon how these individual neurons interact (63). Other neurons, e.g., in the medial NTS, have been found to receive pulmonary afferent inputs (95,172). It has been suggested that these neurons participate in functions such as graded laryngeal adductor inhibition (*see* section 4.2.1., Laryngeal Control of Airway Caliber) (101).

2.1.2. Ventral Respiratory Group

The VRG has been reported to consist of a longitudinal column of both I and E neurons, extending from the midpons to the upper spinal column (62). Several populations of respiratory neurons have been identified within the VRG. These have included (i) the rostral end of the nucleus ambiguus (NA), (ii) the Botzinger complex (BC), (iii) the nucleus paraambigualis (NPA), and (iv) the nucleus retroambigualis (NRA) (62,171,327).

The NA is the motor nucleus of the glossopharyngeal and vagus nerves. The BC (171) at the most rostral aspect of the VRG near nucleus retrofacialis has been found to consist of both I units

Table 1
Classification of Recorded Neurons by Discharge Pattern and
Response to Inflation[a]

| Type | No. | E T F R | | Depth, mm |
		Unit	Phrenic	
I augmenting	69	0.90 ± 0.59	1.12 ± 0.16	
Inflation (0)	19	0.97 ± 0.07	1.13 ± 0.15	1.78 ± 0.38
Inflation (+)	34			
Early onset	22	0.61 ± 0.14	1.08 ± 0.18	1.83 ± 0.37
Late onset	12	0.25 ± 0.28	1.17 ± 0.23	1.72 ± 0.24
Inflation (−)	16	1.70 ± 0.60	1.13 ± 0.18	1.87 ± 0.40
I decrementing	13			1.97 ± 0.34
Pump	8			1.51 ± 0.35
Total	90			

[a]Values are mean ± SD. ETFR, equivalent-time firing ratio; *see* explanation in text (from ref. *63*, with permission of the author and publisher).

and a condensed group of E neurons. In the intermediate VRG, NPA lies alongside NA. The NPA has been found to contain mainly, albeit not exclusively, I neurons (with both augmenting and decrementing patterns), many of which project contralaterally to the PHR and intercostal motoneurons. The NRA has been found to contain E neurons whose axons project contralaterally to intercostal and abdominal motoneurons (*222*). This latter group of E neurons is not believed to be involved in the generation of the basic pattern of the respiratory rhythm. There has been some evidence that E neurons of NRA receive projections from the BC. Projections from DRG to VRG have been reported to be numerous, i.e., late firing I (+) to NRA on contralateral side. Projections from VRG to DRG have also been reported, i.e., Botzinger neurons to the caudal contralateral NS (*199,327*).

2.1.3. Pontine Neurons

In the rostral and lateral pons, immediately caudal to the inferior colliculi, several subgroups of respiratory-related neurons have been identified in (i) the nucleus parabrachialis medialis (NPBM), (ii) the region of the Kölliker-Fuse nucleus (K-F), and (iii) the ventral part of the brachium conjunctivum (*62*). Projections from

NPBM and K-F to VRG, and less so to DRG, have also been identified (49).

The NPBM corresponds to the pneumotaxic center (PneuC) identified by lesion and transection studies (22,62). Respiratory modulation of PneuC neurons has been found to originate from medullary neurons (probably VRG) (62), and to be strongly inhibited by pulmonary afferent activity (106). In fact, few respiratory-related neurons have been identified in animals with intact vagi (62). Without phasic vagal discharge, PneuC neurons have been found to be critical for I–E switching. It has also been suggested that tonic PneuC activity may be important for off-switching (i.e., acts to shorten I) since in vagotomized animals, lesions in, or transactions caudal to, this area have been reported to prolong I (apneusis) (105). Electrical stimulation in the ventral NPBM has been observed to terminate I. On the other hand, stimulation in dorsal NPBM has facilitated PHR discharge (58).

It has been suggested that the input and output traffic of the PneuC may be channeled through the K-F (62). In addition to respiratory-related activities, the K-F has been shown to be involved in autonomic, neuroendocrine, behavioral, and sensorimotor functions (207), and may serve to integrate these different functions, including respiration (101).

In summary (i) the respiratory rhythm appears to originate from structures within the lower brainstem, but a precise anatomic locus has not been identified, (ii) both VRG and DRG neurons are required for normal, periodic rhythm-generation and contain neurons capable of generating a progressive ramp-like increase, if not originating such activity, (iii) the DRG seems to be specialized for transmission of IXth and Xth nerve inputs into the RRG and for control of PHR activity, (iv) the VRG seems more specialized for control of efferent thoracic and vagal (e.g., laryngeal) respiratory neuronal activity projections, (v) input from pulmonary afferents is received by both pontine and medullary respiratory-related neurons, and (vi) the activity of respiratory-related pontine neurons is modulated by medullary neurons, particularly those of the VRG, and strongly inhibited by pulmonary afferent discharge.

The next sections will briefly review some mechanisms thought to be important for generation of the PHR inspiratory ramp, off-switching, and regulation of the expiratory phase.

2.2. Neural Mechanisms

2.2.1. Inspiration

In the analysis of respiratory periodicity, two major questions have arisen: (i) How does the relatively slow oscillation (on a time scale of seconds) of respiratory discharge develop from rapid (time scale of milliseconds) neural changes?, and (ii) How is activity synchronized between large numbers of neurons discharging in a particular portion of the respiratory cycle (59)?

Some insight into these issues has been obtained from analyses of short-term synchronization between inspiratory neurons, i.e., the phenomenon of high-frequency oscillations (HFO) in PHR discharge. HFO has been noted in several species and is considered to reflect a fundamental property of the organization of the RRG, i.e., that there is active excitation of I neurons during I, rather than disinhibition by E neurons.

Both species- and maturation-related differences in the peak frequency of HFO in PHR activity have been observed. In the cat (2 h–8 wk of age), the peak frequency of HFO was reported to progressively increase from 20 to 60 Hz (300). The same increase (20 to 60 Hz) was observed in puppies, but occurred within the first 3 wk of life (300). In swine (125), the peak frequency, detected by power spectral analysis of PHR discharge, has also been observed to increase with age. However, for any given postnatal age, the peak frequency has been found to be greater than that of either the cat or dog (Fig. 6).

Analysis of excitatory (EPSPs) and inhibitory (IPSPs) postsynaptic potentials from intracellular recordings of cat brainstem respiratory neurons has offered further evidence that the generation of the augmenting pattern of I activity (or I ramp) depends upon a recurrent excitatory loop of I neurons, i.e., alpha neurons received an augmenting pattern of EPSPs during I (270). Similar recordings in newborn animals have not as yet been reported.

The rate of increase in I activity or the slope of the I-augmenting ramp have been considered as indexes of "central inspiratory drive" (CID). CID has been found to depend upon the interaction of a number of afferent inputs, e.g., chemical, pulmonary. Other indices of CID have included mean inspiratory flow (V_T/T_I) and mouth or esophageal pressure at 100 ms after end-inspiratory occlusion (P.100). Under eupneic conditions in the adult, V_T/T_I and

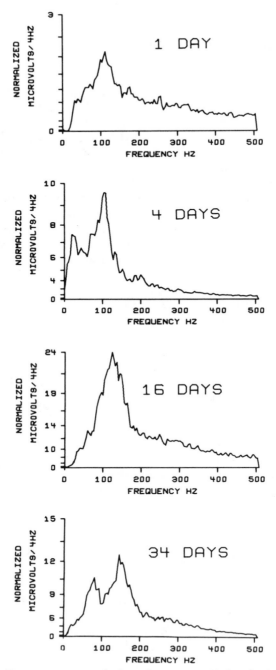

Fig. 6. Power spectra of phrenic nerve activity during inspiration in newborn swine from 1 to 34 d of age. Note increasing frequency of peak power with increasing age (from ref. *125*).

P.100 have been closely correlated to neural activity, i.e., slope of I-ramp (*98,297*). However, as will be discussed later in this chapter, this has not always been true in adults under abnormal conditions, e.g., hypoxia, or in normal newborns (*198*).

2.2.2. Off-Switching

A much-debated issue in respiratory neurophysiology has been how I is switched off. However, it has been agreed upon that the classical view of reciprocal inhibition between I and E neurons was incorrect. Instead, it has appeared that higher-order interneurons are necessary to terminate I. The specific neurons or types(s) of neurons (e.g., higher-order interneurons) that comprise a central I inhibitory (CII) system and effect off-switching remain to be identified (*63,101,270*).

The threshold of the off-switch has been found to depend upon both the progressively augmenting central neuronal activity (acting indirectly via interneurons of an inspiratory inhibitory system), and inputs from pulmonary afferents. Other afferent inputs have also been observed to alter the off-switch threshold; it has been raised (I activity reached a greater intensity before termination) by increased chemoreceptor discharge, and lowered by increased input from NPBM, cerebrum, reticular activating system (RAS), and chest wall proprioceptors (*101*).

2.2.3. Expiration

Expiration (E) begins with off-switching efferent I activity. The control of airflow during this part of the respiratory cycle has been found to be largely governed by elastic recoil forces developed during I. However, under certain conditions active mechanisms have been observed that retard expiratory airflow. These have included (i) postinspiration inspiratory activity (PIIA) of I muscles and laryngeal adductor activity, during early E, and (ii) recruitment of E muscles, during late E. In a mechanical sense, these have been considered to be important for maintaining the resting lung volume or functional residual capacity (FRC) (i.e., they vary inversely with lung inflation *127,227*).

With respect to the functional organization of the RRG, PIIA has been considered important for depressing both inspiratory higher-order neurons and expiratory activities. It has been seen as a powerful mechanism "gating," over a considerable time, the recycling excitation of I ramp-generating neurons, and thus, preventing either apneusis or rapid, shallow breathing. Although PIIA activity was not observed in PHR recordings of neonatal

Table 2

Respiratory Patterns of Newborn Mammals[a]

Species	n	No. of breaths	Age, days	BW, g	V_T, mL	T_I, S	T_E, S	f, breaths/min	V_E, mL/s	T_T, S	T_I/T_T, %	V_T/T_I, mL/s
Mice	5	1,000	2.8	2.5	0.16	0.160	0.289	141.5	2.3	0.453	36.4	0.105
			±1.1	±1.5	±0.010	±0.056	±0.124	±33.3	±1.4	±0.160	±10.1	±0.065
CV					61.3	35.2	42.9	23.5	61.2	35.3	27.8	62.3
Rats	4	800	2	7.2	0.068	0.190	0.427	99.3	6.6	0.640	30.8	0.374
			±0	±0.1	±0.019	±0.069	±0.141	±22.1	±2.0	±0.228	±8.8	±0.113
CV					28.6	36.3	33.0	22.3	30.3	35.6	28.6	30.2
Rabbits	4	800	1.7	79.3	0.83	0.385	0.541	68.6	55.4	0.932	42.0	2.18
			±1	±17.9	±0.22	±0.092	±0.174	±19.7	±15.3	±0.268	±5.7	±0.54
CV					26.2	23.9	32.2	28.7	27.6	28.8	13.6	24.8
Guinea pigs	2	400	3	91.1	0.67	0.215	0.313	115.0	76.1	0.540	41.1	3.17
			±0	±13.4	±0.15	±0.025	±0.070	±18.4	±17.3	±0.227	±5.5	±0.73

	n			BW	V_T	T_I	T_E	f	V_E	T_T	T_I/T_T	V_T/T_I
CV					22.3	11.8	22.3	16.0	22.7	42.0	13.4	23.0
Kittens	4	800	1.7 ± 0.5	118.8 ± 7.1	1.43 ± 0.51	0.344 ± 0.126	0.702 ± 0.479	72.5 ± 32.6	95.0 ± 40.8	1.057 ± 0.581	36.9 ± 10.8	4.29 ± 1.28
CV					35.5	36.6	68.2	45.0	43.0	55.0	29.3	29.8
Puppies	3	600	1 0	297.2 ± 21.5	2.59 ± 1.32	0.193 ± 0.069	0.618 ± 0.569	90.0 ± 39.1	241.8 ± 159.4	0.811 ± 0.569	29.6 ± 13.3	13.74 ± 5.98
CV					51.0	35.8	92.1	43.4	65.9	70.2	44.9	43.5
Piglets	3	600	1 0	1167 ± 307	14.2 ± 7.0	0.622 ± 0.278	1.055 ± 0.633	45.0 ± 20.9	665.9 ± 518.4	1.670 ± 0.874	39.2 ± 9.5	26.4 ± 15.9
CV					49.0	44.7	60.0	46.4	77.8	52.3	24.2	60.2
Infants			3.9 ± 3.5	3388 ± 79	20.0 ± 4.0	0.485 ± 0.007	0.76 ± 0.26	43.1 ± 14.1	863.4 ± 393.9	1.507 ± 0.468	40.5 ± 9.0	42.2 ± 11.5

[a]Values are means ±SD. CV, coefficient of variation; BW, body weight; V_T, tidal volume; T_I, inspiratory time; T_E, expiratory time; T_T, total breath duration; f, respiratory frequency; V_E, minute ventilation; T_I/T_T, inspiratory time-to-total cycle duration ratio; V_T/T_I, mean inspiratory flow (from ref. 227, with permission of the author and publisher).

swine (Fig. 5), it has been demonstrated in several other species under different experimental conditions (*113,148,204,228*). It is believed that PIIA is generated by a different population of interneurons than those that generate the I ramp.

Activation of E neurons has been considered to depend upon disinhibition of CII mechanisms that progressively decay during E. The rate of decay of CII has been found to be influenced by several factors, including lung inflation. Thus, the duration of E (T_E) has been found to depend upon the preceding duration of I (T_I), i.e., when T_I was lengthened (e.g., lung inflation or NPBM stimulation), the succeeding T_E was increased (*57,100,110*). This relationship has been observed soon after birth in sleeping humans (*25,76*) and anesthetized kittens and rabbits (*85,321*).

It is hoped that the above review has been of some assistance to those relatively unfamiliar with respiratory neurophysiology. It should be kept in mind that the anatomic and functional organization of the RRG remains a subject of considerable controversy and active investigation. For more detailed explanation(s) of mechanisms concerning the RRG, the reader is referred to the papers of Cohen (*62*), Cohen and Feldman (*63*), von Euler (*101*), and Richter (*270*). The remainder of this chapter will be concerned primarily with maturation-related characteristics of the respiratory pattern and age-related changes in afferent inputs of the RRG.

3. Maturation-Related Characteristics of the Respiratory Pattern

Metabolic rate (O_2 consumption per unit body weight) has been found to vary inversely with size and maturation. As a consequence, minute ventilation ($\dot{V}$) has been found to be higher in newborns than in adults of the same species. Thus, the newborns of the smallest mammalian species have the highest $\dot{V}$. In addition, the postnatal decrease in $\dot{V}$ has been found to parallel that in O_2 consumption (per unit body weight) in humans (*137,160*) and animals (*214*). These differences in $\dot{V}$ have been reflected mainly in changes in respiratory rate rather than tidal volume (V_T) (Tables 2 and 3).

Although maturation-related, the breathing pattern has not been found to be homogeneous at any age. Considerable variability in both respiratory rate and V_T have been observed. This has been especially true in the newborn; the respiratory pattern has

Table 3
Pulmonary Ventilation[a]

		Infant	Adult	
Respiratory frequency	(f)	34–45	13	BPM
Tidal volume	(V_T)	6–8	7	mL/kg
Alveolar volume	(V_A)	3.8–5.8	4.8	mL/kg
Dead-space volume	(V_D)	2–2.2	2.2	mL/kg
Minute ventilation	(V_E)	200–260	90	mL/kg/min
Alveolar ventilation	(V_A)	100–150	60	mL/kg/min
		2.3	2.4	L/m²/min
Wasted (dead space) ventilation	(V_D)	77–99	30	mL/kg/min
Dead space/tidal volume	(V_D/V_T)	27–37	0.3	
Oxygen consumption	(V_{O_2})	6–8	3.2	mL/kg/min
Ventilation equivalent	(V_A/V_{O_2})	16–23	19–25	

[a]From ref. 234, with permission of publisher.

often appeared irregular, and episodes of apnea (cessation of respiration) (205) and periodic breathing (repetitive periods of brief apnea, 3–10 s, alternating with regular respirations) have been observed frequently (55,164,173–175,268,271,272).

Some investigators have found these dysrhythmias (apnea and periodic breathing) to be of pathophysiologic significance, e.g., apnea of prematurity (219), infant apnea syndrome (304), and sleep apnea syndrome (132). Others have considered the possibility that sudden infant death syndrome (SIDS) might be caused by maturation-related perturbation of respiratory control (96,139,173–175,301). Still others have examined the possibility that the variability of the respiratory pattern reflects the behavioral state, changes in metabolic drive, or employment of the respiratory musculature for nonhomeostatic functions, e.g., speech. Much of what is known about the maturational characteristics of the RRG has come from studies concerned with the issues of apnea, SIDS, and sleep.

3.1. Apnea

Three types of apnea have been identified from polygraphic recordings (e.g., simultaneous measurements of chest excursion, airflow, EKG, EEG) (i) central apnea—cessation of breathing effort (i.e., no chest movement), (ii) obstructive apnea—interruption of inspiratory airflow despite breathing efforts, and (iii) mixed apnea—obstructive and central apnea occurring in close temporal proximity. There has been no consensus as to whether one type, if any, is predominant in the newborn (*48,226, 289,317,318*).

3.1.1. Central Apnea

In the adult, central apnea has been found to occur almost exclusively during sleep (*see* section 3.2, Behavioral State). Prolonged apnea (>20 s), observed in normal subjects during rapid eye movement (REM) sleep, has been attributed to inhibitory mechanisms inherent to that sleep state. On the other hand, in adults with disorders of the brainstem (i.e., infectious, vascular, and neoplastic), central apnea has occurred mainly during non-rapid eye movement (NREM) sleep (*254*). The severity and duration of apnea during REM has been found to be self-limited; the developing hypoxemia has been found to stimulate breathing (*311*) and evoke arousal (*253*), i.e., there was a properly functioning system for metabolic modulation of the RRG by both peripheral and central chemoreceptors (*see* section 4.4, Chemical Drive to the RRG). On the other hand, the vulnerability to apnea during NREM has been attributed to the blunting or absence of metabolic mechanisms that modulate breathing in that state (i.e., abnormalities of peripheral or central chemoreceptors) or injury to brainstem neurons that process metabolic information.

In contrast to the adult, apnea in newborns and infants has been observed not only during sleep, but also during wakefulness and transitional periods (*304*). Prolonged apnea has been found to occur more often in premature infants, especially those of lowest birthweight; 25% at <2500 g (*79,276*), 85% at <1000 g (*4*). It has also been found to be most frequent in newborns presenting with another pathology, such as intracranial hemorrhage or sepsis. However, no associated disorder has been identified for at least 20% of newborns presenting with periods of apnea (*276*).

The postnatal incidence of central apnea has been the subject of several studies (Table 4). According to these reports, in normal subjects (a) apnea of >10 s is not uncommon in the first 2 or 3 mo of life; (b) apnea >15 s is unusual after the first week postterm; and (c) apnea >10 s appears to be more frequent during REM.

A number of mechanisms have been proposed for the pathogenesis of central apnea in newborns and infants. These have included (i) reduced responsiveness to chemical drive, e.g., CO_2

(*121,122*); (ii) immaturity of the reticular formation, or more specifically the reticular activating system, e.g., diminished arousal response (*32*); (iii) increased inhibitory afferent input, e.g., chest wall afferents (*178*); (iv) increased central sensitivity to inhibitory afferent inputs, e.g., laryngeal mechano- and chemoreceptors (*96*); (v) paradoxical depression of the RRG by afferent inputs that are facilitatory later in postnatal development, e.g., hypoxia (*272*); (vi) ventilatory muscle fatigue (*229*); and (vii) increased sleep state-associated depression of the RRG, i.e., newborns spend most of the 24-h day sleeping, particularly in REM (*254*). These various mechanisms will be discussed further in subsequent sections of this chapter.

3.1.2. Obstructive Apnea

Brief periods of obstructive apnea have been observed in normal adults and newborns (*133*). Such episodes have been found to be of longer duration and greater severity (i.e., significant hypoxemia) in subjects with marked structural abnormalities or deformities of the upper airway, e.g., Pickwick syndrome (*120*) and adenoid and tonsillar hypertrophy (*254*). However, obstructive apnea of pathophysiologic consequence has been also reported in otherwise apparently normal patients (*132*). Partial obstruction has been observed to result in snoring and hypoventilation; complete obstruction has been found to cause profound disturbances of the cardiovascular and respiratory systems. Anatomically, the location of obstruction has usually been supraglottic (above the vocal cords). It has been found to occur physiologically when the transpharyngeal pressure exceeded the dilating action of the pharyngeal musculature (*124,267,318*).

In the adult (*254*), obstructive apnea has been found almost always during sleep (both in REM and NREM). In infants (3 wk–6 mo), it has been reported to occur predominantly in REM sleep (*159*). This association of sleep and obstructive apnea has been considered indicative that the critical factor in determining the severity of obstruction is functional rather than structural (*38,50, 220,318*). Thus, the terminology "central" or "obstructive" is somewhat misleading; all apnea actually has a central basis, i.e., inhibition of I or I-related neurons. Postulated pathogenetic mechanisms for obstructive apnea during sleep have included (i) REM-associated depression of pharyngeal musculature, e.g., genioglossus muscle (*38,287,318*); (ii) reduced responsiveness of upper airway motoneurons, e.g., HYP vs. PHR during hypoxia;

Table 4
Postnatal Incidence of Central Apnea in Preterm and Term Infants

Gestation	No. studied	Postnatal age	Method	Duration of study	Significant finding(s)	Ref.
Term	123	0–52 wk	Pneumogram	?	Longest apnea recorded was 14 s	175
Term	52	2 d–13 wk	Polysomnogram	Between 2 feedings	No apnea >10 s after 5th week. Apnea of all duration decreases by end of 2nd mo. Short apnea more common during REM	115
Preterm (30–37 wk)	7	3 d–9 wk	Polysomnogram	1–2 h	Apnea >10 s common during REM. Amplitude of monosynaptic stretch reflex decreased during apnea	289
Preterm and term (33–40 wk)	18	0–9 wk	Polysomnogram	?	Apnea highest at 44 wk postconception and during REM at all ages. Short apneas were the most frequent type. No episodes >20 s were observed. Apnea >10 s was infrequent by 52 wk postconceptional age	126

Population	n	Age	Recording	Duration	Findings	Ref
Preterm (28–34 wk)	8	2–48 d	Polysomnogram	2–4 h	Apnea >10 s frequent in each infant. 65% of these were accompanid by bradycardia. Short apnea more frequent during REM	119
Preterm term	2337 6914	1 and/or 6 wk	Pneumogram	24 h	In 29 infants who died of SIDS, none had apnea >13.2 s	301
Term	110	1–231 d	Pneumogram	22 h	Large variation in duration of apneas in first 15 d. Apneas of >18 s not present after 7 d of age. Incidence of short apnea decreased after 4 wk	268
Preterm (26–35 wk) Term	28	1–66 d	Polysomnogram	1½ h	Apnea more common during NREM in infants 30–33 wk gestation. Apnea (>10 s), most common in less mature infants, none in term infants	188
Term	9	0–6 mo	Polysomnogram	12 h	Apnea most common at earliest age. Apnea >15 s limited to 1-wk-old. Apnea of 6–8 s peaked at 2 mo	159
Term	11	5 d–6 mo	Polysomnogram	6 h	No apnea >12 s. Apnea 3–5 s common in the first 3 mo	51

and (iii) alteration of the normal temporal relationships between muscles of the upper airway and thorax (53). Such mechanisms might underlie obstructive apnea seen in the newborn, and obviously much work needs to be done in this field.

3.1.3. Periodic Breathing

Periodic breathing has been observed in normal newborns, older infants, children, and adults, particularly during sleep (254). In normal adults, the incidence of periodic breathing has been found to increase with aging (330) and with hypoxic exposure, e.g., high altitude (234). It has also been found to be greater in adults with a number of clinical disorders, e.g., chronic pulmonary disease, heart failure, cerebrovascular insufficiency, and bilateral pyramidal tract damage (39,254). In patients with these diseases, periodic breathing has been observed almost exclusively during NREM sleep (254).

In newborns, the reported percentage of infants who have episodes of periodic breathing has varied inversely with (a) birthweight (94.5% at <2500 g, 36.1% at >2500 g), (b) gestation (86.5% at <37 wk, 41% at >37 wk (107), and (c) postnatal age (78% at 0–2 wk, 29% at 39–52 wk) (175). The percentage of time in which infants breathe periodically has also been reported to decrease with age (51,159,174,268). However, it has also been suggested that the proportion of time spent breathing periodically may be characteristic of the individual subject and independent of age: From birth to one year of life, a lesser number of subjects were found to have periodic breathing, but in serially studied subjects, the proportion of time in which periodic breathing was identified was unchanged (175).

As has been reported for the adult, altered chemical drive has been associated with a change in the frequency and duration of periodic breathing in the newborn; i.e., increased by hypoxia (e.g., altitude) and abolished by hyperoxia and hypercapnia (91,107,272). In contrast to the adult, periodic breathing in newborns and infants has been reported to be common both in NREM and REM (151,159). However, this discrepancy might be caused as much by difficulties in determining sleep state as by true maturational differences in those state-dependent mechanisms that might determine the breathing pattern (*see* section 3.2, Behavioral State).

Several mathematical models have been developed to describe periodic breathing, in an effort to explain its (patho)genesis under various circumstances, e.g., sleep, hypoxia, cerebrovascular insufficiency (*52,176,202,223*). In these models, periodic breathing has resulted from (a) aberrations in controller (RRG) gain to peripheral and central chemoreceptor inputs, and/or (b) phase lags in the timing of the controller response to peripheral chemoreceptor inputs. In the former, the gain of the RRG has been considered to be abnormally high (i.e., very steep $\dot{V}/PaCO_2$ curve) (*see* section 4.4.2.2., Response to Carbon Dioxide) resulting in an "overcompensation" (*176*) of the RRG to various perturbations. In the latter, prolonged circulatory delay has been seen to produce "the wrong response at the wrong time" (*176*).

Since both periodic breathing and apnea have been associated with disturbances in metabolic modulation of the RRG, it has been suggested that they might share a common oscillatory mechanism (*328,329*). Some investigators have extended this view even further, claiming that periodic breathing might be a "marker" for babies at greater risk for SIDS (*173–175*). However, this association has not been substantiated by the findings of other laboratories (*139,268*). There has also been support for the position that periodic breathing and apnea might originate from independent pathways or mechanisms (*164*). Based on the available evidence, it seems probable that both oscillatory and non-oscillatory mechanisms may lead to apnea. How this may be possible will be more clear after review of a number of afferent inputs and reflexes of the RRG.

3.2. Behavioral State

One of the most unique characteristics of mammals and birds is the need for sleep, particularly during the period soon after birth (Fig. 7). Cyclical differences in behavior and electroencephalographic (EEG), electrooculographic (EOG), and electromyographic (EMG) patterns have permitted classification of the sleep state into several stages (*90,167*). In the adult, sleep has been found to progress from wakefulness into synchronized or nonrapid eye movement sleep (NREM) and subsequently into active, desynchronized, or rapid eye movement sleep (REM) (*197*).

3.2.1. NREM in the Adult

NREM has been divided by EEG characteristics into four stages. From stages 1–4, there is a progressive decrease in frequency and increase in amplitude. Stages 1 and 2 have been associated with

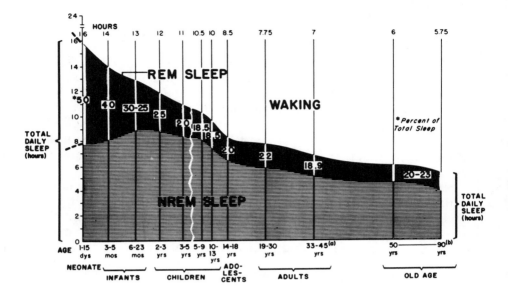

Fig. 7. Graph showing changes (with age) in total amounts of daily sleep, daily REM sleep, and percentage of REM sleep. Note sharp diminution of REM sleep in the early years. REM sleep falls from 8 h at birth to less than 1 h in old age. The amount of NREM throughout life remains more constant, falling from 8 to 5 h. In contrast to the steep decline of REM sleep, the quantity of NREM sleep is undiminished for many years. Although total daily REM sleep falls steadily during life, the percentage rises slightly in adolescence and early adulthood. This rise does not reflect an increase in amount; it is because REM sleep does not diminish as quickly as total sleep. Data for the 33–45- and 50–90-yr groups are taken from Strauch (*139*), Kales et al. (1967), Feinberg et al. (1967), and Kahn and Fisher (1969), respectively (revised by the authors since publication in *Science*, 152: 604–619, 1966, with permission of author and publisher).

drowsiness and light sleep. Stage 2 has also been characterized by the presence in the EEG of spindles (12–14 Hz). Stages 3 and 4 have also been called quiet sleep or slow-wave sleep because of the slow waves (2–7 Hz) and high voltage activity (>100 μV) that have been observed in the EEG (*254*). During NREM, muscular activity has been observed to be diminished, but proprioceptive reflexes remained intact (*197*).

Respiration during NREM appears to be governed by metabolic mechanisms. The usual breathing pattern has been found to be regular. However, periodic breathing has also been observed, usually during stages 1 and 2, and particularly in an hypoxic environment, e.g., high altitude (254). During each NREM epoch, $\dot{V}$, V_T, and respiratory rate have been noted to progressively decrease in parallel with a decline in metabolic rate (251,254). Other ventilatory changes that have been observed during NREM have included a net retention of CO_2 and decrease in both alveolar and arterial O_2 (131,254).

3.2.2. REM in the Adult

REM has been observed to follow NREM about 50–70 min after the onset of sleep, then recurring every 80–90 min. It has been characterized by an EEG of low-voltage oscillations of mixed frequency, resembling beta waves and bursts of phasic muscular activity (e.g., grimaces, athetoid movements of the torso) that accompany rapid eye movements (279). Proprioceptive reflexes have been found to be virtually absent. In addition, the tone of the most limb, neck, and chin muscles has been demonstrated to be profoundly suppressed (i.e., postsynaptic inhibition of alpha-motoneurons and trigeminal motoneurons via the reticulospinal system).

The pattern of breathing in REM has been observed to be typically irregular; it has also been found to be independent of a variety of different afferent inputs, e.g., vagal (116,251,265), aortic and carotid chemoreceptors (131), and chest wall receptors (235). In the cat, pontogeniculooccipital waves in the lateral geniculate body, considered to be part of the REM process, have been correlated with medullary respiratory activity and phasic events such as eye movements (54,305,13). In comparison to NREM, observations have included (i) a greater mean respiratory rate and minute ventilation (10), (ii) a smaller V_T (251,254,265), and (iii) an increased or unchanged arterial O_2 (254). Thus, the respiratory pattern is believed to be characteristic of processes inherent to REM and, therefore, relatively independent of the usual metabolic controls.

3.2.3. Sleep Staging in the Newborn

In the newborn, the sleep state has been more difficult to classify than in the adult. Sleep–wake staging has relied heavily on behavioral and physiologic activity, as well as EEG patterns (8,247,332). Even with modification of adult schema, neither REM nor NREM have always been identifiable. Thus, a transitional or indeterminate state has often been added (69,332).

During the first few months of postnatal life, marked changes have been recognized in both the distribution of time spent sleeping and awake, and in the (electroencephalographic and behavioral) characteristics that identify different behavioral states (*8,247,332*). At birth, nearly two-thirds of the 24-h day has been found to be spent in sleep, with 50% in REM. Subsequently, both total sleep time and REM have been reported to diminish. However, NREM has been found to remain fairly constant (Fig. 5). In addition, the sequence of states in the newborn's sleep cycle has been reported to differ from the sequence in the adult; REM has been observed to directly follow either wakefulness or NREM (*332*). Furthermore, the duration of each state during a particular cycle has been reported to be somewhat shorter in the newborn than in the adult, i.e., 20–25 min, although periods of REM exceeding 50 min have been observed (*279*).

3.2.4. NREM in the Newborn

Until 44–46 wk postconception, the dominant EEG pattern of NREM has been observed to be the trace alternant (long period of low-voltage activity interrupted by irregular delta and theta waves 1–4 s long, which have also been seen in awake infants) (*332*). The low-voltage activity in the premature infant before 32 wk has often appeared as a flat line. Sleep spindles have been observed by the second month after term in stages 2 and 3, and have become restricted to stage 2 after 4 mo of age (*197*). Except for occasional generalized body movements (startles), NREM in the newborn has been found to be essentially devoid of muscle contractions (except, of course, respiratory muscles, e.g., diaphragm, intercostals).

3.2.5. REM in the Newborn

In infants ≤ 30 wk gestation, the EEG during REM (*197*) has been found to consist of some beta activity superimposed on high delta waves. By term, these waves have been observed to disappear and the EEG has been found to be of low amplitude (i.e., appeared flat) with a high percentage of delta and sub-delta waves. Over the next several months, the EEG pattern has been found to be variable until 3–4 mo of age, when mostly theta waves have been observed.

In contrast to the virtual motionless during NREM, the newborn during REM has been observed to have frequent grimaces, facial and limb twitches, and gross positional changes of the limbs (*197,279*). On the other hand, a reduction in resting muscle activity has been observed in conjunction with the onset of each REM period. Thus, the overall picture during REM has been described as one of "suppression of muscle tone punctuated by frequent muscle contractions" (*279*).

3.2.6. Neonatal State-Associated Characteristics of Respiration

State-associated characteristics of the breathing pattern, O_2 consumption (*310*), and $\dot{V}$ (*29,108*) in the term human (birth to 6 mo) have been found to be similar to those of the adult (*11, 29,77,108,137,141,150,151,160,256,257*). In prematures, however, the respiratory rate has appeared to be independent of state (*298*) until >35 wk gestation (*78*). In both premature and term newborns, ventilatory perturbations such as increased work of breathing (*43*) paradoxical chest movements (*109,154*), atelectasis, and reduced thoracic volume (*144,155*) have found to be more frequent during REM than NREM; in contrast to the adult, mean arterial O_2 (*144,218,219*) and CO_2 (*219*) have been reported to be lower during REM.

The effect of maturation on the state-dependent characteristics of respiration have been examined in several nonhuman species. The results of these few studies have differed considerably from those reported in humans. (i) In monkeys (2–21 d), rate was independent of state (*135*); (ii) in puppies (14–29 d), rate was slower and V_T greater during REM (*140*); and (iii) in kittens (1–8 wk), virtually all ventilatory parameters were independent of state (*214*). Thus, considerable research is needed to verify maturational and/or species differences in state-dependent characteristics of respiration.

The interactions of sleep state and other afferent inputs to the RRG have also received considerable attention. When appropriate, these will be discussed in subsequent sections.

4. Afferent Inputs to the RRG

A number of different afferents feed into the DRG (NTS) and other parts of the CNS making up the RRG. These inputs are summarized in Fig. 2.

4.1. Pulmonary Afferents

Afferent activity from receptors within the airways and lungs has been found to be transmitted by vagal myelinated (type A) and nonmyelinated (type C) fibers (*66,241,246*). The nomenclature used to distinguish among different pulmonary afferent endings has come from a number of studies in adult animals. Table 5 summarizes receptor location, distribution, conduction velocities, and known adequate stimuli.

Most of the information about the developmental characteristics of pulmonary afferents has been obtained indirectly; although many have been concerned with the response of the RRG to various perturbations, only a few morphologic investigations of vagal pulmonary fibers have been undertaken (*212,281,290*). In the human, myelination has been found to increase linearly from 24 wk (0.6 term), reaching adolescent values by 40 wk (term) gestation (*281*). In the cat, the degree of vagal myelination in the neonatal period has been less clear: similar degrees between 2 and 35 d (*290*) vs marked increases between birth and 60 d (*212*). Conduction velocities of fibers 1 and 5 μm in diameter have been found to be similar to adult values: at birth, 20 μm/s and 6μm/s; at 60 d, 35 μm/s and 10 μm/s, respectively (*211*).

4.1.1. Pulmonary Stretch Receptors

Pulmonary stretch receptors (PSR) have been so named because their activity has been found to increase rhythmically in phase with I (*246*). They have also been called slowly adapting receptors (SAR), since their discharge has been found to adapt slowly or at various rates to a sudden maintained inflation (*1*). It is the PSR or the SAR that have been considered to be the receptors of the Breuer-Hering reflex (*37,157*).

Most knowledge of the distribution of PSR has come from indirect physiologic measurements rather than from anatomic studies. In adult dogs (*16*), the majority of PSR were found in the extrapulmonary airways (especially the trachea) that represented only a small proportion of the total airway surface area. In the puppy, a similar distribution of receptors was noted (*112*). In contrast, the majority of PSR in kittens were localized to the intrapulmonary airways (*213*).

Although more frequently linked to lung volume (*1*), PSR activity has had a better correlation to transpulmonary pressure (P_{tp}) (*84*). Several investigators have considered the circumferential tension (T) in the airways to be the actual stimulus. This T has

Table 5
Pulmonary Afferent Vagal Fibers

Fiber type	Receptor	Location/distribution	Conduction velocity, m/s				Natural stimuli
			Mean	Range			
A	Stretch (also called slowly adapting receptors) "Tonic" or "low-threshold" "Phasic" or "higher-threshold"	Between muscle fibers of sub-mucosa airways (246)	38	8–62	dog	(221)	Lung Inflation (1)
			53	33–69	dog	(64)	
			36	14–59	cat	(243)	
			39	22–54	cat	(244)	
			37	14–58	cat	(244)	
A	Irritant (also called rapidly adapting receptors)	Superficial ciliary epithelium of tracheobronchial tree (333)	25	16–37	cat	(238)	Mechanical and chemical irritation (225) Rapid lung inflation and deflation (180,240,242)
			13	4–26	rabbit	(225)	
C	Pulmonary J	Interstitial tissues close to the pulmonary capillaries (66)	1.4	0.8–2.4	dog	(65)	Pulmonary congestion (251)
			1.3	0.9–2.1	cat	(9)	
				0.8–8	cat	(245)	
C	Bronchial	Intrapulmonary airways both larger, near hilum, and smaller (66)	(see Pulmonary, above)				Hyperinflation, other (66) Inflamation (66) Lung Inflamation Other

been found to depend upon the length–tension relationship of the airways smooth muscle, the transpulmonary pressure (P_{tp}), and the airways radius (R); [T = P_{tp} × R] (*16,211*).

The PSR have been observed to respond to both the overall degree of inflation (static) and the rate of inflation (dynamic) (*84,241*). They have also been characterized as (i) "high threshold" or "phasic" (modulated only with inflation, i.e., silent during E), or (ii) "low threshold" or "tonic" (active during E, but also modulated with inflation) (*182,213,244,246,269*).

Both species- and age-dependent differences of PSR have been reported concerning the proportion of PSR with tonic characteristics (Table 6). These have been related mainly to functional characteristics (i.e., the length–tension relationship of the airways smooth muscle), rather than to morphologic aspects. In the adult, the percentage of receptors considered tonic has been reported to vary directly with size, i.e., larger mammalian species have a higher proportion of tonic PSR. In puppies (*112*) and kittens (*213*), a relative paucity of tonic receptor activity (in comparison to adults of these species) has been reported. This has been attributed, in part, to the lower functional residual capacity (FRC) and resting P_{tp} of the newborn, i.e., PSR P_{tp} threshold of the newborn exceeds that at FRC of the adult (*2,110,111*). It has also been suggested that different mechanical coupling between the tracheal backwall, site of the receptor (*16*), and the U-shaped cartilages leads to lower tension at the receptor site in the newborn than in the adult (*112*).

4.1.1.1. Responses to the RRG to Lung Inflation and Deflation

It was first recognized in the mid-19th century that lung inflation may inhibit I effort and deflation may augment respiration, i.e.,

Table 6
Percentage of Stretch Receptors
Considered To Be Tonic

Species		Percent	Ref.
Guinea pig	(adult)	27	*182*
Rabbit	(adult)	44	*269*
Cat	(adult)	60	*244*
	(newborn)	9–10	*211,213*
Dog	(adult)	63	*112*
	(newborn)	16	*112*

I-inhibitory and E-facilatory (*37,152*). Subsequently, facilitation of inspiratory activity by lung inflation has been observed under a variety of experimental conditions (*94,152,230*). In addition, it has been recently recognized that lung inflation may facilitate inspiratory motoneuron discharge under normal conditions (*63,270*).

The effect of lung inflation on respiratory timing has been studied in three ways. (i) Altering lung volume, e.g., increasing V_T or FRC; (ii) altering the normal phasic relationships of lung volume and central respiratory discharge, e.g., end-E occlusion of the airway (no I inflation) and end-I occlusion (maintained inflation during E); and (iii) bilateral vagotomy (*68,93*).

CO_2 inhalation has been used to increase V_T in spontaneously breathing subjects. In anesthetized cats, a hyperbolic relationship between V_T and T_I has been described. However, in conscious humans, such a relationship was not achieved unless V_T reached twice the control value (*57*). Similar studies have been undertaken in neonates. In kittens (*322*), there was an age-dependent response; the amplitude of INT PHR increased independently of age, but T_I decreased more in the 1–7 d-olds than in the 15–21 d-olds. In humans, several studies of the I-inhibitory reflex have yielded conflicting results (Table 7). Rather than true differences in the I-inhibitory reflex, the disparities might have been caused by the effects of other uncontrolled afferent inputs, including (i) a direct effect of CO_2 on irritant receptor (*83,292*) or stretch receptor activity during smooth muscle contraction (*112*), (ii) a vagally independent intercostal-to-PHR inhibitory reflex (*178*) (*see* section 4.3, Chest Wall Afferents), and (iii) behavioral state (*108,251*).

Table 7

Breuer-Hering Reflex: Effect of Co_2 Inhalation of Human Infants on Tidal Volume (V_T) and Inspiratory Time (T_I)

Gestation	Postconceptual age, wk	CO_2 concentration, percent	V_T	T_I	Ref.
Term	42–54	2	↑ (but 2× control)	no Δ	*138*
Term	>37	4	marked ↑	↓	*27*
Preterm	32–36	4	no Δ	↑	*27*
Preterm and term	27–36	2 and 4	↑	no Δ	*316*
Preterm and term	29–41	2 and 4	↑	no Δ	*34*

In the adult, removal of phasic vagal afferent activity by end-E occlusion of the airway was reported to result in prolongation of T_I in spontaneously breathing conscious dogs (*249*), Na pentobarbital or Dial-urethane anesthetized cats (*129*) and spontaneously breathing midcollicular decerebrate unanesthetized cats (*60*); thus, the lung-inflation response to V_T acted to decrease T_I. In anesthetized dogs, the prolongation of T_I after airway occlusion fell short of the T_I obtained following bilateral vagotomy (*250*), suggesting "tonic" vagal activity in addition to "phasic" activity. A state-dependent effect has also been reported; in dogs, the response (i.e., increase in T_I) was diminished during REM sleep (*251*).

Conflicting results have been reported concerning responses to tracheal occlusions in neonates. In phenobarbital anesthetized kittens (1–21 d old), end-E occlusions were followed by prolongation of T_I (*323*). On the other hand, T_I decreased following occlusion in several other studies (kittens, *203,305*; rabbits, *303*). The response of unanesthetized human newborns to end-E occlusions have also not been uniform (Table 8). These results probably reflect other factors that may limit the usefulness of the occlusion technique in newborns; possible explanations include: (i) stimulation of other afferent pathways, e.g., intercostal-to-PHR inhibitory reflex (*178*); and (ii) limitations from mechanical factors, e.g., chest wall distortion (*198*).

Phasic vagal afferent activity has been also controlled by the use of the cycle-triggered pump (CTP) (*60,105*) in paralyzed, artificially ventilated animals. In the adult cat, a no-lung-inflation test (i.e., hold in E) resulted in a prolongation of T_I (*63*). In contrast, in piglets lightly anesthetized with halothane, T_I was not prolonged (Fig. 8). This method, too, might have been confounded by other factors (e.g., anesthesia), as has been suggested by some recent observations in piglets anesthetized with Althesin (Glaxo), paralyzed, thoracotomized, and artificially ventilated, but not on a CTP. The PHR response to a no-lung-inflation test (Fig. 9) appeared to depend on the rate of administration of anesthetic: T_I lengthened when the infusion rate of Althesin was 30–40 mg/kg/h; however, T_I was not altered when the rate was much lower, i.e., 3–4 mg/kg/h. The observed change was clearly dependent on vagal activity; it was abolished by bilateral vagotomy (*125*).

In addition to phasic changes in lung volume, the effect on T_I of a sustained increase in FRC by application of either continuous

Table 8
Effects of Airway Occlusion on Inspiratory Timing in Human Infants

Study population	Findings	Ref.
Preterm (mean 29.9 wk) vs term (mean 39.8 wk)	Longer prolongaton of T_I in preterm (53%) than in term (25%)	236
Preterm (29–36 wk) vs term within first 2 wk and preterm (mean 30 wk) tested serial measurement to 40 wk postconceptual age	1. The inspiratory inhibition reflex decreased with maturation 2. Extrauterine development delayed the rate of disappearance of this reflex	177
Term serial measurements from 10–90 min to several days	The ratio of T_I following occlusion (T_I0) vs control (T_IC) decreased with age	113
Preterm (27–36 wk) and term measurements at birth, 3–4 d, 7–10 d, 2–6 wk	1. In 27–32 wk GA, no change with postnatal age from 1–10 d, increase at 5 wk 2. In 33–36 wk GA, no statistical change from 1–10 d (2–6 wk not measured) 3. Extrauterine development yielded at same postconceptional age a greater increase in T_I 4. 29–32 wk GA vs term are no different by term 5. Two infants had a decrease in T_I	316
Preterm (28–41 wk)	1. Longer prolongation of T_I with increasing age 2. 8 of 10 infants (mean 28–34 wk GA) had shortening of T_I	121
Term (17–51 h old)	1. Much greater increase in NREM than REM 2. In both states T_I decreased following occlusion	108

positive (CPAP) or negative airway pressure (CNeg) has been studied. In anesthetized adult animals (*80,125*), the response has been variable. Similarly, in preterm and term infants, T_I has been reported to either increase or decrease attributed to one or more

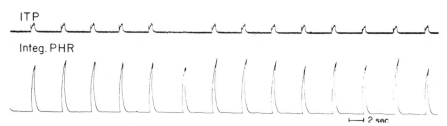

Fig. 8. No-lung-inflation test in newborn pig anesthetized with halothane. Note decrease in integrated phrenic signal (Integ. PHR) when inflation was withheld for one cycle and intratracheal pressure (ITP) was maintained at FRC (from Gootman, unpublished observation).

of the following: (i) elimination of the deflation reflex (*216*), (ii) elimination of chest distortion and the intercostal-PHR inhibitory reflex, particularly during REM (*143*); and (iii) improvement in chest wall mechanics (*143,217*).

The Breuer-Hering deflation reflex (an increase in respiratory rate caused by moderate deflation of the lungs) has been described in the anesthetized human (*215*) and cat (*136*). Deflations applied during E have been found to produce shortening of this

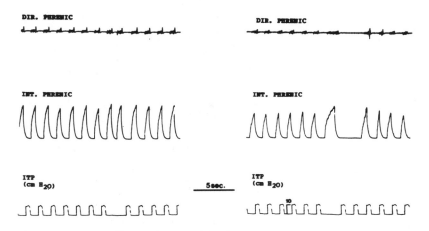

Fig. 9. No-lung-inflation tests in newborn pig anesthetized with Althesin (GLAXO). Left: low althesin (3 mg/kg/h, iv). Note that there is no change in direct phrenic (Dir. Phrenic) nor integrated phrenic (INT. PHRENIC) signals. Right: High Althesin (30 mg/kg/h, iv). Note augmented Dir. Phrenic and Int. Phrenic signals and prolongation of both inspiratory and expiratory durations (from ref. *125*).

phase; the threshold for such premature switching from E to I has been found to decrease with time from the onset of E (*62,181*). This reflex has also been the subject of several neonatal studies. Although no change in rate followed deflations in awake newborn humans (*116*), an increase was observed in anesthetized newborn kittens, monkeys, and rabbits (*85,320*): T_I was variable and T_E has been observed to lengthen in anesthetized kittens (*323*). With the CTP, Gootman et al. (*125*) (Fig. 8) observed no change in T_E with a no-lung-inflation test in halothane-anesthetized piglets. Further studies are needed in decerebrate unanesthetized animals (e.g., piglets), to determine whether these observations were species- or maturation-dependent, or the result of anesthesia.

A number of studies have examined the effect of deflation on respiratory timing by applying inflations during E or by maintaining the inflation achieved during I. Inflations delivered during E have been found to act in an opposite fashion to deflations; T_E has been prolonged such that the later in E that the inflation has been applied, the larger the prolongation (*60,172*). Similarly, T_E in term humans (*76*) has been prolonged by inflations achieved by compressing a rubber bag connected to a face mask or by body decompression. The effect was age-related; T_E was less prolonged in 5 d of age than immediately after birth. Increases in T_E, following end-inspiratory occlusions (maintained I), have also been reported in preterm as well as term newborns (*25*); the increase in the T_E was progressively greater from 32 to 37 wk, but progressively lesser with a further advance in gestational age. In addition to phasic changes in lung volume, increased tonic activity (elevation of FRC) has been associated with an increase in T_E in adult animals (*17,80*), as well as term and preterm human infants (*217,306*).

In addition to the aforementioned effects on respiratory timing, the effect of vagal activity on CID (i.e., the rate of rise of I activity, or slope of I ramp) has also been studied. In paralyzed, midcollicular decerebrate adult cats, the typical response pattern to a no-lung-inflation test has been to show no change in slope for most of I and an increase in slope in late I (*63*). Similar responses have been seen in deeply anesthetized (*14,339*) and spontaneously breathing unanesthetized decerebrate cats (*60*). On the other hand, in the anesthetized dog (*18*) and pig (*162*), the INT PHR slope has been reported to be unchanged in early I and increased in late I, suggesting lung inflation-facilitated PHR discharge.

Augmentation of PHR (or INT PHR) activity and V_T has been observed following deflation in anesthetized newborn kittens, monkeys, rabbits (*85,320*), sheep (*161*), and humans (*76, 315,319,325*). Such facilitation of I activity has also been observed following withholding of inflation in newborn kittens and piglets. However, the efferent response of the RRG has varied, possibly as a result of differences in species or anesthesia. In kittens (birth to 21 d old), INT PHR response to airway occlusion was inhibited by increasing the dosage of phenobarbital; although there was an increase in slope of INT PHR in late I at 20 mg/kg, there was no increase in late I and a decrease in early I at 30 mg/kg (*323*). In contrast, augmentation of INT PHR has been elicited with a no-lung-inflation test in piglets under deep anesthesia (Althesin, 40 mg/kg/h), but not light anesthesia (Althesin, 4 mg/kg/h) (Fig. 9) (*125*). Furthermore, piglets lightly anesthetized with halothane have been observed to have an inhibition of INT PHR discharge when inflation was withheld using a CTP (Fig. 8).

In summary, the PSR have appeared active during the neonatal and postnatal period. However, identification of whether certain maneuvers facilitate or inhibit the efferent output of the RRG has often been confounded by other factors (e.g., anesthetics, state, chest wall afferents). Future studies (such as lung-inflation tests using CTP, single-unit analysis, decerebration) are needed to delineate more clearly the effect of maturation on vagal afferent modulation of the RRG.

4.1.2. Irritant Receptors

Irritant receptors have been found in both extrapulmonary (*333*) and intrapulmonary (*225*) airways of adult animals. They have been so named because their discharge increased with mechanical irritation from mucus, dust, or other particles (*246*). Bronchoconstriction and cough have followed stimulation of extrapulmonary irritant receptors (*230,231,334–336*); stimulation of intrapulmonary receptors has also elicited bronchoconstriction, but not always a cough component (*225*).

Because irritant receptors rapidly adapt to sudden inflation or deflation (*180*), they have also been called rapidly adapting receptors (RAR). Recent evidence (*241*) has suggested that the primary stimulus of these receptors might be transductive, rather than nocioceptive as once was thought. For example, a decrease in T_E has been observed following selective inhibition of PSRs by cold block (*232*) or SO_2 block (*83*). Some investigators have attributed augmented breaths and subsequent rapid shallow breathing following sudden deflation (*123,242,292*) to stimulation of these receptors. Augmentation of PHR nerve activity during rapid inflation has also been attributed to increasing discharge of these endings (*240,242*). It has also been suggested that Head's reflex (paradoxical facilitation of the RRG by lung inflation) was initiated by stimulation of irritant receptors (*152,337*).

Both anatomic and functional immaturity of irritant receptors have been reported (*112,114,183*). The percentage of total myelinated fibers, considered as irritant receptors, have been reported to be both age- and species-related: 4% in the newborn puppy, 15% in the dog (*112*), 10% in the cat (*333*), and 30% in the rabbit (*280*). Maturation-dependent responses to stimulation of the large airways with a fine nylon catheter or fiber have been reported in the conscious human (*114*) and anesthetized kittens (*183*). The effect(s) of maturation on the transductive properties of these endings has not, as yet, been established.

4.1.3. C Fibers and J Receptors

The identification and differentiation of pulmonary and bronchial C fibers has been based upon the response of these fibers to infusion of specific chemicals (i.e., capsaicin or phenyl diguanide) into the pulmonary and bronchial arterial circulations, respectively (*64–66,246*). The responses have included E apnea, followed by rapid, shallow breathing bradycardia and inhibition of myotatic reflexes. In addition, stimulation of C fibers has also been considered by some investigators to be responsible for the tachypnea of pulmonary congestion (*246*), inflammation, and embolism, perhaps secondary to release of serotonin (*66*). Whether C-fiber stimulation elicits apnea or tachypnea has been thought to be "dose-dependent," i.e., tachypnea with low doses of phenyldiguanide, apnea with higher doses (*66,246*). It has also been speculated that the respiratory response to many transductive and nocioceptive stimuli under both normal and pathologic conditions may be caused by stimulation of C fibers. Thus, C fibers have also been considered as the afferents involved in Head's reflex (*66*).

From a developmental point of view, there has been little investigation of the effects of J-receptor or C-fiber stimulation on the RRG. In newborn kittens, there was no response to PDG until 7 d of age; the fully mature response was present by 10 d (*170*). In newborn swine, apnea was seen at 2 d of age (the youngest age studied); the duration of apnea was inversely related to postnatal age (*288*). In view of the sparse myelination in preterm human infants, further information concerning the role of C fibers in respiratory regulation would be of considerable interest.

4.2. The Larynx

Respiration-related studies of laryngeal function have emphasized mechanisms controlling (i) efferent nerve (RL) and muscle activity that regulate airway caliber (*299,302*), and (ii) reflex responses of the RRG and laryngeal muscles to stimulation of laryngeal receptors and/or the superior laryngeal nerve (SLN).

4.2.1. Laryngeal Control of Airway Caliber

Airway caliber has been found to be dependent upon the activity of (i) the posterior cricoarytenoid (PCA), which predominantly regulates I resistance and is facilitated by activation of the cricothyroid (CT) (*147,296*); and (ii) the thyroarytenoid (TA), which mainly regulates E airflow, i.e., adducts during E (*146,148*). Postnatal regulation of the activity of these muscles has been studied in the opossum (*104*) and the lamb (*146–148*).

4.2.1.1. Inspiration

In opossums (24–35 d old), PCA activity was observed to begin during E, preceding the onset of diaphragmatic activity during I. Both PCA and diaphragmatic activity increased during asphyxia (10% O_2 + 7% CO_2) and were depressed by barbituate anesthesia. In lambs, except when rumination began at several weeks of postnatal life, diaphragmatic activity was always accompanied by PCA activity, often facilitated by CT activity (*147*). In adult cats (*237,238*) and rats (*296*), sleep-dependent depression of laryngeal activity has been observed. However, the effect of state on PCA activity has not yet been studied to the best of my knowledge in newborn animals.

4.2.1.2. Expiration

The importance of the larynx in determining E airflow has been clearly demonstrated (*266*). However, the relative contribution of the adductors and abductors has been controversial; TA activity during E was not observed in adult cats (*15*) or opossums (*104*), but was present in lambs (*146–148*). In the latter, maturation-related, state-dependent, and pulmonary afferent-dependent effects on TA activity have been studied. In the first several weeks of life, TA activity was found to increase progressively, being greatest during wakefulness and least during REM (*148*). The TA activity was also found to vary inversely with FRC, e.g., decreased with increasing CPAP or CNeg. In the human, there has been indirect evidence of pulmonary afferent and state-related regulation of laryngeal activity at birth, e.g., "grunting" (E against a closed glottis) in infants with atelectasis caused by respiratory distress syndrome, and reduced thoracic volume during REM (*156*).

4.2.2. Afferent Laryngeal Reflexes

Chemical and mechanical irritation of the laryngeal mucosa has been found to cause a number of reflex changes. In the adult, these have included transient respiratory inhibition, coughing and laryngeal constriction, and swallowing (*30,31*). Such responses have been viewed as protective, i.e., to defend the airways and lungs against aspiration. However, in the newborn, stimulation of the laryngeal region has been observed to elicit prolonged apnea. This effect has proved fatal in neonatal animals, has been implicated as a mechanisms causing SIDS (*96*), and has led to numerous studies concerned with the postnatal development of laryngeal receptors and reflexes (*28,103,191,194, 206,224,308,309,314*).

4.2.2.1. Respiration

Mechanoreceptors sensitive to tactile stimulation (e.g., airflow, flowing water, cotton thread) have been identified from superior laryngeal nerve (SLN) recordings in adult rabbits, cats (*31, 117,283,308*), and newborn lambs (*309*). In kittens and puppies (*5*), the reflex respiratory response (prolongation of T_E) was found to be directly related to age; in the least-mature animals, prolonged apnea was frequently observed.

Chemosensitive receptors have been identified from direct SLN recordings in lambs (*308,309*) and puppies (*28*), as well as from reflex responses following the introduction of fluids into the tracheal region (Table 9). The common ionic basis among these substances is believed to be a reduction or absence of chloride ions in the laryngeal fluid (*28*).

The chemoreflex has been mimicked by electrical stimulation of the superior laryngeal nerve (SLN) in piglets (*103,191,194*), monkeys (*314*), and kittens (*206,224*). At low intensities, such stimulation has resulted in apnea of irregular respiratory movements of

Table 9
Laryngeal Chemoreflex

Species	Gestation	Stimulus	Response	Ref.
Human	Preterm	Water, glucose water	Apnea and swallowing	248
Human	Preterm	Menthol or nasal drops	Transient apnea and coughing	165
Lamb	Term	Water, sugars, acids, salt solutions (lithium, sodium, and calcium chlorides)	Apnea and swallowing	184
Lamb	Preterm	Water	Apnea	209
Puppy	Term	Water, sucrose, urea, potassium chloride, anions of large hydrated size (e.g., acetate)	Apnea	28
		Sodium chloride, small anions (e.g., bromide)	No apnea	
Kitten	Term	Water, sodium bicarbonate, sodium citrate, sodium sulfate	Apnea	206
Piglet	Term	Water, potassium chloride, cow's milk	Apnea	96 187
		Sow's milk	No apnea	188

low tidal volumes. With higher intensities, prolonged apnea (long past removal of the stimulus) has been observed, especially in the youngest animals.

Maturation-related effects on the chemoreflex have included (i) no poststimulus apnea by 28 d in monkeys (314), (ii) apnea of no greater than 10–15 s by 50–60 d in kittens (206), and (iii) no sustained apnea in piglets by 2 mo of age (194). The response and duration of respiratory inhibition has been enhanced by anemia and catecholamine depletion (195), hypocarbia (191), and chloralose anesthesia (194) in piglets, and barbiturates in kittens (224). On the other hand, the chemoreflex was diminished by aminophylline in piglets (194) and terbutaline in lambs (125).

The above-mentioned factors that have altered the strength and/or duration of the respiratory inhibition following laryngeal stimulation suggest that central processing, rather than peripheral mechanisms (e.g., laryngeal receptor properties), underlie maturation-related differences in the chemoreflex. Single-unit analyses of neurons in the vicinity of NTS in the kitten (206) and cat (293) have offered additional support for this view; following SLN stimulation, the percentage of units inhibited varied inversely, whereas the percentage excited varied directly with postnatal age. However, these neurons were not cross-correlated to PHR activity nor definitively identified by histochemical means. Thus, future studies, such as single-unit recordings cross-correlated to whole inspiratory nerve activity by cycle-triggered histograms, are needed to determine possible factors (e.g., age, anesthetic) that might influence how individual respiratory-related neurons respond to afferent laryngeal input and how these responses might be summated.

4.2.2.2. Coughing and Swallowing

Coughing has been observed in premature (27%) and term (49%) newborns during laryngoscopy, and in prematures following mucosal depression by a nylon fiber (25%). It has been also observed following chemostimulation with menthol drops in preterm infants (19–37 wk) (165). However, coughing has not been reported as part of the chemoreflex associated with apnea in most other investigations. Rapid swallowing has been observed following application of water into the laryngeal region in lambs (149,184) and in human newborns (248). A direct relationship was found between the frequency of swallows and duration of respiratory inhibition. From these results, it appears unlikely that prolonged apnea following laryngeal chemostimulation was caused by an inability to remove foreign substances from the laryngeal region or age-related changes of afferent SLN activity. It seems more likely that prolonged apnea was the result of maturation-related characteristics in CNS processing.

4.2.2.3. Laryngeal Constriction

EMG recordings from TA muscles in fetal (0.7 gestation) and newborn lambs (146) have demonstrated prolonged adductor contractions accompanying laryngeal stimulation. In puppies, single-shock SLN stimulation did not elicit adductor responses prior to 2 d of age (286). In preterm and term newborn monkeys, brief glottic adduction (measured as a change in airway resistance) followed SLN stimulation, but was no longer present during poststimulus apnea (*314*). From these studies, it has not as yet become clear as to whether the presence of a laryngeal adduction component of the chemoreflex is maturation related. At present, one can only speculate that an inadequacy of glottic adduction might underlie the greater susceptibility to aspiration seen in the newborn period.

4.2.2.4. Behavioral State

Species differences have also been observed with respect to behavioral state, i.e., arousal or sleep. Data at present are insufficient to lead to any conclusions about maturation. Arousal from sleep by laryngeal stimulation was more difficult to elicit during REM than NREM in adult dogs, whereas inhibition of respiration and bradycardia occurred at a lower threshold during REM (*312*). On the other hand, arousal and respiratory responses in lambs were reported to be independent of sleep state (*209*).

4.2.2.5. Cardiovascular Responses

In lambs (*128,145*) and piglets (*96,194*), reflex apnea was accompanied by bradycardia and hypertension. These responses were attenuated by administration of a beta-adrenergic agonist (terbutaline) and inhibited by a beta-adrenergic antagonist (propranolol) (*128*). Although carotid denervation had no effect on the apnea, the hypertensive response was attentuated and bradycardia was eliminated (*102*). In the monkey, SLN stimulation resulted in a fall in heart rate and blood pressure at stimulus onset, followed at higher stimulus intensities by an increase in blood pressure and fall in heart rate (*314*). Although the newborn has appeared more vulnerable than the adult to adverse consequences of laryngeal stimulation, i.e., prolonged apnea and death (*96,206*), these cardiovascular changes strongly suggest that at least the initial response (apnea, bradycardia, hypertension) to laryngeal stimulation might be part of an oxygen-conserving mechanisms, such as in the diving reflex (*6*).

4.3. Chest Wall Afferents

In the adult, chest wall afferents have been shown to affect the timing and force of I muscle contraction by segmental (*291*), intersegmental, and supraspinal reflexes (*89,97,263,264*). Intercostal-to-intercostal and intercostal-to-PHR reflexes (acting in concert with, or independently of, vagally mediated reflexes) have been demonstrated following mechanical and electrical stimulation of chest wall receptors and intercostal nerves (*70,99,263,294,295*).

During mechanical loading, such as increased resistance to airflow, facilitation of external intercostal muscle activity has been observed; stimulation of muscle spindles has resulted, via thoracic dorsal roots (*70,208,210*) in recruitment of new motor units of intercostal muscles and increased rate of unit activity (*294*). This "load-compensation reflex" has been considered important for providing stability to the rib cage, i.e., preventing collapse of the thorax (*294*), and for maintaining FRC, particularly during periods of diaphragmatic fatigue (*203,204*).

The PHR discharge has been both facilitated and inhibited following intercostal afferent stimulation. Facilitation, mainly via spinal pathways, has been reported following stimulation of caudal thoracic nerves in spinal and decerebrate cats (*89*). On the other hand, inhibition (shortening of I, lengthening of E) has followed mechanical and electric stimulation of midthoracic nerves, e.g., by rapid compression of the rib cage and/or airway occlusion (*263,264*).

There have been several age-related studies, both morphologic (*324*) and physiologic (*324–326*), concerned with chest wall afferents. These have been prompted partly by observations of paradoxical chest wall movements (i.e., collapse during I) and frequent episodes of apnea in premature infants.

Age-related changes have been demonstrated in the degree of myelination and ~aliber of dorsal roots. In kittens, the percentage of fibers that were myelinated doubled over the first 60 d of life. Intercostal nerve myelination lagged behind that of the PHR, suggesting a relative delay in the functional innervation of the intercostal muscles vs the diaphragm (*212*). On the other hand, by staining with succinic dehydrogenase, the intercostal muscles and diaphragm of newborn rabbits and kittens were found to mature at a similar rate. Study of the caliber of the thoracic roots suggested a paucity of gamma fibers before 25 d (*290*).

Although the forementioned studies suggest immaturity of both afferent nerves and respiratory muscles, physiologic studies have found chest wall afferents to be active soon after birth; the load-compensation reflex was observed in vagotomized kittens (<1 wk old) (*325*) and normal newborn humans. However, the strength of the response has been reported to be age-related, i.e., increased with postconceptual age (*26*).

Sleep has also been found to alter load-compensation and chest wall discharge. Although less than during wakefulness, the load-compensation reflex was reported to be greater during NREM than REM (*179,236*). In addition, intercostal nerve activity was observed to be decreased during REM (*258*) and believed to underlie the paradoxical chest movements frequently seen during that state (*109,142,154*).

With respect to respiratory timing, the intercostal-to-PHR inhibitory reflex has been reported following chest wall compression, airway occlusion, and application of vibratory stimuli in preterm human infants (*143,178*). Similarly, in <1 wk-old kittens, electrical stimulation of intercostal nerves was observed to inhibit I and prolong E. An inverse relationship between stimulus strength and T_I was reported. Also, the later in E that stimulation was applied, the longer the prolongation of T_E (*326*).

One can conclude that, although anatomically and functionally immature, soon after birth chest wall afferents may significantly alter efferent respiratory activity of both the intercostal muscles and diaphragm. However, an inability to compensate for respiratory loads and an active intercostal-to-PHR inhibitory reflex (especially during REM) may be important mechanisms leading to apnea (*143,178*).

4.4. Chemical Drive to the RRG

The pattern of respiration may be altered by excitation or inhibition of central (ventrolateral medulla, level of the cranial nerves VII–XII) and peripheral (aortic and carotid bodies) chemoreceptors. A decrease in PaO_2 or pH or an increase in $PaCO_2$ has been found to augment the discharge frequency of these receptors (*53,92*). Stimulation of these receptors has been found to elicit an increase in efferent respiratory discharge; when chemoreceptor activity was decreased, the opposite was observed.

4.4.1. Peripheral Chemoreceptors

In the adult, the carotid chemoreceptors (innervated by the glossopharyngeal or IXth cranial nerve) have been found to be influenced by changes in PaO_2, but also sensitive to breath-to-breath changes in $PaCO_2$ (*14,33*), venous CO_2 (*255*), and pH. In the neonate, most studies have been concerned with the response of the RRG to changes in the concentration of inspired O_2. Thus, little is known about the effect(s) of pH and CO_2 on these receptors soon after birth.

In the adult, the aortic bodies (innervated by the vagus or Xth cranial nerve) have appeared less important than the carotid bodies in modulating the respiratory response to changes in PaO_2 (*67*). However, the relative importance of the different chemoreceptor in the newborn has not as yet been determined because few studies have examined this question. In lambs, aortic body denervation did not alter the ventilatory response to cyanide (*86*). On the other hand, lambs responded with an increase in ventilation during hypoxia, despite bilateral carotid nerve transection (*20*). Whereas sectioning of the carotid sinus nerves had no effect on establishing effective breathing after birth in lambs (*163*), denervation of both carotid and aortic bodies in newborn rats resulted in severe episodic disturbances of respiration (50% of the subjects died) (*158*).

Another peripheral chemoreceptor site, the glomus pulmonale, which has a dual blood supply from the pulmonary artery and aorta, has been identified mainly in neonates (*164*). Its function has not been determined as yet.

4.4.1.1. Response to Hypoxia

When placed in a moderately hypoxic environment (e.g., 15% O_2), the adult mammal has typically responded with a sustained increase in ventilation. In contrast, newborns have responded with a two-phase pattern that has also been referred to as biphasic; transient increase in ventilation followed by a decrease toward (*24,47,72,192,338*), or below (*36,74,190,273,284*) control values. However, neither response has proved to be unique for a particular age group; sustained hyperventilation has been observed in newborn lambs (*46,260*) and a biphasic pattern has been reorted in the adult human, depending upon the degree and duration of hypoxic exposure (*168,284*). The ventilatory responses of newborns of several species to various hypoxic concentrations are summarized in Table 10. Changes in $\dot{V}$ have been primarily in amplitude, i.e., V_T, (*46,72,30,338*) diaphragmatic EMG (*24*), or INT PHR (*192*); respiratory rate has been variable, i.e., exceeded control values in 2-d-old monkeys (*189,190,338*), but declined in piglets (some animals became apneic) (*192*), kittens (*24*), and preterm humans (*273*).

This inability of newborns to sustain an increase in ventilation during hypoxia has been mainly attributed to "central depression." Several mechanisms have been proposed and/or investigated, as listed in Table 11. Perhaps most intriguing is the idea of

Table 10
Ventilatory Responses of Several Species to an Hypoxic Environment

Species	Age, days	F_IO_2	Response	Ref.
Term human	Newborn	0.12	Biphasic	70
	<1–5.5	0.13	Biphasic or decrease only	69
Preterm human	≤8	0.15	Biphasic	284
	8 (mean)	0.15	Biphasic	273
	18 (mean)	0.15	Sustained increase	
Piglet	1–5	0.15	Biphasic	192
Lamb	2	0.12	Biphasic	47
	2	0.07	Biphasic	
	7	0.12	Sustained increase	
	7	0.07	Biphasic	
	2–90	Progressive to 0.05	Sustained increase	46
Monkey	2	0.14	Biphasic	338
	2	0.12	Biphasic	
	7–8	0.14	Biphasic	
	7–8	0.12	Sustained increase at 6 min	
	19–21	0.14	Sustained increase at 10 min	
Kitten	5–34	0.12	Biphasic	24
		0.06	Biphasic	
Rabbit	1–15	0.05	Biphasic	130

suprapontine inhibition (92), which has also appeared to underlie hypoxia-induced depression of breathing movements in fetal lambs. In intact fetal lambs, or those with rostral brainstem transactions (caudal hypothalamus or anterior commissure), hypoxia inhibited fetal breathing; when transection was more caudal (midcollicular level), fetal breathing was facilitated by hypoxia (87). These findings suggest that respiratory inhibition during hypoxia may be part of an oxygen conservation mechanism. However, similar to the RRG response to SLN stimulation, respiratory inhibition might become prolonged (apnea) and deleterious.

The degree of arousal, or state, has been shown to affect the response of the RRG to hypoxia. In the adult, little difference has been found between wakefulness and NREM. The arousal re-

Table 11
Possible Mechanisms for Hypoxic Ventilatory Depression

Mechanism	Postulated or supported	Ref.	Refuted	Ref.
"Supra brain stem" inhibition	Piglet	200		
Fall in metabolism	Human	36	Piglet	192
	Puppy	140		
Immature excitatory neurotransmitter	Piglet	192		
Inhibitory transmitter				
GABA	Rabbit	153		
Adenosine	Piglet	81	Piglet	201
Endorphins	Rabbit	130	Piglet	81,201
	Human (in part)	88		
Failure to process peripheral chemoreceptor afferent input			Piglet	193
Immature or "weak" peripheral chemoreceptor	Lamb	46,47	Lamb	23
			Kitten	24
			Rabbit	290
			Piglet	193
Altered mechanics	Human	185		
	Monkey	189,190		

sponse remained intact during REM and appeared to be important for preventing severe hypoventilation and for termination of apneic periods (254).

In newborns of several species, state-dependent changes in ventilation have been observed, as summarized in Table 12. In contrast to the adult, the arousal threshold (critical level of O_2 desaturation) has been found to be state-dependent, i.e., REM>NREM in calves (166) and lambs (155). Although some species and individual differences have been reported, a relative impairment in the ability of the newborn to respond to hypoxia is evident especially during REM sleep (283). This vulnerability could be an important mechanism underlying sleep hypoventilation, prolonged apneic episodes, and possibly even sudden death.

Table 12
Effects of Sleep State on Neonatal Response to Hypoxia

Species	Finding(s)	Ref.
Human (premature)	Late decline in REM, biphasic in wakefulness sustained increase in NREM	*278*
Lambs	↑ in NREM, little change in REM	*155*
Calves	↑ in NREM, little change in REM	*166*
Monkeys	Biphasic in NREM	*338*
Puppies	Similar in NREM and REM but greater than degree of hypoxia required during REM	*155*
Puppies (14 d)	Minute ventilation: ↑ in REM, ↓ caused by change in rate. V_T actually increased in NREM and ↓ REM	*140*
Kittens	Decrease rate in NREM, but not REM	*13*

4.4.1.2. Response to Hyperoxia

The activity of the peripheral chemoreceptors has also been examined by chemical denervation by administration of O_2. Ventilatory depression has been observed following a single breath of O_2 in the adult, but not in human newborns <3 d old (*122*). However, this finding might not have been indicative of chemoreceptor depression in the immediate period after birth. Instead, it might have been the result of an insufficient increase in PaO_2 caused by persistance of fetal right-to-left shunting (*259*).

With longer exposure (minutes) to O_2, ventilatory depression has been observed. In newborn lambs, the response was age-dependent: 2-d-olds <30-d-olds (*46*). In the human, a decrease in ventilation has been noted in both preterm and term infants as early as 1 h after birth. With even more prolonged periods of hyperoxia, adults have been reported to show sustained hypoventilation (*35,74,271*). In contrast, the newborn has exhibited a biphasic pattern: transient hypoventilation (sometimes apnea), followed by a sustained increase (*73–75,125,192,239,262,273*). This response has been found to be independent of state (*29,102,278*), gestational age, and postnatal age (*35*). Recent evidence has suggested that the response might have been partly dependent on the baseline PaO_2; when PaO_2 was identical, the adult and newborn responses were qualitatively and quantitatively similar (*284*).

In summary, at birth the pattern of the RRG may be altered by changes in peripheral chemoreceptor discharge. Under certain conditions, respiration may be depressed by hypoxia, even leading to apnea. Further investigations are needed to determine the mechanism(s) underlying such ventilatory depression.

4.4.2. Central Chemoreceptors

4.4.2.1. Response to Changes in pH

In the neonate, the effects of application of acidic and alkaline solutions to the ventral medullary surface have been reported (*331*). Also, a mild systemic metabolic acidosis (by intravenous infusion of hydrochloric acid) has been found to facilitate respiration, but not progressively, as arterial and CSF pH fell (*44*). With a more severe acidosis, ventilation increased further in older lambs (30 and 90 d old), but decreased toward preacidotic levels in younger lambs (2 and 10 d old) (*45*). However, this inhibition might not have been caused by decreased chemoreceptor sensitivity, but by age-related characteristics in CSF acid base regulation (*40,233*). Such a conclusion is suggested by the findings that lambs did not develop paradoxical alkalinization of the CSF (*45*), which has occurred in adults because of rapid loss of CO_2 from CSF to plasma.

4.2.2.2. Response to Carbon Dioxide

Newborns, either animal or human, and preterm or term, have been observed to increase their ventilation in response to CO_2 (*12,56,71,134,169,196,237,261,274*). Responses to both a progressive increase and to an increase in the steady-state level of CO_2 have been investigated (Table 13). Although the results of earlier studies suggested little age-dependent difference in response to hypercapnia (*12,55*), more recent studies have indicated maturational differences: an increase with advancing postnatal age in monkeys (*134,135*), and with both gestational and postnatal ages in humans (*118,138,186,187,274*) and dogs (*140*).

More often than not, the change in V̇ has been caused by a change in amplitude (V_T) rather than rate. However, whether amplitude and/or rate is altered by administration of CO_2 appears to depend upon the preexisting eucapnic respiratory pattern; CO_2 inhalation during periodic or irregular breathing has been associated with a change in frequency, whereas during regular breathing, a change in V_T has been observed (*169,277*).

Table 13

Ventilatory Responses to CO_2 Inhalation

Subjects	Gestational age, wk	Postnatal age	Method, conc. of CO_2	Finding(s)	Ref.
Preterm	32 vs 37	2–27 d	1. Steady state, 2 and 4% CO_2 vs room air 2. Nosepiece 3. Measured mean slope of CO_2 response curve $\Delta\dot{V}/P_ACO_2$	1. $\Delta\dot{V}/P_ACO_2$ highest for oldest subjects; 37, 32 wk 27, 2 d	274
Preterm	32 ± 1	28 ±5 d	1. Steady state, 0.5–1.5% CO_2 vs 21% O_2 2. Nosepiece 3. Polysomnography	1. $\dot{V}$ Increased in both REM and NREM 2. Increase in V_T when control breathing regular 3. Increase in rate when control breathing periodic 4. V_T/T_I not altered under any condition, i.e., state or breathing pattern	169

Group	Gestational age	Age	Methods	Results	Ref
Term		<14–120 d	1. Steady state, 2% CO_2 vs room air 2. Barometric 3. Polysomnography	1. V_T increased at all ages independent of state 2. No effect on timing 3. V_T/T_I increased at all aages in both REM and NREM	138
Newborns	Term	<1–21 d	1. Rebreathing expired air for 4–6 min (P_ACO_2 = 50–60 mm/Hg) 2. Face mask 3. Measured $\Delta\dot{V}/P_ACO_2$	1. No difference between newborns and adults	12
Preterm	28–36	1–50 d	1. Rebreathing 5% CO_2/40% O_2 2. Face mask 3. Measured $\Delta\dot{V}/P_ACO_2$	1. Sensitivity increased with gestational and postnatal ages; body size, lung volume, and compliance	186

Preterm and term	26–40	1–25 d	1. Rebreathing 5% CO_2/40% O_2
			2. Face mask
			3. $\Delta\dot{V}/P_ACO_2$
				1. Identified two types nonresponsive to CO_2:
				  1. Those with decreased compliance but attempt to increase V, i.e., unable to increase respiratory work
				  2. No response to CO_2 at all
					187

Whether the effect of CO_2 inhalation on CID might be age-related has not been clarified; calculated $\dot{V}/T_I$ has been reported to increase with maturation (*138*), but age-dependent differences were not observed using the occlusion method (*71*). It is possible that both of these indices of CID might have been affected by hypercapnia-induced changes in respiratory mechanics (*186,187*). Further elucidation of this issue by separation of neural and nonneural factors might be possible if a more direct measurement of neural activity were employed, e.g., PHR nerve recording during artificial ventilation with the CTP.

In the adult, it has been suggested that the preexisting state of arousal may affect the response of the RRG to hypercapnia, i.e., depressed during sleep, with REM > NREM (*251,252,254*). In the neonate, the magnitude of the ventilatory response to CO_2 has been reported to be both dependent (*41*) and independent of state (*7,82,134,138,169*). In addition, a greater variability in the RRG response to CO_2 has been noted during REM.

In adult dogs, a higher CO_2 during REM than NREM has been necessary for arousal (*32*). Arousal during inhalation of 4% CO_2 has been observed in human infants (*138*), but it was not reported whether this was state-dependent. Further study of the arousal response to CO_2 is needed in order to determine whether arousal might be a critical mechanism for interrupting episodes of hypoventilation or apnea in newborns.

4.4.3. Peripheral and Central Chemoreceptor Interaction

In the human, the newborn response at the peripheral chemoreceptor level to combined hypercapnia and hypoxia has been similar to that of the adult. However, at the level of the central chemoreceptors, responses have differed (*189*); the immediate response to hypoxia (15% O_2) was enhanced by the addition of hypercapnia (4%) (*3*), but the response to 2% CO_2 was depressed by hypoxia in the newborn (*275*). In lambs, the response has differed from that in humans; CO_2 was facilitated by hypoxia, varying inversely with the PaO_2 (*261*).

5. Conclusions

At birth, the newborn appears capable of sustaining regular respiration and maintaining metabolic homeostasis under normal conditions. However, under conditions of stress, the newborn's ability to appropriately increase and maintain efferent respiratory activity appears limited (e.g., hypoxia, inspiratory loading). Reflexes that are part of oxygen-conserving mechanisms during fetal life, or protect the airways against noxious substances during adulthood, have been observed postnatally to have deleterious and sometimes catastrophic effects (e.g., apnea from laryngeal chemostimulation). Hopefully, future studies will offer answers to many questions raised by these observations.

Acknowledgments

The author's research is supported by LIJ-HMC research grant 3-401 and by grant HL 20864 awarded to Dr. Phyllis M. Gootman by NIH. The author sincerely thanks Dr. Nancy Buckley for her considerate review of this manuscript and most helpful suggestions, and Ms. Irene Barling for her secretarial assistance in its preparation.

References

1. Adrian, E. D. (1933) Afferent impulses in the vagus and their effect on respiration. *J. Physiol.* **79,** 332–358.
2. Agostini, E. (1959) Volume–pressure relationships of the thorax and lung in the newborn. *J. Appl. Physiol.* **14,** 909–913.
3. Albersheim, S., Boychuk, R. B., Seshia, M. M. K., Cates, D., and Rigatto, H. (1976) Effects of CO_2 on immediate ventilatory response to O_2 in preterm infants. *J. Appl. Physiol.* **41,** 609–611.
4. Alden, E. R., Mandelhorn, T., Woodrum, D. E., Wennberg, R. P., Parks, C. R., and Hodson, A. (1972) Morbidity and mortality of infants weighing less than 1000 grams in an intensive care nursery. *Pediatr.* **50,** 40–49.
5. Al-Shway, S. and Mortola, J. P. (1982) Respiratory effects of airflow through the upper airways in newborn kittens and puppies. *J. Appl. Physiol.* **53,** 805–814.
6. Anderson, H. T. (1966) Physiologic adaptations in diving vertebrates. *Physiol. Rev.* **46,** 212–243.
7. Anderson, J. V., Jr., Martin, R. J., Abboud, E. F., Dyme, I. Z., and Bruce, E. N. (1983) Transient ventilatory response to CO_2 as a function of sleep state in full term infants. *J. Appl. Physiol.* **54,** 1482–1488.

8. Anders, T. F., Ende, R., and Parmalee, A. H. (1971) A Manual of Standardized Terminology, Techniques and Criteria for the Scoring of States of Sleep and Wakefulness in Newborn Infants. UCLA Brain Inf. Serv./BRI Pub. Off., Los Angeles.

9. Armstrong, D. J. and Luck, J. C. (1974) A comparative study of irritant and type J receptors in the cat. *Respir. Physiol.* **21,** 47–60.

10. Aserinsky, E. and Kleitman, N. (1953) Regularly occurring periods of eye motility and concomitant phenomena during sleep. *Science* **118,** 273–274.

11. Ashton, R. and Connelly, K. (1971) The relation of respiration rate and heart rate to sleep states in the human newborn. *Dev. Med. Child. Neurol.* **13,** 180–187.

12. Avery, M. E., Chernick, V., Dutton, R. E., and Permutt, S. (1963) Ventilatory response to inspired carbon dioxide in infants and adults. *J. Appl. Physiol.* **18,** 895–903.

13. Baker, T. L. and McGinty, D. J. (1977) Reversal of cardiopulmonary failure during active sleep in hypoxic kittens: Implications for sudden infant death. *Science* **199,** 419–421.

14. Band, D. M., McClelland, M., Phillips, D. L., Saunders, K. R., and Wolff, C. B. (1978) Sensitivity of the carotid body to within-breath changes in arterial CO_2. *J. Appl. Physiol.* **45,** 768–777.

15. Bartlett, D., Jr., Remmers, J. E., and Gautier, H. (1973) Laryngeal regulation of respiratory airflow. *Resp. Physiol.* **18,** 194–204.

16. Bartlett, D., Jr., Jeffrey, P., Sant'Ambrogio, G., and Wise, J. C. M. (1976) Location of stretch receptors in the trachea and bronchi of the dog. *J. Physiol.* (Lond.) **258,** 409–420.

17. Bartoli, A., Bystrzycka, Guz, A., Jain, S. K., Noble, M. I. M., and Trenchard, D. (1973) Studies of the pulmonary vagal control of central respiratory rhythm in the absence of breathing movements. *J. Physiol.* (Lond.) **230,** 449–465.

18. Bartoli, A., Cross, B. A., Guz, A., Huszczuk, A., and Jeffries, R. (1975) The effect of varying tidal volume on the associated phrenic motorneurone output: Studies of vagal and chemical feedback. *Respir. Physiol.* **25,** 133–155.

19. Baumgarten, R. von and Kanzow, E. (1958) The interaction of two types of inspiratory neurons in the region of the tractus solitarius of the cat. *Arch. Ital. Biol.* **96,** 361–373.

20. Belenky, D. A., Standaert, T. A., and Woodrum, d. e. (1979) Maturation of hypoxic ventilatory response of the newborn lamb. *J. Appl. Physiol.* **47,** 927–930.

21. Berger, A. J. (1977) Dorsal respiratory group neurons in the medulla of cat: Spinal projections, responses to lung inflation and superior laryngeal nerve stimulation. *Brain Res.* **135,** 231–254.

22. Bertrand, F., Hugelin, A., and Vibert, J. F. (1973) Quantitative study of anatomical distribution of respiration related neurons in the pons. *Exp. Brain Res.* **16,** 383–399.

23. Biscoe, T. J. and Purves, M. J. (1967) Carotid body chemoreceptor activity in the new-born lamb. *J. Physiol.* (Lond.) **190**, 443–454.
24. Blanco, C. E., Hanson, M. A., Johnson, P., and Rigatto, H. (1984) Breathing pattern of kittens during hypoxia. *J. Appl. Physiol.* **56**, 12–17.
25. Bodegard, G., Schwieler, G. H., Skoglund, S., and Zetterstrom, R. (1969) Control of respiration in newborn babies. I. The development of the Hering-Breuer inflation reflex. *Acta Paediatr. Scand.* **58**, 567–571.
26. Bodegard, G. and Schwieler, G. H. (1971) Control of respiration in newborn babies. II. The development of the thoracic reflex response to an added respiratory load. *Acta Paediatr. Scand.* **60**, 181–186.
27. Bodegard, G. (1975) Control of respiration in newborn babies. III. Developmental changes in respiratory depth and rate responses to CO_2. *Acta Paediatr. Scand.* **64**, 684–692.
28. Boggs, D. F. and Bartlett, D., Jr. (1982) Chemical specificity of a laryngeal reflex in puppies. *J. Appl. Physiol.* **53**, 455–462.
29. Bolton, D. P. G. and Herman, S. (1974) Ventilation and sleep state in the newborn. *J. Physiol.* (Lond.) **240**, 67–77.
30. Boushey, H. A., Richardson, P. S., and Widdicombe, J. G. (1972) Reflex effects of laryngeal irritation on the pattern of breathing and total lung resistance. *J. Physiol.* (Lond.) **224**, 501–513.
31. Boushey, H. A., Richardson, P. S., Widdicombe, J. G., and Wise, J. C. M. (1974) The response of laryngeal afferent fibers to mechanical and chemical stimuli. *J. Physiol.* (Lond.) **240**, 153–175.
32. Bowes, G., Woolf, G. M., Sullivan, C. E., and Phillipson, E. A. (1980) Effect of sleep on ventilatory response and arousal of sleeping dogs to respiratory stimuli. *Am. Rev. Resp. Dis.* **122**, 899–908.
33. Bowes, G., Andrey, S. M., Kozar, L. F, and Phillipson, E. A. (1983) Carotid chemoreceptor regulation of expiratory duration. *J. Appl. Physiol.* **554**, 1195–1201.
34. Boychuk, R. B., Rigatto, H., and Seshia, M. M. K. (1977) The effect of lung inflation on the control of respiratory frequency in the neonate. *J. Physiol.* (Lond.) **270**, 653–660.
35. Brady, J. P., Cotton, E. C., and Tooley, W. H. (1964) Chemoreflexes in the newborn infant: Effects of 100% oxygen on heart rate and ventilation. *J. Physiol.* **172**, 332–341.
36. Brady, J. P. and Cerutti, E. (1966) Chemoreceptor reflexes in t he newborn infant: Effects of varying degrees of hypoxia on heart rate and ventilation in a warm environment. *J. Physiol.* (Lond.) **1884**, 631–645.
37. Breuer, J. (1868) Die Selbststeuerung der Atmungdurch den Vervus Vagus. *Sitzber. Akad. Wiss. Wien. Abt.* **2**, **58**, 909–937.
38. Brouillette, R. T. and Thach, B. T. (1980) Control of genioglossus muscle inspiratory actvity. *J. Appl. Physiol.* **49**, 801–808.

39. Brown, H. W. and Plum, F. (1961) The neurologic basis of Cheynes-Stokes respiration. *Am. J. Med.* **30,** 849–860.

40. Bruce, E. N. (1981) Control of breathing in the newborn. *Ann. Biomed. Eng.* **9,** 425–437.

41. Bryan, M. H., Hagan, R., Gulston, G., and Bryan, A. C. (1976) CO_2 response and sleep state in infants (Abst.). *Clin. Res.* **24,** 689.

42. Bryan, A. C. and Bryan, M. H. (1978) Control of respiration in the newborn. *Clin. Perinatol.* **5,** 269–281.

43. Bryan, M. H. (1979) The work of breathing during sleep in newborns. *Am. Rev. Respir. Dis.* **119,** 137–138.

44. Bureau, M. A., Begin, R., and Berthiaume, Y. (1979) Central chemical regulation of respiration in term newborn. *J. Appl. Physiol.* **47,** 1212–1217.

45. Bureau, M. A. and Begin, R. (1982) Depression of respiration induced metabolic acidosis in newborn lambs. *Biol. Neonate* **42,** 277–283.

46. Bureau, M. A. and Begin, R. (1982) Postnatal maturation of the respiratory response to O_2 in awake newborn lambs. *J. Appl. Physiol.* **52,** 428–433.

47. Bureau, M. A., Zinman, R., Foulon, P., and Begin, R. (1984) Diphasic ventilatory response to hypoxia in newborn lambs. *J. Appl. Physiol.* **56,** 84–90.

48. Butcher-Peuch, Z. M., Holley, D., and Henderson-Smart, D. J. (1984) Apnea Type, Bradycardias, and Neurological Outcome in Preterm Infants, in *Perinatal Physiology and Behavior,* Melbourne: IUPS Sat *Symp. Aust. Paed. J.* **19,** 4.

49. Bystrzycka, E. K. (1980) Afferent projections from the dorsal and ventral respiratory nuclei in the medulla oblongata of the cat studied by the horseradish peroxidase technique. *Brain Res.* **185,** 59–66.

50. Carlo, W. A., Martin, R. J., Abboud, E. F., Bruce, E. N., and Strohl, K. P. (1983) Effect of sleep state and hypercapnia on alae nasi and diaphragm EMGs in preterm infants. *J. Appl. Physiol.* **554,** 1590–1596.

51. Carse, E. A., Wilkinson, A. R., Whyte, P. L., Henderson-Smart, D. J., and Johnson, P. (1981) Oxygen and carbon dioxide tensions, breathing and heart rate in normal infants during the first six months of life. *J. Dev. Physiol.* **3,** 85–100.

52. Cherniack, N. S. and Longobardo, G. S. (1973) Cheyne-Stokes breathing—an instability in physiologic control. *N. Eng. J. Med.* **288,** 952–957.

53. Cherniack, N. S. and Longobardo, G. S. (1981) The chemical control of respiration. *Ann. Biomed. Eng.* **9,** 395–407.

54. Cherniack, N. S. (1981) Respiratory dysrhythmias during sleep. *N. Eng. J. Med.* **305,** 325–330.

55. Chernick, V., Heldrich, F., and Avery, M. E. (1964) Periodic breathing of premature infants. *J. Pediatr.* **64**, 330–340.
56. Chernick, V. and Avery, M. E. (1966) Response of premature infants with periodic breathing to ventilatory stimuli. *J. Appl. Physiol.* **21**, 434–440.
57. Clark, F. J. and von Euler, C. (1972) On the regulation of depth and rate of breathing. *J. Physiol.* (Lond.) **222**, 267–295.
58. Cohen, M. I. (1971) Switching of the respiratory phases and evoked phrenic responses produced by rostral pontine electrical stimulation. *J. Physiol.* (Lond.) **217**, 133–158.
59. Cohen, M. I. (1973) Synchronization of discharge, spontaneously evoked, between inspiratory neurons. *Acta Neurobiol. Exp.* **33**, 189–218.
60. Cohen, M. I. (1975) Phrenic and recurrent laryngeal discharge patterns and the Hering-Breuer reflex. *Am. J. Physiol.* **228**, 1489–1496.
61. Cohen, M. I. and Feldman, J. L. (1977) Models of respiratory phase-switching. *Fed. Proc.* **36**, 2367–2374.
62. Cohen, M. I. (1979) Neurogenesis of respiratory rhythm in the mammal. *Physiol. Rev.* **59**, 1105–1173.
63. Cohen, M. I. and Feldman, J. L. (1984) Discharge properties of dorsal medullary inspiratory neurons: Relation to pulmonary afferent and phrenic efferent discharge. *J. Neurophysiol.* **51**, 753–776.
64. Coleridge, H. M., Coleridge, J. C. G., and Luck, J. C. (1965) Pulmonary afferent fibers of small diameter stimulated by capsaicin and by hyperinflation of the lungs. *J. Physiol.* (Lond.) **179**, 248–262.
65. Coleridge, H. M. and Coleridge, J. C. G. (1977) Impulse activity in afferent vagal C-fibres with endings in intrapulmonary airways of dogs. *Respir. Physiol.* **29**, 125–142.
66. Coleridge, J. C. G. and Coleridge, H. M. (1984) Afferent vagal C fiber innervation of the lungs and airways and its functional significance. *Rev. Physiol. Biochem. Pharmacol.* **99**, 1–110.
67. Comroe, J. H., Jr. and Mortimer, L. (1964) The respiratory and cardiovascular responses of temporally sparated aortic, and carotid bodies to cyanide, nicotine, phenyl diguanide, and serotonin. *J. Pharmacol. Exp. Ther.* **146**, 33–41.
68. Coombs, H. C. and Pike, F. H. (1930) The nervous control of respiration in kittens. *Amer. J. Physiol.* **95**, 681–693.
69. Coons, S. and Guilleminault, C. (1982) Development of sleep-wake patterns and non-rapid eye movement sleep stages during the first six months of life in normal infants. *Pediatrics* **69**, 793–798.
70. Corda, M., Eklund, G., and von Euler, C. (1965) External intercostal and phrenic motor responses to changes in respiratory load. *Acta Physiol. Scand.* **63**, 391–400.
71. Cosgrove, J. F., Neunburger, N., Bryan, M. H., Bryan, A. C., and

Levison, H. (1975) A new method of evaluating the chemosensitivity of the respiratory center in children. *Pediatrics* **56,** 972–980.

72. Cotton, E. K. and Grunstein, M. M. (1980) Effects of hypoxia on respiratory control in neonates at high altitude. *J. Appl. Physiol.* **48,** 587–595.

73. Cross, K. W. and Warner, P. (1951) The effect of inhalation of high and low oxygen concentrations on the respiration of the newborn infant. *J. Physiol.* (Lond.) **114,** 283–295.

74. Cross, K. W., and Oppe, T. E. (1952) The effect of inhalation o f high and low concentrations of oxygen on the respiration of the premature infant. *J. Physiol.* (Lond.) **117,** 38–55.

75. Cross, K. W. (1954) Respiratory control in the neonatal period. *Cold Spring Harbor Symp. Quant. Biol.* **19,** 126–132.

76. Cross, K. W., Klaus, M., Tooley, W. H., and Weisser, K. (1960) The response of the newborn baby to inflation of the lungs. *J. Physiol.* (Lond.) **151,** 551–565.

77. Curzi-Dascolova, L., Gaudebout, C., and Dreyfus-Brisac, C. (1981) Respiratory frequencies of sleeping infants during the first six months of life: Correlations between values in different sleep states. *Early Hum. Develop.* **5,** 39–54.

78. Curzi-Dascolova, L., Lebrun, F., and Korn, G. (1983) Respiratory frequency according to sleep states and age in normal premature infants: A comparison with full term infants. *Pediatr. Res.* **17,** 152–156.

79. Daily, W. J. R., Klaus, M., and Meyer, H. B. P. (1969) Apnea in premature infants: Monitoring, incidence, heart rate changes, and an effect of environment temperature. *Pediatrics* **43,** 510–518.

80. D'Angelo, E. and Agostini, E. (1975) Tonic vagal influences on inspiratory duration. *Respir. Physiol.* **24,** 287–302.

81. Darnall, R. A. (1983) The effect of opioid and adenosine on hypoxic ventilatory depression in the newborn piglet. *Pediatr. Res.* **17,** 374A.

82. Davi, M., Sankaran, K., MacCallum, M., Cates, D., and Rigatto, H. (1979) Effect of sleep state on chest wall distortion and on the ventilatory response to CO_2 in neonates. *Pediatr. Res.* **13,** 982–986.

83. Davies, A., Dixon, M., Callahan, D., Huszczuk, A., Widdicombe, J. G., and Wise, J. C. M. (1978) Lung reflexes in rabbits during pulmonary stretch receptor block by sulfur dioxide. *Respir. Physiol.* **34,** 83–101.

84. Davis, H. L., Fowler, W. S., and Lambert, E. H. (1956) Effect of volume and rate of inflation and deflation on transpulmonary pressure and response of pulmonary stretch receptors. *Am. J. Phys.* **187,** 558–566.

85. Dawes, G. S. and Mott, J. C. (1959) Reflex respiratory activity in the newborn rabbit. *J. Physiol.* (Lond.) **145Z,** 85–97.

86. Dawes, G. S., Lewis, B. V., Milligan, J. E., Roach, M. R., and Talner, N. S. (1968) Vasomotor responses in the hind limbs of fetal and newborn lambs to asphyxia and aortic chemoreceptor stimulation. *J. Physiol.* (Lond.) **195**, 55–81.

87. Dawes, G. S., Gardner, W. N., Johnston, B. M., and Walker, D. W. (1983) Breathing in fetal lambs: The effect of brain stem section. *J. Physiol.* (Lond.) **335**, 535–553.

88. DeBoeck, C., Van Reempts, P., Rigatto, H., and Chernick, V. (1984) Naloxone reduces decrease in ventilation induced by hypoxia in newborn infants. *J. Appl. Physiol.* **56**, 1507–1511.

89. Decima, E. E., von Euler, C., and Thodem, V. (1969) Intercostal-to-phrenic reflexes in spinal cat. *Acta Physiol. Scand.* **75**, 568–579.

90. Dement, W. C. and Kleitman, N. (1957) Cyclic variations in EEG during sleep and their relation to eye movements, body motility and dreaming. *Electroencephalogr. Clin. Neurophysiol.* **9**, 673–690.

91. Deming, J. and Washburn, A. H. (1935) Respiration of infancy. I. A method of studying rates, volume and character of respiration with preliminary report of results. *Am. J. Dis. Child.* **49**, 108–124.

92. Dempsey, J. A. and Forster, H. V. (1982) Mediation of ventilatory adaptations. *Physiol. Rev.* **62**, 262–346.

93. Denisova, M. P. and Figeurin, N. L. (1926) Phenomenes Periodiques au Cours du Sommeil des Infants, *Nouveautes de la Reflexologie de la Physiologie du Systeme Nerveux.* **2**, 338–345.

94. DiMarco, A. F., von Euler, C., Romaniuk, J. R., and Yamamoto, Y. (1981) Positive feedback facilitation of external intercostal and phrenic inspiratory activity by pulmonary stretch receptors. *Acta Physiol. Scand.* **113**, 375–386.

95. Donoghue, S., Garcia, M., Jordan, D., and Spyer, K. M. (1982) The brain-stem projections of pulmonary stretch afferent neurones in cats and rabbits. *J. Physiol.* (Lond.) **322**, 353–363.

96. Downing, S. E. and Lee, J. C. (1975) Laryngeal chemosensitivity: A possible mechanism for sudden infant death. *Pediatrics* **55**, 640–649.

97. Downman, C. R. B. (1955) Skeletal muscle reflexes of splanchnic and intercostal nerve origin in acute spinal and decerebrate cats. *J. Neurophysiol.* **18**, 217–235.

98. Eldridge, F. L. (1975) Relationship between respiratory nerve, muscle activity and muscle force output. *J. Appl. Physiol.* **39**, 567–574.

99. Euler, C. von and Fritts, J. W., Jr. (1963) Quantitative aspects of respiratory reflexes from the lungs and chest walls of cats. *Acta Physiol. Scand.* **57**, 284–300.

100. Euler, C. von and Trippenbach, T. (1976) Excitability changes of the inspiratory "off-switch" mechanism tested by electrical stimulation in the nucleus parabranchialis in the cat. *Acta Physiol. Scand.* **977**, 175–188.

101. Euler, C. von (1983) On the central pattern generator for the basic breathing rhythmicity. *J. Appl. Physiol.* **55,** 1647–1659.
102. Fagenholz, S. A., O'Connell, K., and Shannon, D. C. (1976) Chemoreceptor function and sleep state in apnea. *Pediatrics* **58,** 31–36.
103. Fagenholz, S. A., Lee, J. C., and Downing, S. E. (1979) Laryngeal reflex apnea in the chemodenervated newborn piglet. *Am. J. Phys.* **237,** R10–R14.
104. Farber, J. P. (1978) Laryngeal effects and respiration in the suckling opposum. *Resp. Physiol.* **35,** 189–200.
105. Feldman, J. L. and Gautier, H. (1976) Interaction of pulmonary afferents and pneumotaxic center in control of respiratory pattern in cats. *J. Neurophysiol.* **39,** 31–44.
106. Feldman, J. L., Cohen, M. I., and Wolotsky, P. (1976) Powerful inhibition of pontine respiratory neurons by pulmonary afferent activity. *Brain Res.* **104,** 341–346.
107. Fenner, A., Schalk, U., Hoenicke, H., Weedenberg, A. and Roehling, T. (1973) Periodic breathing in premature and neonatal babies: Incidence, breathing pattern, respiratory gas tensions, response to changes in the composition of ambient air. *Pediatr. Res.* **7,** 171–183.
108. Finer, N., Abroms, I. F., and Taeusch, H. W., Jr. (1976) Ventilation and sleep states in newborn infants. *J. Pediatr.* **89,** 100–108.
109. Finkel, M. L. (1972) The character of respiration in mature neonates during diurnal sleep. *Vopr. Okhr. Mat.* **1,** 47–52.
110. Fisher, J. T. and Mortola, J. P. (1980) Statics of the respiratory system in newborn mammals. *Respir. Physiol.* **41,** 155–172.
111. Fisher, J. T. and Mortola, J. P. (1981) Statics of the respiratory system and growth: An experimental and allometric approach. *Am. J. Physiol.* **241,** R336–R341.
112. Fisher, J. T. and Sant'Ambrogio, G. (1982) Location and discharge of properties of respiratory vagal afferents in the newborn dog. *Respir. Physiol.* **50,** 209–220.
113. Fisher, J. T., Mortola, J. P., Fox, G. S., and Weeks, S. (1982) Development of the control of breathing. *Am. Rev. Resp. Dis.* **125,** 650–657.
114. Fleming, P. J., Bryan, A. C., and Bryan, M. H. (1978) Functional immaturity of pulmonary irritant receptors in newborn preterm infants. *Pediatrics* **61,** 515–518.
115. Flores-Guevera, R., Plouin, P., Curzi-Dascaalova, L., Radvanyi, M. F., Guidasci, S., Pajot, N., and Monod, N. (1982) Sleep apneas in normal neonates and infants. *Neuropediatrics* **13** (suppl.), 21–28.
116. Foutz, A. S., Netick, A., and Dement, W. C. (1979) Sleep state effects on breathing after spinal cord transection and vagotomy in the cat. *Respir. Physiol.* **37,** 89–100.

117. Frankenhauser, B. (1948) Sensory impulses in large nerve fibers from the epiglottis of the rabbit. *Acta Physiol. Scand.* (suppl.) **53**, 24.

118. Frantz, I. D., III, Adler, S. M., Thach, B. T., and Tauesch, W., Jr. (1978) Maturational effects on respiratory responses to carbon dioxide in premature infants *J. Appl. Physiol.* **41**, 41–45.

119. Gabriel, M., Albani, M., and Schulte, F. J. (1976) Apneic spells and sleep states in preterm infants. *Pediatrics* **57**, 142–147.

120. Gastaut, H., Tassinari, C. A., and Duron, B. (1966) Polygraphic study of the episodic diurnal and nocturnal (hypoxic and respiratory) manifestations of Pickwick syndrome. *Brain Res.* **2**, 167.

121. Gerhardt, T. and Bancalari, E. (1981) Maturational reflexes influencing inspiratory timing in newborns. *J. Appl. Physiol.* **50**, 1282–1285.

122. Gerhardt, T. and Bancalari, E. (1984) Apnea of prematurity. 1. Lung function and regulation of breathing. *Pediatrics* **74**, 58–62.

123. Girard, F., Lacaisse, A., Lemaitre, R., and Houlon, J. P. (1960) Le stimulus O_2 ventilatoirea la periode neonate chez l'homme. *J. Physiol.* (Paris) **52**, 108–109.

124. Glogowska, M., Richardson, P. S., Widdicombe, J. G., and Wimming, A. J. (1972) The role of the vagus nerves, peripheral chemoreceptors and other afferent pathways in the genesis of augmented breaths in cats and rabbits. *Resp. Physiol.* **16**, 179–196.

125. Gootman, P. M., Steele, A. M., and Cohen, H. L. (1985) Postnatal Maturation of the Respiratory Rhythm Generator, in *Physiologic Development of the Fetus and Newborn.* (Jones, C. T., ed.), Academic, in press.

126. Gould, J. B., Lee, A. F. S., James, O., Sander, L., Teager, H., and Fineberg, N. (1977) The sleep state characteristics of apnea during infancy. *Pediatrics* **59**, 182–194.

127. Griffiths, G. B., Nowaraj, A., and Mortola, J. P. (1983) End-expiratory leval and breathing pattern in the newborn. *J. Appl. Physiol.* **55**, 243–249.

128. Grogaard, J. and Sundell, H. (1983) Effect of beta-adrenergic agonists on apnea reflexes in newborn lambs. *Pediatr. Res.* **17**, 213–219.

129. Grunstein, M. M., Younes, M., and Milic-Emili, J. (1973) Control of tidal volume and respiratory frequency in unanesthetized cats. *J. Appl. Physiol.* **35**, 463–476.

130. Grunstein, M. M., Hazinski, T. A., and Schlueter, M. A. (1981) Respiratory control during hypoxia in newborn rabbits: Implied action of endorphins. *J. Appl. Physiol.* **51**, 122–130.

131. Guazzi, M. and Freis, E. D. (1969) Sinoaortic reflexes and arterial pH, PO_2, and PCO_2 in wakefulness and sleep. *Am. J. Physiol.* **217**, 1623–1627.

132. Guilleminault, C., Eldridge, F. L., Simmons, F. B., and Dement, W. C. (1976) Sleep apnea in eight children. *Pediatrics* **58**, 23–31.

133. Guilleminault, C., Ariagno, R., Korobkin, R., Nagel, L., Baldwin, R., Coons, S., and Owen, M. (1979) Mixed and obstructive sleep apnea and near-miss for sudden infant death syndrome. 2. Comparison of near miss and normal control infants by age. *Pediatrics* **64**, 882–891.

134. Guthrie, R. D., Standaert, T., Hodson, W. A., and Woodrum, D. (1980) Sleep maturation of eucapnic ventilation and CO_2 sensitivity in the premature primate. *J. Appl. Physiol.* **48**, 347–354.

135. Guthrie, R. D., Standaert, T. A., Hodson, W. A., and Woodrum, D. (1981) Development of CO_2 sensitivity: Effects of gestational age, postnatal age, and sleep state. *J. Appl. Physiol.* **50**, 956–961.

136. Guz, A., Noble, N. I. M., Eisele, J. H., and Trenchard, D. (1970) The Role of Vagal Inflation Reflexes in Man and Other Animals, in *Breathing: Hering-Breuer Centenary Symposium* J. & A. Churchill, London.

137. Haddad, G. G., Epstein, R. A., Epstein, A. F., Leistner, H. L., Marino, P. A., and Mellins, R. B. (1979) Maturation of ventilation and ventilatory pattern in normal sleeping infants. *J. Appl. Physiol.* **46**, 990–1002.

138. Haddad, G. G., Leistner, H. L., Epstein, R. A., Epstein, M. A. F., Grodin, W. K., and Mellins, R. B. (1980) CO_2-induced changes in ventilation and ventilatory pattern in normal sleeping infants. *J. Appl. Physiol.* **48**, 684–688.

139. Haddad, G. G., Leistner, H. L., Lai, T. L., and Mellins, R. B. (1981) Ventilation and ventilatory pattern during sleep in aborted Sudden Infant Death Syndrome. *Pediatr. Res.* **15**, 879–883.

140. Haddad, G. G., Gandhi, M. R., and Mellins, R. B. (1982) Maturation of ventilatory response to hypoxia in puppies during sleep. *J. Appl. Physiol.* **52**, 309–314.

141. Haddad, G. G., Lai, T. L., and Mellins, R. B. (1982) Determination of the ventilatory pattern in REM sleep in normal infants. *J. Appl. Physiol.* **53**, 52–56.

142. Hagan, R., Bryan, A. C., Bryan, M. H., and Gulston, G. (1976) The effect of sleep states on intercostal muscle activity and rib cage motion (abstract). *Physiologist* **19**, 214.

143. Hagan, R., Bryan, M. H., Bryan, A. C., and Gulston, G. (1977) Neonatal chest wall afferents and regulation of respiration. *J. Appl. Physiol.* **42**, 362–367.

144. Hanson, N. and Okken, A. (1980) Transcutaneous oxygen tension of newborn infants in different behavioral states. *Pediatr. Res.* **14**, 911–915.

145. Harding, R., Johnson, P., Johnston, B. M., McClelland, M. E., and Wilkinson, A. R. (1975) Cardiovascular changes in newborn

lambs during apnea induced by stimulation of laryngeal receptors with water. *J. Physiol.* (Lond.) **256,** 35–36P.

146. Harding, R., Johnson, P., McClelland, M. E., McCleod, C. N., and Whyte, P. L. (1977) Laryngeal function during breathing and swallowing in fetal and newborn lambs. *J. Physiol.* (Lond.) **272,** 14–15P.

147. Harding, R., Johnson, P., and McClelland, M. E. (1980) Respiratory function of the larynx in developing sheep and the influence of sleep state. *Resp. Physiol.* **40,** 165–179.

148. Harding, R. (1980) State-related and developmental changes in laryngeal function. *Sleep* **3,** 307–322.

149. Harned, H. S., Jr., Myracle, J., and Ferreiro, J. (1978) Respiratory suppression and swallowing from introduction of fluids into the laryngeal region of the lamb. *Pediatr. Res.* **12,** 1003–1009.

150. Hathorn, M. K. S. (1974) The rate of depth of breathing in newborn infants in different sleep states. *J. Physiol.* (Lond.) **243,** 101–113.

151. Hathorn, M. K. S. (1978) Analysis of periodic changes in ventilation in newborn infants. *J. Physiol.* (Lond.) **285,** 85–99.

152. Head, H. (1889) On the regulation of respiration. *J. Physiol.* (Lond.) **10,** 1–70.

153. Hedner, J., Hedner, T., Bergman, B., and Lundberg, D. (1980) Respiratory depression of GABA-ergic drugs in the preterm rabbit. *J. Dev. Physiol.* **2,** 401–407.

154. Henderson-Smart, D. J. and Read, D. J. C. (1976) Depression of respiratory muscles and defective responses to nasal obstruction during active sleep in the newborn. *Austr. Pediatr. J.* **12,** 261–266.

155. Henderson-Smart, D. J. and Read, D. J. C. (1979) Ventilatory responses to hypoxaemia during sleep in the newborn. *J. Dev. Physiol.* **1,** 195–208.

156. Henderson-Smart, D. J. and Read, D. J. C. (1979) Reduced lung volume during behavioral active sleep in the newborn. *J. Appl. Physiol.* **46,** 1081–1085.

157. Hering, E. (1868) Die Selbststeuerung der Atmung durch den Nervus Vagus. *Sitzber. Akad. Wiss. Wein. Abt.* **2, 57,** 672–677.

158. Hofer, M. A. (1984) Lethal respiratory disturbance in neonatal rats after arterial chemoreceptor denervation. *Life Sci.* **34,** 489–496.

159. Hoppenbrouwers, T., Hodgman, J. E., Harper, R. M., Hoffman, E., Sternman, M. B., and McGinty, D. J. (1977) Polygraphic studies of normal infants during the first six months of life. III. Incidence of apnea ans periodic breathing. *Pediatrics* **60,** 418–425.

160. Hoppenbrouwers, T., Harper, R. M., Hodgman, J. E., Sternman, M. B., and McGinty, D. J. (1978) Polygraphic studies of normal infants during the first six months of life. II. Respiratory rate and variability as a function of state. *Pediatr. Res.* **12,** 120–125.

161. Hughes, D. T. D., Parker, H. R., and Williams, J. V. (1967) The response of fetal sheep and lambs to pulmomary inflation. *J. Physiol.* (Lond.) **189**, 177–187.
162. Huszczuk, A., Jankowska, L., Kulesza, J., and Ryba, M. (1977) Studies on reflex control of breathing in pigs and baboons. *Acta Neurobiol. Exp.* **37**, 275–298.
163. Jansen, A. H., Ioffe, S., Russell, B. J., and Chernick, V. (1981) Effect of carotid chemoreceptor denervation on breathing in utero and after birth. *J. Appl. Physiol.* **51**, 630–633.
164. Jansen, A. H. and Chernick, V. (1983) Development of respiratory control. *Physiol. Rev.* **63**, 437–483.
165. Javorka, K., Tomori, Z., and Zavarska, L. (1980) Protective and defensive airways reflexes in premature infants. *Physiol. Bohem.* **29**, 29–36.
166. Jeffrey, H. E. and Read D. J. C. (1980) Ventilatory responses of newborn calves to progressive hypoxia in quiet and active sleep. *J. Appl. Physiol.* **48**, 892–895.
167. Jouvet, M. (1967) Neurophysiology of the states of sleep. *Physiol. Rev.* **47**, 117–177.
168. Kagawa, S., Stafford, M. J., Waggener, T. B., and Severinghaus, J. W. (1982) No effect of naloxone on hypoxia-induced ventilatory depression in adults. *J. Appl. Physiol.* **52**, 1030–1034.
169. Kalepsi, Z., Durand, M., Leahy, F. N., Cates, D. B., MacCallum, M., and Rigatto, H. (1981) Effect of periodic breathing on regular respiratory pattern and the ventilatory response to low inhaled CO_2 in preterm infants during sleep. *Am. Rev. Respir. Dis.* **123**, 8–11.
170. Kalia, M. (1976) Visceral and somatic reflexes produced by J pulmonary receptors in newborn kittens. *J. Appl. Physiol.* **41**, 1–6.
171. Kalia, M. (1981) Anatomical organization of central respiratory neurons. *Ann. Rev. Physiol.* **43**, 105–120.
172. Kalia, M. and Sullivan, J. M. (1982) Brainstem projections of sensory and motor components of the vagus nerve in the rat. *J. Comp. Neurol.* **211**, 248–264.
173. Kelly, D. H. and Shannon, D. C. (1979) Periodic breathing infants with near miss sudden infant death syndrome. *Pediatrics* **63**, 355–360.
174. Kelly, D. H., Walker, A. M., Cahen, L., and Shannon, D. C. (1980) Periodic breathing in siblings of sudden infant death syndrome victims. *Pediatrics* **66**, 515–520.
175. Kelly, D. H., Kaitz, E., and Shannon, D. C. (1984) Incidence of apnea and periodic breathing in 124 normal infants 0–12 months at home. *Am. Rev. Resp. Dis.* **129**, A208.
176. Khoo, M. C. K., Kronauer, R. E., Strohl, K. P., and Slutsky, A. S. (1982) Factors inducing periodic breathing in humans: A general model. *J. Appl. Physiol.* **53**, 644–659.

177. Kirkpatrick, S. M. L., Olinsky, A., Bryan, M. H., and Bryan, A. C. (1976) Effect of premature delivery on the maturation of the Hering-Breuer inspiratory inhibitory reflex in human infants. *J. Pediatr.* **88**, 1010–1014.

178. Knill, R. and Bryan, A. C. (1976) An intercostal-phrenic inhibitory reflex in human newborn infants. *J. Appl. Physiol.* **40**, 352–356.

179. Knill, R., Andrews, W., Bryan, A. C., and Bryan, M. H. (1976) Respiratory load compensation in infants. *J. Appl. Physiol.* **40**, 357–361.

180. Knowlton, G. C. and Larabee, M. G. (1946) A unitary analysis of pulmonary volume receptors. *Am. J. Physiol.* **147**, 110–114.

181. Knox, C. K. (1973) Characteristics of inflation and deflation reflexes during expiration in the cat. *J. Neurophysiol.* **36**, 284–295.

182. Koller, E. A. and Ferrer, P. (1970) Studies on the role of the lung deflation reflex. *Respir. Physiol.* **110**, 172–183.

183. rKorpas, J. and Kalocsayova, G. (1973) Mechanoreception of the cat respiratory tract on the first days of post-natal life. *Physiol. Bohemoslov* **22**, 365–373.

184. Kovar, I., Selstram, V., Catterton, W. Z., Stahlman, M. T., and Sundell, H. W. (1979) Laryngeal chemoreflex in newborn lambs: Respiratory response to salt, acids, and sugars. *Pediatr. Res.* **13**, 1144–1149.

185. Krauss, A. N., Tori, C. A., Brown, J. M., Soodalter, J., and A uld, P. A. M. (1973) Oxygen chemoreceptors in low-birth weight infants. *Pediatr. Res.* **7**, 569–574.

186. Krauss, A. N., Klain, D. B., Waldman, S., and Auld, P. A. M. (1975) Ventilatory response to carbon dioxide in newborn infants. *Pediatr. Res.* **9**, 46–50.

187. Krauss, A. N., Waldman, S., and Auld, P. A. M. (1976) Diminished response to carbon dioxide in premature infants. *Biol. Neonate* **30**, 216–223.

188. Krauss, A. N., Solomon, G. E., and Auld, P. A. M. (1977) Sleep state, apnea and bradycardia in preterm infants. *Dev. Med. Child Neurol.* **9**, 160–168.

189. LaFramboise, W. A., Standaert, T. A., Woodrum, D. E., and Guthrie, R. D. (1981) Occlusion pressures during the ventilatory response to hypoxemia in the newborn monkey. *J. Appl. Physiol.* **51**, 1169–1174.

190. Guthrie, R. D., Standaert, T. A., and Woodrum, D. E. (1983) Pulmonary mechanics during the ventilatory response to hypoxemia in the newborn monkey. *J. Appl. Physiol.* **55**, 1008–1014.

191. Lawson, E. E. (1982) Recovery from central apnea: Effect of stimulus duration and end-tidal Co_2 partial pressure. *J. Appl. Physiol.* **53**, 105–109.

192. Lawson, E. E. and Long, W. A. (1983) Central origin of biphasic breathing pattern during hypoxia in newborns. *J. Appl. Physiol.* **55**, 483–488.

193. Lawson, E. E. and Long W. A. (1984) Central neural response to carotid sinus nerve stimulation in newborns. *J. Appl. Physiol.* **56,** 1614–1620.

194. Lee, J. C., Stoll, B. J., and Downing, S. E. (1977) Properties of the laryngeal chemoreflex in neonatal piglets. *Am. J. Physiol.* **233,** R30–R36.

195. Lee, J. C. and Downing, S. E. (1980) Laryngeal chemoreflex inhibition of breathing in piglets: Influences of anemia and catecholamine depletion. *Am. J. Physiol.* **239,** R25–R30.

196. Lees, M. H., Olsen, G. D., McGilliard, K. I., Newcomb, J. D., and Sunderland, C. O. (1982) Chloral hydrate and the carbon dioxide chemoreceptor response: A study of puppies and infants. *Pediatrics* **70,** 447–450.

197. Lenard, H. G. (1970) Sleep studies in infancy. *Acta Paediatr. Scand.* **59,** 572–581.

198. LeSouef, P. N., Lopes, J. M., England, S. J., Bryan, M. H., and Bryan, A. C. (1983) Effect of chest wall distortion on occlusion pressure and the preterm diaphragm. *J. Appl. Physiol.* **55,** 359–364.

199. Lipski, J. and Merrill, E. G. (1980) Electrophysiological demonstration of the projection from respiratory neurones in rostral medulla to contralateral dorsal respiratory group. *Brain Res.* **197,** 521–524.

200. Long, W. A. and Lawson, E. E. (1982) Supra-brainstem inhibition of respiration during carotid sinus nerve stimulation in piglets. *Pediatr. Res.* **16,** 297A.

201. Long, W. A. and Lawson, E. E. (1984) Neurotransmitters and biphasic respiratory response to hypoxia. *J. Appl. Physiol.* **57,** 213–22.

202. Longobardo, G. S., Cherniack, N. S., and Fishman, A. P. (1966) Cheyne-Stokes breathing produced by a model of the human respiratory system. *J. Appl. Physiol.* **21,** 1839–1846.

203. Lopes, J. M., Muller, N. L., Bryan, M. H., and Bryan, A. C. (1981) Synergistic behavior of inspiratory muscles after diaphragmatic fatigue in the newborn. *J. Appl. Physiol.* **51,** 547–551.

204. Lopes, J., Muller, N. L., Bryan, M. H., and Bryan, A. C. (1981) Importance of inspiratory muscle tone in maintenance of FRC in the newborn. *J. Appl. Physiol.* **51,** 830–834.

205. Lucey, J. F. (1977) The Sleeping Dreaming (?) Fetus Enters the Intensive Care Nursery, in *Apnea of Prematurity.* 71st Ross Conference on Pediatric Research, p. 13–18.

206. Lucier, G. E., Storey, A. T., and Sessle, B. J. (1979) Effects of upper respiratory tract stimuli on neonatal respiration: Reflex and single neuron analysis in the kitten. *Biol. Neonate* **35,** 82–89.

207. Lydic, R. and Orem, J. (1979) Respiratory neurons of the pneumotaxic center during sleep and wakefulness. *Neurosci. Lett.* **15,** 187–192.

208. Lynne-Davies, P., Couture, J., Pengelly, L. D., West, D., Bromagee, P. R., and Milic-Emili, J. (1971) Partitioning of immediate ventilatory stability to added elastic loads in the cat. *J. Appl. Physiol.* **30**, 814–819.
209. Marchal, F., Corke, B. C., and Sundell, H. (1982) Reflex apnea from laryngeal chemo-stimulation in the sleeping premature lamb. *Pediatr. Res.* **16**, 621–627.
210. Margaria, L. E., Iscoe, S., Pengelly, L. D., Couture, J., Don, H., and Milic-Emili, J. (1973) Immediate ventilatory response to elastic loads and positive pressure in man. *Resp. Phhysiol.* **18**, 347–369.
211. Marlot, D. and Duron, B. (1979) Postnatal development of vagal control of breathing in the kitten. *J. Physiol.* (Paris) **75**, 891–900.
212. Marlot, D. and Duron, B. (1979) Postnatal maturation of phreic, vagus, and intercostal nerves in the kitten. *Biol. Neonate* **35**, 264–272.
213. Marlot, D., Mortola, J. P., and Duron, B. (1982) Functional localization of pulmonary stretch receptors in the tracheobronchial tree of the kitten. *Can. J. Physiol. Pharmacol.* **60**, 1073–1077.
214. Marlot, D., Bonora, M., Gautier, H., and Duron, B. (1984) Postnatal maturation of ventilation and breathing pattern in kittens; Influence of sleep. *J. Appl. Physiol.* **56**, 321–325.
215. Marshall, R. and Widdicombe, G. J. (1958) The weakness of the Hering-Breuer reflexes in man. *J. Physiol.* (Lond.) **140**, 36P.
216. Martin, R. J., Nearman, H. S., Katona, P. G., and Klaus, M. H. (1977) The effect of a low continuous positive airway pressure on the reflex control of respiration in the preterm infant. *J. Pediatr.* **90**, 976–981.
217. Martin, R. J. Okken, A., Katona, P. G., and Klaus, M. H. (1978) Effect of lung volume on expiratory time in the newborn infant. *J. Appl. Physiol.* **45**, 18–23.
218. Martin, R. J., Okken, A., and Rubin, D. (1979) Arterial oxygen tension during active and quiet sleep in the normal neonate. *J. Pediatr.* **94**, 271–274.
219. Martin, R. J., Herrell, N., and Pultusker, M. (1981) Transcutaneous measurement of carbon dioxide tension. Effect of sleep state in term infants. *Pediatrics* **67**, 622–625.
220. Matthew, O. P., Roberts, J. L., and Thach, B. T. (1982) Pharyngeal airway obstruction in preterm infants during mixed and obstructive apnea. *J. Pediatr.* **100**, 964–968.
221. Mayou, R. A. (1973) Efferent effects on visceral receptors (BSc. Thesis) Oxford Univ. Lab. of Physiol. Library, 1963 (see Paintal, 1973).
222. Merril, E. G. (1970) The lateral respiratory neurons of the medulla: Their associations with nucleus ambiguus, nucleus retroambigualis, the spinal accessory nucleus and the spinal cord. *Brain Res.* **24**, 11–28.

223. Milhorn, H. T., Jr., and Guyton, A. C. (1965) An analog computer analysis of Cheyne-Stokes breathing. *J. Appl. Physiol.* **20**, 328–333.

224. Miller, A. J. and Dunmire, C. R. (1976) Characterization of the postnatal development of the superior laryngeal nerve fibres in the postnatal kitten. *J. Neurobiol.* **7**, 483–494.

225. Mills, J. E., Sellick, H., and Widdicombe, J. G. (1969) Activity of lung irritant receptors in pulmonary microembolism, anaphylaxis and drug induced bronchoconstriction. *J. Physiol.* (Lond.) **203**, 337–357.

226. Milner, A. D., Boon, A. W., Saunders, R. A., and Hopkins, I. E. (1980) Upper airways obstruction and apnea in preterm infants. *Arch. Dis. Child.* **55**, 22–25.

227. Mortola, J. P. (1984) Breathing pattern in newborns. *J. Appl. Physiol.* **56**, 1533–1540.

228. Mortola, J. P., Milic-Emili, J., Nowaraj, A., Smith, B., Fox, G., and Weeks, S. (1984) Muscle pressure and flow during expiration in infants. *Am. Rev. Dis.* **129**, 49–53.

229. Muller, N., Gulston, G., Cade, D., Whitton, J., and Froese, A. B. (1979) Diaphragmatic muscle fatigue in the newborn. *J. Appl. Physiol.* **46**, 688–695.

230. Nadel, J. A. and Widdicombe, J. G. (1963) Reflex control of airway size. *Ann. N.Y. Acad. Sci.* **109**, 712–722.

231. Nadel, J. A., Salem, H., Tamplin, B., and Tokiwa, Y. (1965) Mechanisms of bronchoconstriction during inhalation of sulfur dioxide. *J. Appl. Physiol.* **20**, 164–167.

232. Nadel, J. A., Phillipson, E. A., Fishman, N. H., and Hickey, R. F. (1973) Regulation of respiration by bronchopulmonary receptors in conscious dogs. *Acta Neurobiol. Exp.* **33**, 33–50.

233. Nattie, E. E. and Edwards, W. H. (1981) CSF acid-base regulation and ventilation during acute hypercapnia in the newborn dog. *J. Appl. Physiol.* **50**, 566–574.

234. Nelson, N. M. (1976) Respiratin and Circulation After Birth, in *Physiology of the Newborn Infant*. 4th ed. (Smith, C. A. and Nelson, N. M., eds.), Charles C. Thomas, Springfield, Illinois.

235. Nettick, A., Foutz, A. S., and Dement, W. C. (1977) Sleep state effects upon respiration during behavioral sleep following vagotomy and cord transection in the cat (abstract). *Soc. Neurosci. Annu. Meeting* **3**, 407.

236. Olinsky, A., Bryan, M. H., and Bryan, A. C. (1974) Influence of lung inflation on respiratory control in neonates. *J. Appl. Physiol.* **36**, 426–429.

237. Orem, J., Nettick, A., and Dement, W. C. (1977) Increased upper airway resistance to breathing during sleep in the cat. *EEG Clin. Neurophysiol.* **43**, 14–22.

238. Orem, J., Norris, P., and Lydic, R. (1978) Laryngeal abductor activity during sleep. *Chest* **73(2)** (suppl.), 300–301.
239. Orem, J. (1980) Neuronal mechanisms of respiration in REM sleep. *Sleep*, **3**, 251–267.
240. Pack, A. I., DeLanay, R. G., and Fishman, A. P. (1981) Augmentation of phrenic neural activity by increase rates of lung inflation. *J. Appl. Physiol.* **50**, 149–161.
241. Pack, A. I. (1981) Sensory inputs to the medulla. *Ann. Rev. Physiol.* **43**, 73–90.
242. Pack, A. I. and DeeLaney, R. G. (1983) Response of pulmonary rapidly adapting receptors during lung inflation. *J. Appl. Physiol.* **50**, 955–963.
243. Paintal, A. S. (1953) The conduction velocities of respiratory and cardiovascular afferent fibers in the vagus nerve. *J. Physiol.* (Lond.) **121**, 341–359.
244. Paintal, A. S. (1966) Re-evaluation of respiratory reflexes. *Quart. J. Exp. Physiol.* **51**, 151–163.
245. Paintal, A. S. (1969) Mechanism of stimulation of type J pulmonary receptors. *J. Physiol.* (Lond.) **203**, 511–532.
246. Paintal, A. S. (1973) Vagal sensory receptors and their reflex effects. *Physiol. Rev.* **53**, 159–227.
247. Parmalee, A. H., Jr., Wenner, W. H., Akiayuma, Y., Schultz, M., and Stern, E. (1967) Sleep states in premature infants. *Dev. Med. Child. Neur.* **9**, 70–77.
248. Perkett, E. A. and Vaughan, R. L. (1982) Evidence for a laryngeal chemoreflex in some human preterm infants. *Acta Paed. Scand.* **71**, 969–972.
249. Phillipson, E. A. and Murphy, E. (1973) Vagal control of respiratory rate and depth independent of lung inflation in conscious dogs. *Fed. Proc.* **32**, 355.
250. Phillipson, E. A. (1974) Vagal control of breathing pattern independent of lung inflation in conscious dogs. *J. Appl. Physiol.* **37**, 183–189.
251. Phillipson, E. A., Murphy, E., and Kozar, L. F. (1976) Regulation of respiration in sleeping dogs. *J. Appl. Physiol.* **40**, 688–693.
252. Phillipson, E. A., Kozar, L. F., Rebuck, A. S., and Murphy, E. (1977) Ventilatory and waking responses to CO_2 in sleeping dogs. *Am. Rev. Respir. Dis.* **115**, 251–259.
253. Phillipson, E. A., Sullivan, C. E., Read, D. J. C., Murphy, E., and Kozar, L. F. (1978) Ventilatory and waking responses to hypoxia in sleeping dogs. *J. Appl. Physiol.* **44**, 512–520.
254. Phillipson, E. A. (1978) Control of breathing during sleep. *Am. Rev. Resp. Dis.* **118**, 909–939.
255. Phillipson, E. A., Bowes, G., Townsend, E. R., Duffin, J., and

Cooper, J. D. (1981) Carotid chemoreceptors in ventilatory responses to changes in venous CO_2 load. *J. Appl. Physiol.* **151,** 1398–1403.

256. Prechtl, H. F. R. and Lenard, H. G. (1967) A study of eye movements in sleeping newborn infants. *Brain Res.* **5,** 477–493.

257. Prechtl, H. F., Weinmann, H., and Akiama, Y. (1969) Organization of physiological parameters in normal and neurologically abnormal infants. *Neuropadiatrie* **1,** 101–129.

258. Prechtl, H. F. R., VanEykern, L. A., and O'Brien, M. J. (1977) Respiratory muscle EMG in newborns: A non-intrusive method. *Early Hum. Dev.* **1,** 265–283.

259. Purves, M. J. (1966) Respiratory and circulatory effects of breathing 100% O_2 in the newborn lamb before and after denervation of the carotid chemoreceptors. *J. Physiol.* (Lond.) **185,** 42–59.

260. Purves, M. J. (1966) The effects of the hypoxia in the newborn lamb before and after denervation of the carotid chemoreceptors. *J. Physiol.* (Lond.) **185,** 60–77.

261. Purves, M. J. (1966) The respiratory response of the newborn lamb to inhaled CO_2 with and without accompanying hypoxia. *J. Physiol.* (Lond.) **185,** 78–94.

262. Reinstorff, D. and Fenner, A. (1972) Ventilatory response to hyperoxia in premature and newborn infants during the first three days of life. *Respir. Physiol.* **15,** 159–165.
Inhibition of respiratory activity by intercostal muscle afferents. *Respir. Physiol.* **10,** 358–383.

264. Remmers, J. E. and Martilla, I. (1975) Action of intercostal muscle afferents on the respiratory rhythm of the anesthetized cat. *Respir. Physiol.* **24,** 31–41.

265. Remmers, J. E., Bartlett, D., and Putnam, M. D. (1976) Changes in the respiratory cycle associated with sleep. *Resp. Physiol.* **28,** 227–238.

266. Remmers, J. E. and Bartlett, D., Jr. (1977) Reflex control of expiratory airflow and duration. *J. Appl. Physiol.* **42,** 80–87.

267. Remmers, J. E., DeGrott, W. J., Sauerland, E. K., and Anch, A. M. (1978) Pathogenesis of upper airway occlusion during sleep. *J. Appl. Physiol.* **44,** 931–938.

268. Richards, J. M., Alexander, J. R., Shinebourne, E. A., de Swiet, M., Wilson, A. J., and Southall, D. P. (1984) Sequential 22-hour profiles of breathing patterns and heart rate in 110 full-term infants during the first 6 months of life. *Pediatrics* **74,** 763–777.

269. Richardson, P. J., Sant'Ambrogio, G., Mortola, J. P., and Bianconi, R. (1973) The activity of lung afferent nerves during tracheal occlusion. *Resp. Physiol.* **18,** 273–283.

270. Richter, D. W. (1982) Generation and maintenance of the respiratory rhythm. *J. Exp. Biol.* **100,** 93–107.

271. Rigatto, H. and Brady, J. P. (1972) Periodic breathing and apnea in

preterm infants. I. Evidence for hypoventilation possibly due to central respiratory depression. *Pediatrics* **50**, 202–218.

272. Rigatto, H. and Brady, J. P. (1972) Periodic breathing and apnea in preterm infants. II. Hypoxia as a primary event. *Pediatrics* **50**, 219–228.

273. Rigatto, H., Brady, J. P., and Verduzco, R. T. (1975) Chemoreceptor reflexes in preterm infants. I. The effect of gestational age and postnatal age on the ventilatory response to inhalation of 100% and 15% oxygen. *Pediatrics* **55**, 604–613.

274. Rigatto, H., Brady, J. T., and de la Torre Verduzco, R. (1975) Chemoreceptor reflexes in preterm infants. II. The effect of gestational age on the ventilatory response to carbon dioxide. *Pediatrics* **55**, 614–620.

275. Rigatto, H., de la Torre Verduzco, R., and Cates, D. B. (1975) Effects of O_2 on the ventilatory response to CO_2 in preterm infants. *J. Appl. Physiol.* **39**, 896–899.

276. Rigatto, H. (1977) Apnea and periodic breathing. *Sem. Perinatal.* **1**, 375–381.

277. Rigatto, H., Kalepsi, Z., Leahy, F. N., Durand, M., MacCallum, M., and Cates, D. (1980) Chemical control of respiratory frequency and tidal volume during sleep in preterm infants. *Respir. Physiol.* **41**, 117–125.

278. Rigatto, H., Kalepsi, Z., Leahy, F. N., MacCallum, M., and Cates, D. B. (1982) Ventilatory response to 100% and 15% O_2 during wakefulness and sleep in preterm infants. *Early Hum. Dev.* **7**, 1–18.

279. Roffwarg, H. P., Muzio, J. N., and Dement, W. C. (1966) Ontogenetic development of the human sleep-dream cycle. *Science* **152**, 604–619.

280. Roomy, M. and Leitner, L. M. (1980) Localization of stretch receptors and deflation receptors in the airways of the rabbit. *J. Physiol.* (Lond.) **76**, 67–70.

281. Sachis, P. N., Armstrong, D. L., Becker, L. E., and Bryan, A. C. (1982) Myelination of the human vagus nerve from 24 weeks post conceptional age to adolescence. *J. Neuropathol. Exp. Neurol.* **41**, 466–472.

282. Salmoiraghi, G. C. and Burns, B. D. (1960) Notes on mechanisms of rhythmic respiration. *J. Neurophysiol.* **23**, 14–26.

283. Sampson, S. and Eyzaguirre, C. (1964) Some functional characteristics of mechanoreceptors in the larynx of the cat. *J. Neurophysiol.* **27**, 464–480.

284. Sankaran, K., Wiebe, H., Seshia, M. M. K., Boychuk, R. B., Cates, D., and Rigatto, H. (1979) Immediate and late ventilatory response to high and low O_2 in preterm infants and adult subjects. *Pediatr. Res.* **13**, 875–878.

285. Sankaran, K., Leahy, F., Cates, D., MacCallum, M., and Rigatto,

H. (1981) Effect of lung inflation on ventilation and various phases of the respiratory cycle in preterm infants. *Biol. Neonate* **40**, 160–166.

286. Sasaki, C. T., Suzuki, M., and Horiuchi, M. (1977) Postnatal development of laryngeal reflexes in the dog. *Arch. Otolaryngol.* **103**, 138–143.

287. Sauerland, E. K., Orr, W. C., and Hairston, L. E. (1981) EMG patterns of oropharyngeal muscles during respiration in wakefulness and sleep. *Sleep ELectromyogr. Clin. Neurophysiol.* **21**, 307–316.

288. Schleman, M., Gootman, N., and Gootman, P. M. (1979) Cardiovascular and respiratory responses to right atrial injections of phenyl diguanide in pentobarbital anesthetized newborn piglets. *Pediatr. Res.* **13**, 1271–1273.

289. Schulte, F. J–, Busse, C., and Eichhorn, W. (1977) Rapid eye movements sleep, motoneurone inhibition and apneic spells in preterm infants. *Pediatr. Res.* **11**, 709–713.

290. Schweiler, G. H. (1968) Respiratory regulation during postnatal development in cats and rabbits and some of its morphological substrate. *Acta Physiol. Scand.* (suppl.) **304**, 49–63.

291. Sears, T. A. (1964) Investigations on spinal motoneurones of the thoracic spinal cord. *Progr. Brain Res.* **12**, 259–272.

292. Sellick, H. and Widdicombe, J. G. (1970) Vagal deflation and inflation reflexes mediated by lung irritant receptors. *Quart. J. Exp. Physiol.* **55**, 153–163.

293. Sessle, B. J., Greenwood, L. F., Lund, J. P., and Lucier, G. E. (1978) Effect of upper respiratory tract stimuli on respiration and single respiratory neurons in the adult cat. *Exp. Neurol.* **61**, 245–259.

294. Shannon, R. and Zechman, F. W. (1972) The reflex and mechanical repsonse of the inhibitory muscles to an increased airflow resistance. *Respir. Physiol.* **16**, 51–69.

295. Shannon, R. and Freeman, D. (1981) Nucleus retroambigualis respiratory neurons. Responses to intercostal and abdominal muscle afferents. *Respir. Physiol.* **357–375**.

296. Sherry, J. H. and Merrigan, D. (1980) Respiratory EMG activity of the posterior cricoarytenoid, cricothyroid. and diaphragm muscles during sleep. *Resp. Physiol.* **139**, 355–365.
R., Bonora, M., Gautier, H., Duron, B., and Milic-Emili, J. (1981) Phrenic activity, respiratory pressures, and volume changes in cats. *J. Appl. Physiol.* **51**, 109–121.

298. Siassi, B., Hodgmaan, J. E., Cabal, L., and Hon, E. H. (1979) Cardiac and respiratory activity in relation to gestational and sleep state in newborn infants. *Pediatr. Res.* **13**, 1163–1166.

299. Sica, A. L., Cohen, M. I., Donnelly, D. F., and Zhang, H. (1984) Hypoglossal motoneuron responses to pulmonary and superior laryngeal afferent inputs. *Respir. Physiol.* **56**, 399–357.

300. Sothers, G. K., Henderson-Smart, D. J., and Read, D. J. C. (1977) Postnatal changes in the rate of firing of high frequency bursts of inspiratory activity in cats and dogs. *Brain Res.* **132,** 537–540.

301. Southall, D. P. (1983) Identification of infants destined to die unexpectedly during infancy: Evaluation of predictive importance of prlonged apnea and disorders of cardiac rhythm or conduction. *Brit. Med. J.* **286,** 1092–1096.

302. Spann, R. W. and Hyatt, R. E. (1971) Factors affecting upper airway resistance in conscious man. *J. Appl. Physiol.* **31,** 708–712.

303. Speck, F. D. and Webber, C. L., Jr. (1981) Time course of intercostal afferent termination of the inspiratory process. *Respir. Physiol.* **43,** 133–145.

304. Spitzer, A. R. and Fox, W. W. (1984) Overview of Infant Apnea and Approach to Clinical Management, in *Clinical Management of Infantile Apnea Monograph Series* (Spitzer, A. R. and Fox, W. W., eds.), DSA Communications, Port Washington, New York.

305. Spreng, L. F., Johnson, L. C., and Lubin, A. (1968) Autonomic correlates of eye movement bursts during stage REM sleep. *Psychophysiol.* **4,** 311–323.

306. Stark, A. R. and Frantz, I. D., III (1979) Prolonged expiratory duration with elevated lung volume in newborn infants. *Pediatr. Res.* **13,** 261–264.

307. Steele, A. M., Gootman, P. M., Cohen, H. L., Eberle, L. P., and Rudell, A. P. (1984) Postnatal maturaton of the respiratory rhythm generator in neonatal swine. *Ped. Res.* **18,** 406A.

308. Storey, A. T. (1968) A functional analysis of sensory units innervating epiglottis and larynx. *Exp. Neurol.* **20,** 366–383.

309. Storey, A. T. and Johnson, P. (1975) Laryngeal water receptors initiating apnea in the lamb. *Exp. Neur.* **47,** 42–55.

310. Stothers, J. K. and Warner, R. M. (1978) Oxygen consumption and neonatal sleep states. *J. Physiol.* (Lond.) **278,** 435–440.

311. Sullivan, C. E., Kozar, L. F., Murphy, E., and Phillipson, E. A. (1978) Primary role of respiratory afferents in sustaining breathing rhythm. *J. Appl. Physiol.* **45,** 11–17.

312. Sullivan, C. E., Murphy, E., Kozar, L., and Phillipson, E. A. (1978) Waking and ventilatory responses to laryngeal stimulation in sleeping dogs. *J. Appl. Physiol.* **45,** 681–689.

313. Sullivan, C. E., Murphy, E., Kozar, L. F., and Phillipson, E. A. (1979) Ventilatory responses to CO_2 and lung inflation in tonic versus phasic REM sleep. *J. Appl. Physiol.* **47,** 1304–1310.

314. Sutton, D., Taylor, E. M., and Lindeman, R. C. (1978) Prolonged apnea in infant monkeys resulting from stimulation of superior laryngeal nerve. *Pediatrics* **61,** 519–527.

315. Thach, B. T. and Taeusch, H. W., Jr. (1976) Sighing in newborn infants: The role of the inflation augmenting reflex. *J. Appl. Physiol.* **41,** 502–507.

316. Thach, B. T., Frantz, I. D., III, Adler, S. M., and Taeusch, H. W., Jr. (1978) Maturation of reflexes influencing inspiratory duration in human infants. *J. Appl. Physiol.* **45,** 203–211.
317. Thach, B. T., Abroms, I. F., Frantz, I. D., III, Sotrel, A., Bruce, E. N., and Goldman, M. D. (1980) Intercostal muscle reflexes and sleep breathing patterns in the human infant. *J. Appl. Physiol.* **48,** 139–146.
318. Thach, B. T. (1983) The Role of Pharyngeal Airway Obstruction in Prolonging Infantile Apneic Spells, in *Sudden Infant Death Syndrome* (Tildon, J., Roeder, L. M., Steinschneider, A., eds.), Academic Press, New York.
319. Thibeault, D. W., Wong, N. M., and Auld, P. (1967) Thoracic gas volume changes in premature infants. *Pediatrics* **40,** 403–411.
320. Trippenbach, T., Zinman, R., Mozes, R., and Murphy, L. (1979) Differences and Similarities in the Control of Breathing Pattern in the Adult and Neonate, in *Central Nervous Control Mechanisms in Breathing* (von Euler, C. and Langercrantz, H., eds.), Pergamon, Oxford.
321. Trippenbach, T., Zinman, R., and Milic-Emili, J. (1980) Caffeine effect on breathing pattern and vagal reflexes in newborn rabbits. *Respir. Physiol.* **40,** 211–225.
322. Trippenbach, T. (1981) Laryngeal, vagal, and intercostal reflexes during the early postnatal period. *J. Dev. Physiol.* **3,** 133–159.
323. Trippenbach, T., Zinman, R., and Mozes, R. (1981) Effects of airway occlusion at functional residual capacity in pentobarbital-anesthetized kittens. *J. Appl. Physiol.* **51,** 143–147.
324. Trippenbach, T., Gaultier, C., and Cooper, L. (1982) Effects of chest wall compression in kittens. *Can. J. Physiol. Pharmacol.* **60,** 1241–1246.
325. Trippenbach, T. and Kelly, G. (1983) Phrenic activity and intercostal muscle EMG during inspiratory loading in newborn kittens. *J. Appl. Physiol.* **54,** 496–501.
326. Trippenbach, T., Kelly, G., and Marlot, D. (1983) Respiratory effects of stimulation of intercostal muscles and saphenous nerve in kittens. *J. Appl. Physiol.* **54,** 1736–1744.
327. Trippenbach, T. (1983) Effect of drugs on the respiratory control system in the perinatal period and during postnatal development. *Pharmacol. Ther.* **20,** 307–340.
328. Waggener, T. B., Frantz, I. D., III, Stark, A. R., and Kronauer, R. E. (1982) Oscillatory breathing pattern leading to apneic spells in infants. *J. Appl. Physiol.* **52,** 1288–1295.
329. Waggener, T. B., Stark, A. R., Conlan, B. A., and Frantz, I. D., III. (1984) Apnea duration is related to ventilatory oscillation characteristics in newborn infants. *J. Appl. Physiol.* **57,** 536–544.
330. Webb, P. (1974) Periodic breathing during sleep. *J. Appl. Physiol.* **37,** 899–903.

331. Wennergren, G. and Wennergren, M. (1980) Respiratory effects elicited in newborn animals via the central chemoreceptors. *Acta Physiol. Scand.* **108**, 309–311.
332. Werner, S. S., Stocklard, J. E., and Bickford, R. G. (1977) *Atlas of Neonatal Electroencephalography*. Raven, New York.
333. Widdicombe, J. G. (1954) Receptors in the trachea and bronchi o f the cat. *J. Physiol.* (Lond.) **123**, 71–104.
334. Widdicombe, J. G., Kent, D. C., and Nadel, J. A. (1962) Mechanism of bronchoconstriction during inhalation of dust. *J. Appl. Physiol.* **17**, 613–616.
335. Widdicombe, J. G. (1963) Regulation of tracheobronchial smooth muscle. *Physiol. Rev.* **43**, 1–37.
336. Widdicombe, J. G. and Nadel J. A. (1963) Reflex effects of lung inflation and tracheal volume. *J. Appl. Physiol.* **18**, 681–686.
337. Widdicombe, J. G. (1967) Head's paradoxical reflex. *Q. J. Exp. Physiol.* **52**, 44–50.
338. Woodrum, D. E., Standaert, T. A., Maycock, D. E., and Guthrie, R. D. (1981) Hypoxic ventilatory response in newborn monkey. *Pediatr. Res.* **15**, 367–370.
339. Younes, M. K., Remmers, J. E., and Baker, J. (1978) Characteristics of inspiratory inhibition by phasic volume feedback in cats. *J. Appl. Physiol.* **45**, 80–86.

APPENDIX

List of Abbreviations

Ac	Adenylate cyclase
AC	Commisura anterior
ACh	Acetylcholine
AChE	Acetylcholinesterase
$a^i K^+$	Intracellular potassium activity
$a^i Na^+$	Intracellular sodium activity
ANOVA	Analysis of variance
ANS	Autonomic nervous system
AoP	Aortic blood pressure
ATP	Adenosine triphosphate
AV	Atrioventricular
BC	Botzinger complex
BCCO	Bilateral common carotid occlusion
BP	Aortic pressure
C	Nucleus cuneatus
C1	Cervical section 1 of the spinal cord
CA	Catecholamine
cAMP	Cyclic 3′,5′-adenosine monophoshate
CarF	Carotid arterial blood flow
CarR	Carotid resistance
CAT	Choline acetyltransferase
CC	Corpus callosum truncus
Cc	Crus cerebri
CF	Corpus fornicus
CG	Substancia grisea centralis
ChAT	Choline acetyltransferase
CI	Capsula interna
CID	Central inspiratory drive
CII	Central inspiratory inhibition
CM	Corpus mamillare
CNeg	Continuous negative end-expiratory pressure
CNS	Central nervous system
CPAP	Continuous positive airway pressure
CRL	Crown-rump length; an estimate of fetal development
CT	Cricothyroid muscle

CTP	Cycle-triggered pump
CVPM	Cardiac vagal preganglionic motoneurons
d	Day
DA	Dopamine
DBH	Dopamine beta hydroxylase
DCN	Dorsal column nuclei
DDC	Dopa decarboxylase
DEX	Dexamethasone
^{3}H-DHA	Dihydroalprenolol
DLRF	Dorsolateral medullary reticular formation
DMN	Dorsal motor nucleus of the X nerve
DrG	Dorsal root ganglia
DRG	Dorsal respiratory group
DTV	Decussatio tegmenti ventralis
Epi	Epinephrine
E	Expiration
EEG	Electroencephalogram
EKG	Electrocardiogram
EMG	Electromyogram
EOG	Electrooculogram
F	Columna fornicis
FBF	Femoral blood flow
FemF	Femoral arterial blood flow
FemR	Femoral resistance
FRC	Functional residual capacity
FVR	Femoral vascular resistance
G	Nucleus gracilis
μg	Microgram
H	Hydralazine
Hab	Habenula
h	Hour
HA	Nucleus hypothalamicus anterior
HL	Nucleus hypothalamicus lateralis
HP	Nucleus hypothalamicus posterior
HR	Heart rate
HRP	Horseradish peroxidase
HVM	Nucleus hypothalamicus ventromedialis
HYP	Hypoglossal nerve
Hz	Hertz
I	Inspiratory
Integr.	Integrated
IntPHR	Integrated phrenic nerve
IP	Nucleus iterpeduncularis
ISO	Isoproterenol
ITP	Intratracheal pressure
K-F	Kölliker-Fuse nucleus

kg	Kilogram
K^+_i	Intracellular potassium concentration
K^+_o	Extracellular potassium concentration
L-Dopa	L-3,4-Dihydroxyphenylalanine
LG	Corpus geniculatum laterale
LNS	Lumbar nerve stimulation
LR	Lateral reticular nucleus
mA	Microampere
MBF	Mesenteric blood flow
MesF	Mesenteric arterial blood flow
MesR	Mesenteric resistance
MG	Corpus geniculatum mediale
mg	Milligram
MLF	Fasciculus longitudinalis medialis
MNS	Mesenteric nerve stimulation
mo	Month
ms	Millisecond
MT	Fasciculus mamillothalamicus
MVR	Mesenteric vascular resistance
NA	Nucleus ambiguus
NCA	Nucleus cuneatus accessorius
NE	Norepinephrine
NGF	Nerve growth factor
NLF	Nucleus fasciculi lateralis
NO	Nucleus olivaris
NPA	Nucleus paraambigualis
NRA	Nucleus retroambigualis
NREM	Non-rapid eye movement sleep
NTS	Nucleus tractus solitorious
NVII	Nucleus nervi facialis
NVIII M	Nucleus vestibularis medialis
NVIII S	Nucleus vestibularis inferior
NX	Nucleus dorsalis nervi vagi
NXII	Nucleus nervi hypoglossi
ODC	Ornithine decarbonylase
6-OHDA	6-Hydroxydopamine
OPT	Tractus opticus
P	Pyramis
Phen	Phenoxybenzamine
PC	Commissura posterior
PneuC	Pneumotaxic center
PCA	Posterior cricoarytenoid muscle
PDG	Phenyl diguanide
PHR	Phrenic nerve
PIIA	Postinspiration inspiratory activity
P_K	Membrane permeability to potassium

P_{Na}	Membrane permeability to sodium
PNMT	Phenylethanolamine N-methyltransferase
PRT	Pretectum
PSR	Pulmonary stretch receptor
P_{tp}	Transpulmonary pressure
PTU	Propylthiouracil
^{3}H-QNB	Quinuclidynl benzoate
r	Radius
R	Nucleus ruber
RAR	Rapidly adapting receptor
RAS	Reticular activating system
REM	Rapid eye movement sleep
RenF	Renal arterial blood flow
RenR	Renal resistance
RF	Formatio reticularis
RL	Recurrent laryngeal nerve
RNS	Renal nerve stimulation
RRG	Respiratory rhythm generator
RVR	Renal venal resistance
s	Second
SA	Sinoatrial
SAR	Slowly adaping receptor
SCG	Superior cervical ganglion
SHR	Spontaneously hypertensive rat
SIDS	Sudden infant death syndrome
SLN	Superior laryngeal nerve
SN	Substantia nigra
SNS	Sympathetic nervous system
SPL	Splanchnic nerve
SPV	Spinal trigeminal complex
stim	Stimulation
T	Tension
TA	Thyroartenoid muscle
T3	Triiodothyronine
T_E	Expiratory time
TEA	Tetraethylammonium chloride
TH	Tyrosine hydroxylase
THAL	Thalamus
T_I	Inspiratory time
TMA	Tetramethylammonium chloride
TS	Tractus solitarius
TTX	Tetrodotoxin
$\dot{V}$	Minute ventilation
V	Nucleus tractus spinalis nervi trigemini
VIP	Vasoactive intestinal polypeptide
V_K	Membrane reversal potential for potassium

VLNA	Ventrolateral nucleus ambiguus
$\dot{V}_{max}$	Maximum velocity of phase O depolarization of the cardiac transmembrane action potential
VMRF	Ventromedial medullary reticular formation
V_{Na}	Membrane reversal potential for sodium
VRG	Ventral respiratory group
V_T	Tidal volume
V_T/T_I	Mean inspiratory flow
VVA	Veratrum viride alkaloids
WKY	Wistar Kyoto rat
ZI	Zona incerta

INDEX